Calibration in Analytical Science

Calibration in Analytical Science

Methods and Procedures

Paweł Kościelniak

WILEY-VCH

Author

Professor Dr. Paweł Kościelniak
Jagiellonian University
Department of Analytical Chemistry
Gronostajowa St. 2
30-387 Krakow
Poland

Cover Design and Image: Wiley

Library of Congress Card No.: applied for

British Library Cataloguing-in-Publication Data
A catalogue record for this book is available from the British Library.

Bibliographic information published by the Deutsche Nationalbibliothek
The Deutsche Nationalbibliothek lists this publication in the Deutsche Nationalbibliografie; detailed bibliographic data are available on the Internet at <http://dnb.d-nb.de>.

Print ISBN: 978-3-527-34846-6
ePDF ISBN: 978-3-527-83110-4
ePub ISBN: 978-3-527-83112-8
oBook ISBN: 978-3-527-83111-1

Typesetting Straive, Chennai, India
Printing and Binding CPI Group (UK) Ltd, Croydon, CR0 4YY

C109761_100223

Contents

Preface

Analytical chemistry is an exceptionally beautiful scientific area. Behind such a description stands not only the author's undoubtedly subjective view but also completely objective observations. In probably no other chemical discipline is the purpose of theoretical and experimental work so clearly and unambiguously defined as in analytical chemistry. This aim is simply to look deep into the matter and to determine the type or amount of components contained in it. Taking into account that the guiding principle of all scientific research is the pursuit of truth, it can be said that every chemical analysis (and there are many thousands of such analyses carried out every day in the world) is in fact the fulfillment of this principle, and every analytical chemist can have the feeling of getting as close to the "truth" as possible during his work, if only he does it correctly and carefully. This is certainly a motivating and rewarding aspect.

The specificity of analytical chemistry also lies in the fact that the scientific principles, rules, and methods developed over the years can be used in practice extremely directly, rapidly, and usefully, which on the other hand promotes the development of new theoretical and instrumental concepts. This coupling is the reason why, especially in recent decades, analytical chemistry has developed rapidly and become increasingly important in all areas of business and society. Through the application of new analytical methods and techniques, innovative chemical and biochemical materials, as well as specialized, high-tech apparatus, analysts are able to penetrate deeper and deeper into matter, detecting and determining the components contained in it in ever smaller quantities and in a variety of chemical forms.

In the current of this progress, however, it is easy to succumb to the fascination of its technical and instrumental aspects, gradually forgetting that the mere creation of new analytical methods and inventions – that is, the search for paths to "truth" – is insufficient, even if these paths are the most ingenious and innovative. It is equally important that the analytical results obtained by these routes should, as far as possible, have the characteristics of the "true," which, in analytical language, means above all their maximally high accuracy and precision.

No one needs to be convinced of the importance of high-quality chemical analysis. Several years ago, it was calculated that repetitions of analyses performed in industrial laboratories in the USA, which are necessary due to incorrect analytical results, generate losses of several billion dollars a year. But even more important is the fact that only on the basis of reliable results of various types of analysis, in particular clinical, pharmaceutical, environmental or forensic analysis, it is possible to make a reliable diagnosis, which consequently determines our health and living conditions today and in the future.

What is the meaning and role of analytical calibration in this context? The answer to this question can be given in one word – enormous. One need only realize that a calibration process must accompany almost every analytical proceeding regardless of whether the analysis is qualitative or quantitative in nature. In other words, without this process, achieving the analytical goal – or, if you prefer, getting closer to the analytical truth – is simply impossible. Moreover, the proper choice of the calibration path and its correct adaptation to the different stages of the analytical procedure can contribute significantly to the maximum approximation of the true result. Against this background, it seems obvious that, among the various analytical issues, the subject of calibration requires special attention and interest.

Unfortunately, the reality contradicts this thesis – interest in calibration issues among analysts is relatively low both scientifically and practically. First of all, there are no book positions entirely devoted to this topic, except perhaps for multivariate calibration, which, however, is not widely used in analytical laboratories. In analytical chemistry textbooks, little is usually said about calibration methods, usually limiting themselves to basic, customary approaches and solutions. On the other hand, over the years many articles have appeared in which new calibration solutions can be found, testifying to progress in this analytical field as well. Mostly, however, these reports are usually treated as purely academic and are generally not applicable in laboratory practice.

It must also be said that in the field of calibration there is an extremely large nomenclatural chaos, concerning not only the nomenclature and classification of calibration methods but also the concept of the analytical calibration process as such. This state of affairs obviously has negative consequences. Above all, it is not conducive to teaching purposes, since it is difficult to reliably convey specific analytical knowledge using a language that is not standardized and generally accepted. The lack of a common ground for communication in this area can also become a source of misunderstandings and ambiguities leading to erroneous and incorrect analytical procedures. And yet, no one but an analyst should be particularly sensitive to "order" and "purity" in his work.

The main purpose of this book is to fill, at least to some extent, these gaps and backlogs. It collects and describes a variety of calibration methods and procedures for determining the nature and quantity of sample components in different ways. These approaches are tailored to the specific chemical and instrumental conditions of the qualitative and quantitative analyses performed, as well as to the specific objectives the analyst wishes to achieve in addition to the overarching goal. Based on the calibration properties of these methods, their nomenclature and classification are proposed. It is also shown how calibration approaches can be combined and integrated mainly to diagnose, evaluate, and eliminate analytical errors and thus achieve results with increased precision and accuracy.

The contents of this book are largely based on the author's many years of experience. This is to some extent the result of both the layout of the book and the detailed selection of the issues covered, which certainly does not exhaust the entire calibration subject matter. For the same reason, one can find here original, authorial approaches to this subject, which – although previously published in scientific articles and thus verified – may be further debatable. Therefore, I apologize in advance to those who have slightly different views on the issues raised in the book. I understand this and at the same time invite opponents to such a discussion. I believe, however, that despite all possible reservations and doubts, the book will be a useful source of information on analytical calibration and, for many, a valuable addition to analytical knowledge and a helpful tool in scientific and laboratory work.

As mentioned, the calibration process is inextricably linked to the analytical process. Wandering through the various avenues of performing calibrations is thus also an opportunity to learn or recall various analytical methods and general problems related to analytical chemistry and chemical analysis. With this in mind, the author also sees this book as a supplement to general analytical knowledge delivered in a slightly different way and from a different angle than typical analytical science textbooks.

Finally, I would like to express my warm gratitude to Professor Andrzej Parczewski for "infecting" me many years ago with the subject of calibration. I would also like to thank my colleagues from the Department of Analytical Chemistry of the Jagiellonian University in Krakow for accompanying me on exciting analytical adventure and for providing me with many of their research results for it.

But I am most grateful to my beloved Wife – for motivation, words of support, and time which, at the expense of being with her, I could devote to this work. Without you, Ania, this book would not have been written.

Kraków, March 2022 *Paweł Kościelniak*

1

Calibration Fundamentals

The general understanding of the term "calibration" is far from what applies to the concept in an analytical sense. Leaving aside colloquial connotations, such as calibrating a weapon, the term is generally associated with the adjustment of specific parameters of an object to fixed or desired quantities, and in particular with the adjustment of a specific instrument to perform a correct function. It is, therefore, understood more as a process of instrumental standardization or adjustment. This is reinforced by publicly available nomenclatural sources. For example, in the Cambridge Advanced Learner's Dictionary [1] calibration is defined as " … the process of checking a measuring instrument to see if it is accurate," and in the http://Vocabulary.com online dictionary as "the act of checking or adjusting (by comparison with a standard) the accuracy of a measuring instrument" [2]. Even in a modern textbook in the field of instrumental analysis, you can read: "In analytical chemistry, calibration is defined as the process of assessment and refinement of the accuracy and precision of a method, and particularly the associated measuring equipment…" [3].

The ambiguity of the term "calibration" makes it difficult to understand it properly in a purely analytical sense. To understand the term in this way, one must of course take into account the specificity of chemical analysis.

1.1 Analytical Context

The analyst aims to receive **the analytical result**, i.e. to identify **the type** (in qualitative analysis) or to determine the **quantity** (in quantitative analysis) of a selected component (**analyte**) in the material (**sample**) assayed. To achieve this goal, he must undertake a series of operations that make up **the analytical procedure**, the general scheme of which is shown in Figure 1.1.

When starting an analysis, the sample must first be prepared for measurement in such a way that its physical and chemical properties are most suitable for measuring the type or amount of analyte in question. This step consists of such processes as, e.g. taking the sample from its natural environment and then changing its aggregate state, diluting it, pre-concentrating it, separating the components, changing the temperature, or causing a chemical reaction.

Calibration in Analytical Science: Methods and Procedures, First Edition. Paweł Kościelniak.

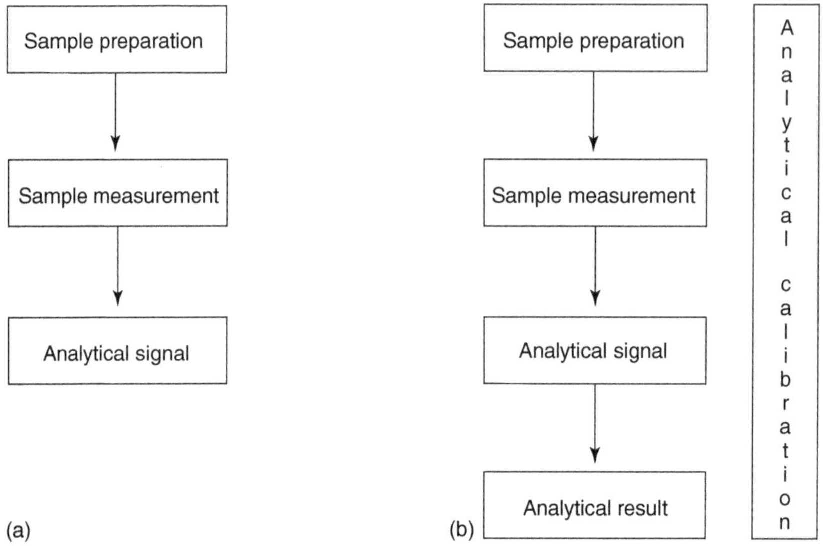

Figure 1.1 Analytical procedure alone (a) and supplemented by analytical calibration (b).

The measurement is generally performed by the chosen using an instrument that operates on the principle of a selected **measurement method** (e.g. atomic absorption spectrometry, potentiometry, etc.). The instrument should respond to the presence of the analyte studied in the form of **measurement signals**. From a calibration point of view, the most relevant signal is the so-called **analytical signal**, i.e. the signal corresponding to the presence of analyte in the sample.

An analytical procedure carried out in a defined manner by a specific measurement method forms an **analytical method**.

The basic analytical problem is that **the analytical signal is not a direct measure of the type and amount of analyte in the sample**, but only information indicating that a certain component in a certain amount is present in the sample. To perform a complete analysis, it is necessary to be able **to transform the analytical signal into the analytical result** and to perform this transformation. This is the role of analytical calibration. As seen in Figure 1.3, the analytical calibration process is an integral part of the analytical procedure and without analytical calibration, qualitative and quantitative analysis cannot be performed. **Realizing this aspect allows one to look at the subject of calibration as a fundamental analytical issue.**

1.2 Principles of Analytical Calibration

However, there is still the question of what the process of transforming an analytical signal to an analytical result consists of, i.e. how analytical calibration should be defined. In this regard, there is also no unified approach, so it is best to rely on official recommendations.

The process of analytical calibration is largely concerned with the making of measurements and the interpretation of measurement data and therefore falls within the scope of metrology. In the Joint Committee for Guides in Metrology (JCGM) document on basic and general terms in metrology, calibration is defined as "... operation that, under specified conditions, in a first step, establishes a relation between the quantity values with measurement uncertainties provided by measurement standards and corresponding indications with associated measurement uncertainties and, in a second step, uses this information to establish a relation for obtaining a measurement result from an indication" [4]. At the same, the document makes it clear that "calibration should not be confused with adjustment of a measuring system ...".

The metrological term, although it allows for a deeper understanding of the concept of calibration, is still rather general because it is inherently applicable to different measurement systems and different types of results obtained. The concept of calibration in the analytical sense is more closely approximated by publications issued by the International Union of Pure and Applied Chemistry (IUPAC). In the paper [5], the IUPAC definition is aligned with the JCGM definition in that it defines analytical calibration as "... the set of operations which establish, under specified conditions, the relationship between value indicated by the analytical instrument and the corresponding known values of an analyte," and in a subsequent IUPAC publication [6] we find an express reference of analytical calibration to both quantitative and qualitative calibration: **"Calibration in analytical chemistry is the operation that determines the functional relationship between measured values (signal intensities at certain signal positions) and analytical quantities characterizing types of analytes and their amount (content, concentration)."**

Such a purely theoretical approach is too general, even abstract, and unrelated to analytical practice. In particular, it does not provide guidance on how the functional relationship (calibration model) should be formulated in different analytical situations and how it relates to the different types of methods used in qualitative and quantitative analysis. Nor does it say anything about the relative nature of the calibration process that the term "measurement standard" gives to the concept in metrological terms.

To extend the definition of analytical calibration, the author proposes to introduce the concept of three functions that relate the signal to the analytical result: the true function, the real function, and the model function [7]. This approach is illustrated in Figure 1.2.

If a sample that an analyst takes for qualitative or quantitative analysis contains a component (analyte) of interest, then before any action is taken with the sample, the type of analyte and its quantity in the sample can be referred to as the **true value** (type or quantity), x_{true}, of the analyte. If it were possible to measure the analytical signal for that analyte at that moment, then the relationship between the resulting signal and its true type or quantity, $Y_{true} = T(x_{true})$ could be called the **true function**.

However, the determination of the true function and the true value of the analyte is not possible in practice because it requires the analyst's intervention in the

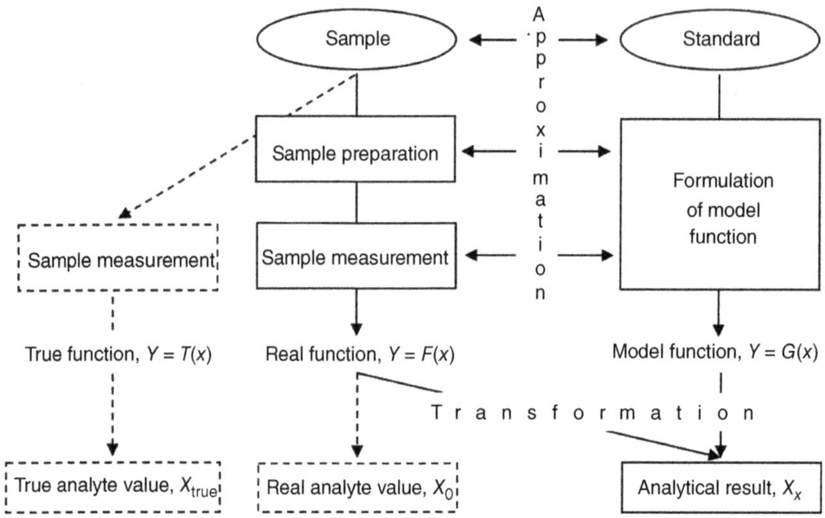

Figure 1.2 Concept of analytical calibration based on the terms of true, $Y = T(x)$, real, $Y = F(x)$, and model, $Y = G(x)$, functions (virtual analytical steps and terms are denoted by dotted lines; for details see text).

form of preparing the sample for measurement and performing the measurement. The initiation of even the simplest and shortest analytical steps results in a change of the true analyte concentration in the sample that continues until the analytical signal is measured. Thus, the concepts of true function and true analyte value are essentially unrealistic and impossible to verify experimentally or mathematically.

When the sample is prepared for analysis, the type or amount of analyte in the sample to be analyzed takes on a **real value, $\boldsymbol{x_0}$**. The relationship between the analytical signal and the type or amount of analyte is described at this point by the **real function**, $Y = F(x)$, which takes the value Y_0 for the value x_0:

$$Y_0 = F(x_0) \tag{1.1}$$

Although the value of Y_0 is measurable, the exact form of the real function is unknown because it depends on a number of effects and processes that led to the current state of this relationship during the preparation of the sample for measurement. Consequently, the determination of the real result x_0 by means of the real function is impossible.

This situation forces the formulation of an additional auxiliary **model function**, $Y = G(x)$. The role of this function is to replace the real function in the search for the true value, x_0. It should therefore meet two basic conditions: to be known and well-defined and to be the most accurate approximation of the real function ($G(x) \leftrightarrow F(x)$). To fulfill these conditions, a **calibration standard** (one or more) should be used, which should be similar to the sample and properly prepared for measurement.

Assuming that the approximation of the real function by the model function, $Y = G(x)$, is accurate, then the inverse form of the model function, $x = G^{-1}(Y)$, has to be created, which is called the evaluation function [6]. Theoretically, it allows the value of Y_0 to be transformed into the real result, x_0:

$$x_0 = G^{-1}(Y_0) \tag{1.2}$$

In practice, the approximation of the real function by the model function is never accurate because the real function is essentially unknown. Therefore, transformation (1.2) leads to a certain value x_x:

$$x_x = G^{-1}(Y_0) \tag{1.3}$$

which is an approximate measure of the real result, x_0. This result can also be considered as the final **analytical result**.

The processes of creating a model function and its approximation and transformation are fundamental, integral, and necessary elements of analytical calibration. Thus, it can be said that **analytical calibration consists of approximating the real relationship between the signal, *Y*, and the type, *b*, or amount, *c*, of an analyte in a sample by means of a model function, and then applying this function to transform the signal obtained for the analyte in the sample to the analytical result.**

Note the natural logic of the above description of analytical calibration. Such quantities as "sample" (considered as a collection of unknown chemical constituents), "real function" and "real type or amount of analyte" have their counterparts in the terms of "standard", "model function" and "obtained type or amount of analyte", which are associated with analytical calibration. The former are largely hypothetical, unknown in fact to the analyst, while the latter are known and are approximations of the former. Just as the composition and properties of a sample can never be faithfully reproduced in a standard, the form of the real function cannot be accurately approximated by a model function, and the real type or amount of analyte in the sample at the time the analytical signal is measured can only be approximated by the analytical result obtained.

1.3 Calibration Standards and Models

Depending on the type of univariate model function used, analytical calibration can be broadly divided into **empirical calibration** and **theoretical calibration** [7]. In some cases, the calibration is also of a complex nature to varying degrees (empirical–theoretical or theoretical–empirical) when, to better represent the real function, empirical information is supported by theoretical information or vice versa.

An essential part of any calibration process is the use of calibration standards, which can be of different nature: **chemical, biological, physical, or mathematical** [7]. A common feature of calibration standards is that they directly or indirectly enable the assignment of a measurement signal to a known, well-defined type or amount of analyte. These standards are therefore used to formulate a model function. According to the principle of analytical calibration, a standard should be able to formulate a model function that approximates the true function as closely as possible.

In empirical calibration, the model function is formulated on the basis of the performed experiment, sensory perception, or observation. The sources of information needed to create this type of **empirical model function**, $Y = G(x)$, are measurements of analytical signals obtained directly or indirectly for chemical, biological, or physical standards. In this case, the analyst does not go into the theoretical aspects of the dependence of the analytical signal on the type or amount of analyte (although in some cases the laws and rules underlying this dependence, e.g. Nernst's or Lambert Beer's law, may be helpful).

A widely recognized and used method of analytical calibration is the empirical calibration performed with a **chemical standard**. This is a synthetic or (less commonly) natural material, single or multicomponent, containing an analyte of known type or amount. In special cases, a chemical standard contains a known type or amount of a substance that reacts with the analyte or a known type or amount of an isotope of the element being determined. Calibration with chemical standards is a universal procedure in the sense that it does not depend on the chosen measurement method. The model function formulated is mathematically usually simple and its graphical form is called a calibration graph.

In theoretical calibration, the model function is formulated on the basis of a mathematical description of physicochemical phenomena and processes occurring during the analysis using a given analytical and measurement method. Such a description includes phenomenological quantities based on physical or chemical measurements (electrochemical potentials, diffusion coefficients, etc.), universal quantities (molar mass, atomic number, stoichiometric factors), and/or fundamental physical constants (Faraday constant, Avogadro constant, etc.). The individual elements of the mathematical description act as **mathematical standards**, and the function created with them, $Y = G(x)$, is a **theoretical model function**.

In analytical chemistry, there are relatively few cases of well-defined theoretical models of relatively simple mathematical form. However, in the literature, one can find many new proposals of such functions formulated for various measurement methods. As a rule, they have a very complex mathematical structure, which results from the desire to approximate the real function as accurately as possible. A strong motivation for these scientific efforts is that the theoretical model allows the calculation of the analytical result without the need to prepare chemical standards and perform measurements for the analyte in these standards.

As mentioned, other types of calibration standards can be found in chemical analysis, as well as model functions of a different nature formulated with them, as discussed in Chapter 2 of this chapter. It can be hypothesized that **analytical calibration is inherently connected with the use of standards and the creation of model functions with their help**.

The implications of this approach to analytical calibration are interesting. Qualitative or quantitative analysis performed on the basis of a theoretical model function is often referred to in the literature as calibration-free analysis or absolute analysis. From the point of view of the accepted definition of analytical calibration, this term is misleading, because the formulation of the theoretical model function, like the empirical model, is part of the full calibration procedure. Thus, the questions arise:

can chemical analysis be performed in practice without analytical calibration and what conditions must an analytical method meet to be called a "absolute method"? The discussion of this issue will be the subject of Chapter 2 of this book.

1.4 Calibration Procedures and Methods

The concept of analytical calibration presented above perhaps do not yet give a clear picture of this process. How, then, does the full empirical and theoretical calibration procedure look in general?

As already stated, the calibration process is essential to the performance of chemical analysis – both qualitative and quantitative – and is an integral, inseparable part of any analytical method. What the calibration process contributes to the analytical procedure is the handling of the calibration standard necessary to formulate the model function and use it to transform the measurement signal to the analytical result. Thus, the calibration procedure consists of three steps: preparative, measurement, and transformation.

The **preparative step** consists in preparing the sample and the standard in such a suitable way that the true function, $Y = F(x)$, and the model function, $Y = G(x)$, are similar to each other as much as possible. In the case of empirical z-calibration, there are two main routes to this goal:

- **the sample and standard are prepared separately**, taking care that the chemical composition of the standard is similar to that of the sample and that the preparation of the sample and standard for measurement is similar,
- **the standard is added to the sample** prior to measurement (less frequently prior to sample processing).

In the case of theoretical calibration, separate treatment of the sample and the standard is obvious and natural. Appropriate preparation of the standard in relation to the sample consists in introducing such mathematical standards to the theoretical model that most adequately describe the state of the sample and the phenomena and processes that the sample undergoes under the conditions of the specific measurement method.

In the measurement stage, signal measurements are made using a selected measurement method. If the calibration is empirical, measurements are related to the sample and standard or on the sample and sample with the addition of the standard (depending on their preparation at the preparative stage). In either case, the measurements involving the standard are used to formulate an empirical model function. In the case of a theoretical calibration, measurements are made only for the sample and the formulated theoretical model is considered as the model function.

In the **transformation step**, the value of the signal obtained for the sample is entered into an empirical or theoretical model function and thus the final analytical result (type or amount of analyte in the sample) is determined.

Referring to the formulated extended definition of analytical calibration, it can be noticed that the preparative and measurement stages are used to approximate the

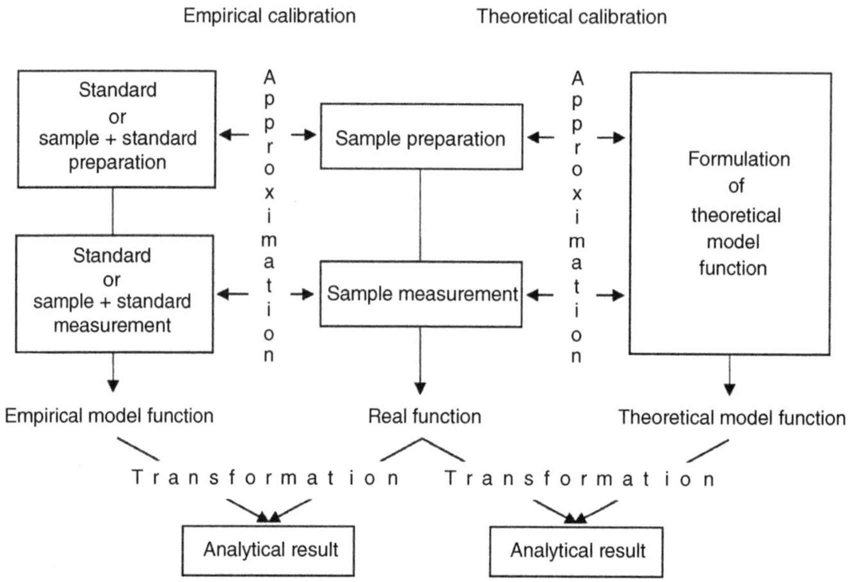

Figure 1.3 General scheme of empirical and theoretical calibration.

model function to the real function, and the key, transformational calibration process takes place at the last stage. A schematic diagram of the procedures of empirical and theoretical calibration is shown in Figure 1.3.

Calibration procedures with specific preparation of sample and standard for measurement form **calibration methods**. In general, therefore, two groups of methods can be distinguished in analytical calibration, which can be called **comparative methods** (when the sample and standard are treated separately) and **additive methods** (when the standard is added to the sample). Within each of these two groups, it is possible to distinguish methods that differ more specifically on the preparative side (e.g. external standard method, internal standard method, standard addition method, etc.). These names are mostly customary and do not always correspond to the specifics of the individual methods. Therefore, another, more essential criterion for the division of the calibration methods in terms of the mathematical way of transforming the measurement signal into the analytical result will also be proposed.

1.5 Calibration in the Context of Measurement Errors

The role of analytical calibration is not only to make it possible to identify or determine an analyte in a sample, but also to do so with as much **accuracy and precision** as possible. The measure of accuracy is the statistically significant difference between the analytical result obtained, x_x, and the true type or amount of analyte, x_{true}, in the sample before it was subjected to any analytical process. The measure of precision is the random difference in analytical results obtained in so-called parallel analyses, that is, performed in the same way and under the same experimental conditions. The accuracy and precision of an analytical result are thus determined by any systematic and random changes in the true function before it becomes, at the time of measurement, the true function, and then by the systematic and random difference between the true function and its representation, the model function.

Changes in the analytical signal that occur both during sample preparation for measurement and during measurement, resulting in the transformation of the true function to the model function, can be called analytical effects [7]. They can be controllable and uncontrollable. Controlled analytical effects include, for example, changes caused by a targeted action by the analyst to decrease or increase the concentration of an analyte in a sample by dilution or concentration, respectively. Effects of this type can usually be calculated and corrected at the stage of analytical result calculation.

During qualitative and quantitative analysis, however, there are also such changes in the analytical signal that are partially or completely out of the analyst's control. These **uncontrolled analytical effects** can be both random and systematic. Although the analyst is usually aware of the risk of their occurrence and usually tries to prevent them accordingly, he or she may overlook or even neglect them while performing the analysis. As a result, control over the entire analytical process is lost in a sense. Uncontrolled effects manifest themselves by changing the position and intensity of the analytical signal, i.e. they are important in both qualitative and quantitative analysis.

1.5.1 Uncontrolled Analytical Effects

Uncontrolled effects can be caused by many factors manifesting themselves at different stages of the analytical process. The classification of these effects covering all possible factors is, of course, a matter of convention. The division presented below is the author's proposal [7].

Uncontrolled effects are primarily caused by the analyst himself (the so-called **human factor**) as a result of incorrect or careless behavior at various stages of the analytical process. The magnitude of these changes depends primarily on the analyst's knowledge and skills, that is, on his or her professional abilities and qualifications. Personal factors such as tiredness, nervousness, and hurry play a large role. The minimization of the human factor is also not favored by a routine, "automatic" approach to individual analytical activities, resulting, for example, from performing analyses according to a single, unchanging analytical method over a long period of time.

The basic effects include **preparative effect**. Under this term, we understand signal changes caused by such sample processing that results in uncontrolled change (loss, less often gain) of analyte amount in the sample. The analyte can be partially lost e.g. when changing the aggregate state of the sample (to make it suitable for the given measurement method) or when separating its components. The process of changing the amount (or rarely the type) of sample can also take place outside the purposive, controlled action of the analyst as a result of e.g. an induced chemical reaction. A preparative effect is also involved when the change in signal results directly from physical changes in the sample or standard (e.g. solution viscosity), or from changes in environmental conditions under which the sample and standard are processed (e.g. temperature, humidity, illumination, etc.).

The **instrumental effect** is caused by the action of various instrumental components used in the analytical process to process the sample prior to measurement. In this case, the source of random changes in the analytical signal is all natural imperfections in the design and operation of these instruments, including the measurement systems that characterize the measurement method. However, as a result of instrument malfunction, signal changes can also be systematic.

The instrumental measurement system is the source of separate specific measurement changes occurring in the detection system. This phenomenon can therefore be called a **detection effect**. These changes are manifested, for example, by the limited ability of the system to respond proportionally to the analyte concentration, which is natural for each detector. Another phenomenon is the so-called measurement trend, which consists of a successive increase or decrease in signal intensity over time. In spectrometric methods, there is sometimes the problem of baseline, which varies more or less randomly between spectra. The detection effect can also be related to natural phenomena underlying the measurement method (a typical example is the phenomenon of self-absorption of radiation emitted in the emission spectrometry method, causing a change in analytical signal intensity out of proportion to the amount of analyte in the sample).

The signal measured for a specific type or amount of analyte can also be affected by other components both naturally present in the sample (native) and introduced during sample preparation for measurement. These components then take on the role of interferents, and the signal change caused by them is the so-called **interference effect**. If the effect comes exclusively from the native components of the sample, then it is called a **matrix effect**, while if the interferents are components added to the sample during sample processing, then the induced changes are called a **blank effect**. The interference effect can originate at the stage of sample preparation for measurement (e.g. due to added reagents), but can also be induced during measurement of the analytical signal as a result of phenomena and processes occurring at this stage.

Finally, a specific effect is the **speciation effect**. It occurs when an analyte contained in a sample unexpectedly for the analyst changes its chemical form and at the same time changes its measurement sensitivity. As with the interference effect, this change can occur before measurement (e.g. as a result of a chemical reaction) or at the time of measurement, when it involves a change in that form that is responsible for causing the analytical signal in the detection system (e.g. a change from atoms to analyte ions in atomic absorption spectrometry).

Uncontrolled effects are revealed by a change in the analytical signal either directly or indirectly by changing the type or amount of analyte, as illustrated in Figure 1.4.

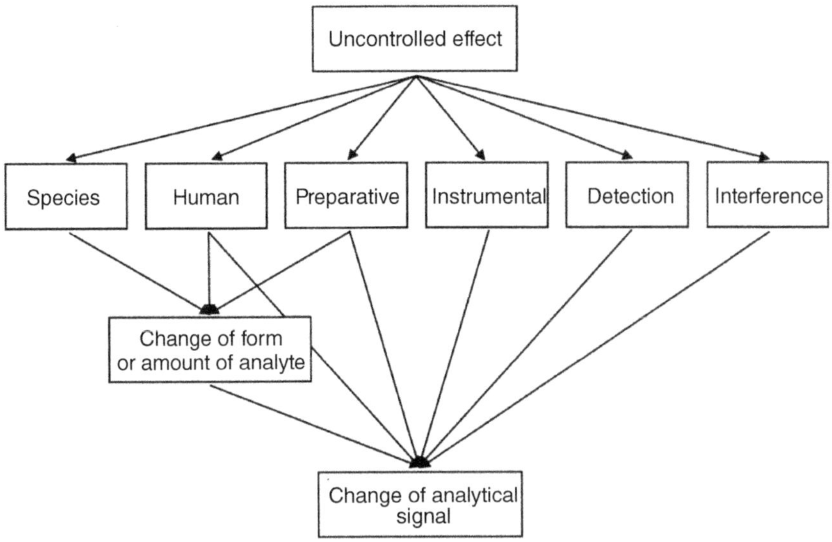

Figure 1.4 Pathways of the various uncontrolled effects. Source: Kościelniak [7]/ Elsevier/CC BY 4.0.

1.5.2 Elimination and Compensation of Uncontrolled Effects

The natural way to avoid uncontrolled effects revealed during sample handling is to employ various means of **eliminating** them. Effectiveness of these actions largely depends on proper identification of the type of these effects and their sources, which are differently situated on the analytical procedure plan. This is shown in Figure 1.5.

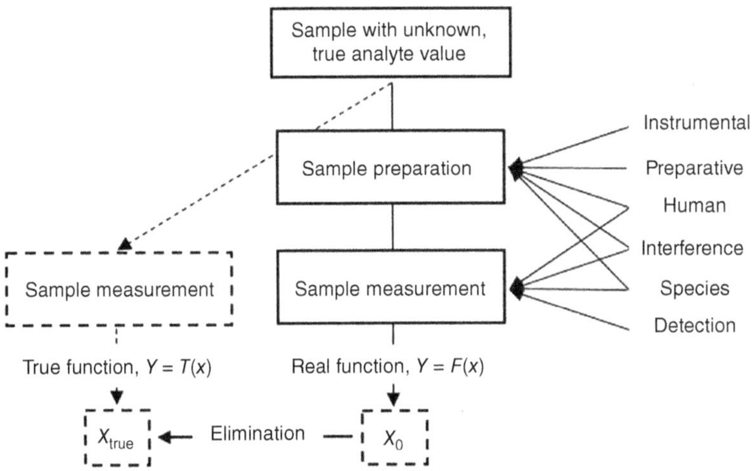

Figure 1.5 Impact of uncontrolled effects on an analyte in the sample during its preparation and measurement; due to elimination of effects the real analyte value approaches the true value ($x_0 \approx x_{true}$). Source: Kościelniak [7]/Elsevier/CC BY 4.0.

The prerequisite for reducing the influence of the human factor is that analyses should be performed only by qualified staff, with a high level of knowledge and skills, maintaining care and caution during the work. Instrumental and detection effects may not be a major problem if the instruments used are of high quality, proven reliability, and low maintenance. In special cases where there are, for example, strong time trends or baseline shifts, special correction procedures are used [8].

In contrast to instrumental and detection effects, speciation effects can be difficult to eliminate if the analytical procedure is relatively complex and involves the use of different types of chemical reactions. The preparative effect can also be difficult to eliminate effectively. This is because no sample processing is in practice free from partial loss of analyte. The degree of this phenomenon should in each case be well recognized by preliminary experiments and then reduced as much as possible. The amount of analyte lost can also be quantified (e.g. by the recovery method, which is discussed later in Chapter 10) and the final analytical result can be corrected on this basis.

The interference effect can be eliminated in basically two ways. The universal way is to remove the interferents from the sample or to isolate the analyte from the sample matrix by appropriately selected laboratory techniques (e.g. by extraction, crystallization, gaseous diffusion, etc.). Another approach is to add an appropriately selected reagent to the sample to eliminate interferents by chemical means.

Progressive elimination of uncontrolled effects causes the two analyte values, true, x_{true}, and real, x_0, to become increasingly similar, as can be seen in Figure 1.5. **When the effects are completely eliminated, the true analyte value becomes an accurate (within random error) measure of the true analyte value in the sample, i.e. $x_0 \approx x_{true}$.**

When proceeding with an analytical calibration, the analyst is forced to use a standard. Importantly, however, this constraint simultaneously provides an opportunity to make the standard similar to the sample. If the sample and standard are similar, then all uncontrolled effects occurring during the analytical procedure should, in theory, manifest themselves in the same way and with appropriate strength with respect to both the sample and the standard. As a result, **compensation** for these effects occurs. Note that effect compensation differs from the process of elimination in that it does not eliminate the uncontrolled effects, but merely involves equalizing them in the sample and standard.

In an empirical calibration performed using a chemical standard, it is easiest to compensate for instrumental effects because it is sufficient to maintain instrumental conditions at the same optimum level during sample and standard preparation for measurement. The detection effect is compensated for just as easily by using the same instrument for both sample and standard measurements and keeping the conditions of the measurements the same.

Compensating for preparative effects is more difficult, although it can be achieved to some extent by subjecting the standard to the same preparative treatments to which the sample was subjected. However, it must be taken into account that the analyte in the standard may be subject to this effect to a different degree than the native analyte due to the different chemical environment and potentially different chemical form. For speciation effects, it is very important that the chemical form of the analyte remains the same in the sample and in the standard during the calibration procedure. If the analyte is present in several chemical forms in the sample, the analyte in the standard need not take all of these forms but should be in the form in which the analyte is to be determined in the sample.

The most difficult effect to compensate for effectively is the interference effect. It is only relatively simple to compensate for the blank effect by adding the reagents used in sample preparation to the standards. The effect from native sample components requires that the composition of the sample in the standard is accurately reproduced (which is very difficult or even impossible in practice) and that this condition be maintained until measurements are made. However, there are various ways to make the sample and the standard at least partially similar in chemical composition or to compensate for the effect by using an appropriate calibration method. These solutions will be shown and discussed in Chapter 6.

The compensation of effects is offered by the calibration process and is therefore closely related to the representation of the real function by the model function. The more accurate the approximation of the two functions is, the more complete the compensation process is. Progressive compensation of effects promotes a progressive approximation of the analytical result, x_x, to the real result, x_0, as well as the real result, x_0, to the true result, x_{true}. Thus, **after complete compensation, the analytical result becomes an accurate (within random error) estimate of the true value of the analyte in the sample, i.e. $x_x \approx x_{true}$.** This is illustrated schematically in Figure 1.6.

In theoretical calibration, compensating for uncontrolled effects involves describing them adequately by means of a mathematical standard, i.e. including in this description the effects of various factors on the signal measured for the analyte in the sample. However, while a chemical standard can be made similar to a sample due to its similar nature, making a mathematical standard similar to a sample is extremely difficult. Thus, when deciding to use a theoretical calibration, it is

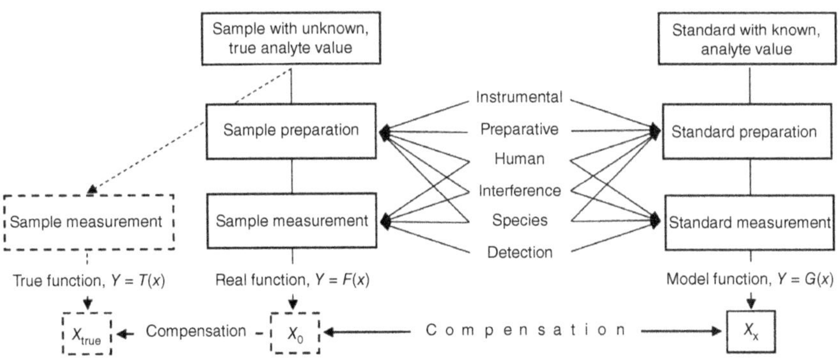

Figure 1.6 Impact of uncontrolled effects on an analyte in both the sample and standard during its preparation and measurement; due to compensation for effects the analytical result value approaches the true value ($x_x \approx x_{true}$). Source: Kościelniak [7]/CC BY 4.0.

important to eliminate, as much as possible, uncontrolled effects affecting the analyte in the sample.

Analytical calibration thus leads to an accurate analytical result either by complete elimination of uncontrolled effects or by their complete compensation. Elimination of an effect thus does not require its compensation (e.g. once an interference effect has been eliminated with a special reagent, there is no need to reconstruct the composition of interferents in the standard), although if it is known that the elimination of an effect may be incomplete, it should be compensated. Similarly, compensation for effects (e.g. instrumental effects) does not require their elimination, although any small reduction increases the chance of their complete compensation. The processes of elimination and compensation of uncontrolled effects are thus complementary activities in the sense that, taken together, they provide the best chance of achieving an accurate assessment of the true value of the analyte in the sample from the analytical result obtained.

So how should the analytical calibration process be evaluated in the context of errors made during the analytical procedure? Certainly, calibration is a potential source of its own random and systematic analytical errors. This is primarily due to the need to use a standard. The empirical standard, like the sample, is subject to uncontrolled effects that may be of a different type than those found in the sample and therefore not compensable. Furthermore, the sample is always more or less different from the standard either because of properties and composition (in empirical calibration) or because of mathematical approximations and corrections (in theoretical calibration). From the imperfection of the calibration standard comes the imperfection of the model function and the added difficulty of accurately approximating the true function.

On the other hand, it should be noted that if it were possible to determine the true value of an analyte in a sample without the contribution of any standard, the analytical procedure used would have to be completely free of uncontrolled effects, or these effects would have to be completely eliminated, and both are impossible in practice. The participation of a calibration standard, i.e. the performance of an analytical calibration, is therefore not only a necessary condition for obtaining an analytical result, but also offers an additional opportunity to improve the quality of this result by compensating for uncontrolled effects.

1.6 Calibration in Qualitative Analysis

Analytical calibration applies equally to qualitative and quantitative analysis [6]. However, in both cases the form of the real function is different, the basis for the formulation of the model function is different, and the accuracy of the results of analyte identification and determination is also evaluated differently. It is therefore worth taking a closer look at these calibration aspects in both types of chemical analysis.

When proceeding with a qualitative analysis, the analyst generally wants to identify the analyte, that is, to find out what component is present in the sample being analyzed or what chemical components the sample is composed of. In some

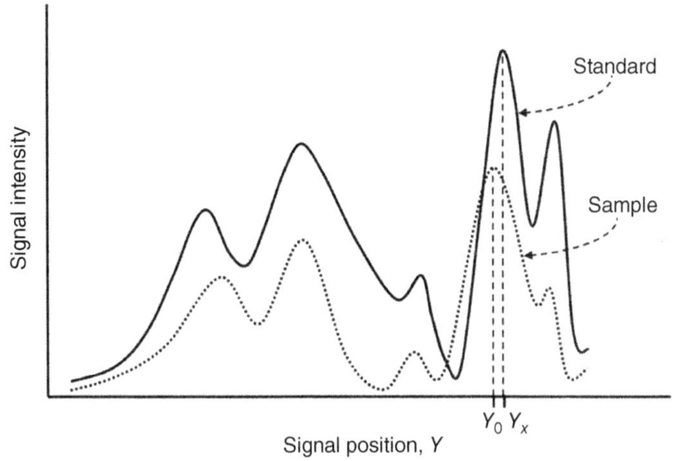

Figure 1.7 Measurement images of the sample and standard used in qualitative analysis: the analyte is identified from the position of the Y_0 and Y_x signals obtained for the unknown analyte in the sample and the known analyte in the standard, respectively.

cases, he asks whether a specific component or several components are present in the sample. In other situations, he may also be interested in questions such as: what is the kind of the whole sample, whether the sample under study is similar to another sample, or whether the sample under study belongs to a particular group of samples.

The relationship between measurement signal and analyte type can be illustrated by the measurement images shown in Figure 1.7. They are created by subjecting a multicomponent sample and a standard of similar chemical composition to the sample to measurements under identical conditions with a specific instrument in such a way that a change in signal intensity is recorded as the specific quantity characteristic of the measurement method used (e.g. wavelength, time, etc.) changes. These signals, when significantly larger than the measurement noise, correspond to the presence of unknown components in the sample and at least one known component, b_x, present in the standard (solid line). **The type of component is indirectly indicated by the signal position** on the abscissa axis, that is, the value of the specific quantity corresponding to the maximum intensity of its signal.

Empirical calibration in qualitative analysis usually involves comparing the signal position value Y_0 obtained for the sample with a similar value Y_x obtained for the standard (see Figure 1.4).[1] Since the value of Y_x obtained for the standard corresponds to the known component b_x, it can be said that both values form a model function, $Y = G(b)$, at a point with coordinates $[Y_x, b_x]$. Because of the similarity of the values of Y_x and Y_0, the real function, $Y = F(b)$, can also be considered as well approximated by the model function at this point. In such a situation, the value of Y_0 is assigned a component b_x using the evaluation function: $b_x = G^{-1}(Y_0)$, and it is

1 This calibration approach refers to a specific comparative calibration method, most commonly used in quantitative analysis.

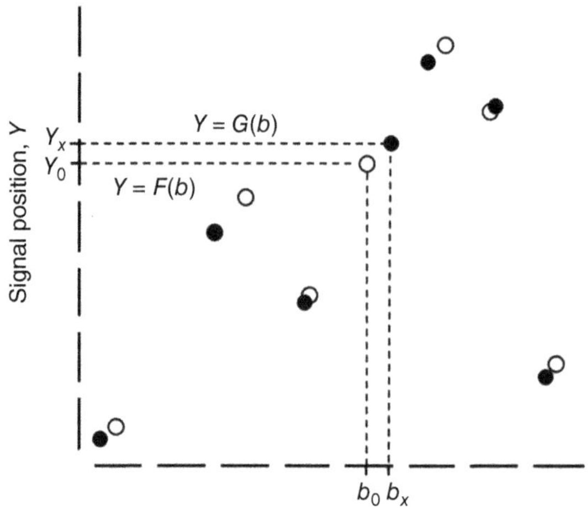

Figure 1.8 Analytical calibration in qualitative analysis: analyte b_0 in a sample (empty points) is identified as analyte b_x in a standard (full points) on the basis of mutual signal positions, Y_0 and Y_x.

claimed that the component b_0 present in the sample is probably the component b_x present in the model. This procedure is illustrated in Figure 1.8.

Theoretical calibration involves the mathematical formulation of a model function, $Y = G(b)$. It should best approximate the real function, $Y = F(b)$, at least in one point with coordinates $[Y_x, b_x]$. When the signal obtained for the sample, Y_0, is substituted into the formula of the inverse function of the formulated model function, the signal position value, b_x, is obtained, indicating the true type, b_0, of the analyte sought.

Mathematically, the real function and the model function are discrete functions in qualitative analysis, as shown in Figure 1.5. When the real function is mapped sufficiently accurately with the model function, any signal of a particular position obtained for known components of the standard is theoretical evidence for the presence or absence of those components in the sample. Some components may be identified by two or more signals,[2] which are analytical signals for them. In many cases, a model function can be used to identify multiple components of a sample, that is, to perform a multicomponent analysis.

When applying the chosen analytical method and recording the signal under appropriately established optimum conditions, the analyst should have at his disposal at least one signal corresponding to a specific type of analyte. It is most advantageous if he has a measurement image of the type shown in Figure 1.7, which covers a wide range of magnitude characterizing the type of constituent. Such an image, obtained under specific experimental conditions, reflects the chemical composition of the entire sample and is characteristic of it. The presence of the desired analyte in the sample can then be indicated not only by the corresponding positions of the analytical signals, but additionally by other parameters, such as the number of these signals, their absolute and relative heights, and even the shape of the entire signal recorded in a given measurement range (some measurement

2 Nevertheless, the relationships $Y = F(b)$ and $Y = G(b)$ can be called functions because, due to the natural random errors of the obtained measured and calculated values, different signals cannot represent perfectly the same measure of a particular sample component.

methods also offer their own specific identification parameters). All these parameters can act as auxiliary identification parameters supporting the basic parameter in the calibration process. A common feature of qualitative analysis is therefore its multi-parametric nature.

Since sample and standard identification parameters are naturally correlated with each other and are highly characteristic of a particular analytical method, the effectiveness of using auxiliary parameters to increase the accuracy of the analytical result with them is limited. Therefore, if there is a need to be more certain about the presence or absence of a sample component (similarity or dissimilarity of samples), then a qualitative analysis of a given sample can be performed by another analytical method (the so-called reference method), preferably as different as possible from the previous one due to the sample processing and measurement method used. In this way, a new range of identification parameter values can be obtained that are not correlated with the previous ones.

A specific aspect of qualitative analysis is the very concept of accuracy of the analytical result. It is clear that if the analyte sought is in the sample or the sample tested is the sample sought (+), and the analytical result confirms this (+), then it is consistent with the actual result, i.e. it is "accurate" ($b_x = b_0$). Similarly, if in such a situation the result obtained is negative (−), it is in fact a false negative, i.e. inaccurate ($b_x \neq b_0$). However, it is, of course, also possible that the analyte sought is not present in the sample or the sample tested is the one sought (+). Then a positive result (+) means that it is in fact inaccurate (false positive, $b_x \neq b_0$), and a negative result (−) means that it is accurate ($b_x = b_0$), although negative. To clearly illustrate these eventualities, they are shown in Table 1.1.

As shown in Figure 1.5, in some cases the degree of similarity of the model function to the real function may raise doubts as to the presence (or absence) of a specific component in the sample, and even this presence (or absence) may possibly rule out. This uncertainty is obviously due to the occurrence of random and systematic uncontrolled effects. Consequently, the accuracy and inaccuracy of an analytical result in a qualitative analysis are always determined with a certain probability, and never with certainty, however certain this certainty may seem to be. By the same token, it cannot be said that a sample and a standard or the two samples

Table 1.1 Estimation of the accuracy and inaccuracy of an analytical result obtained in qualitative analysis.

Analytical result		Evaluation of the obtained result	Accuracy of the result
Obtained, b_x	**Real, b_0**		
+	+	Truly positive	Accurate
+	−	False positive	Inaccurate
−	+	False negative	Inaccurate
−	−	Truly negative	Accurate

being compared are "the same" but at most "the same," and that they are "certainly different" but "possibly different" with a certain probability.

In qualitative analysis, uncontrolled effects usually manifest themselves as shifts in the position of the analytical signal due to random or systematic changes in instrumental parameters. A frequently occurring problem is also the additive interference effect, consisting of overlapping of signals coming from the analyte and the interferent. The analyte signal may also, under the influence of various factors, so reduce its intensity so that the presence of the analyte may go unnoticed. All these effects will be shown by experimental examples in Chapter 3.

The analytical result in qualitative analysis is nonmeasurable (qualitative) and therefore the assessment of its accuracy is more subjective than in quantitative analysis. This assessment comes down to determining whether and to what extent the differences in measurement information provided by the sample and the standard are statistically significant, i.e. are caused by systematic factors, or are insignificant in comparison with the differences resulting from random errors. However, the multiparameter nature of qualitative analysis means that the use of simple statistical tools (also applicable in quantitative analysis) may be unreliable.

Various chemometric methods that are commercially available in the form of computational packages come to the rescue. These methods are used both to match the model function to the real function as accurately as possible and to assess the accuracy of the result of the identification analysis, which is, therefore, more objective than the analyst's intuitive assessment. However, it should be remembered that it is only up to the analyst to choose the chemometric method and its detailed parameters and criteria on the basis of which it works, and all these factors affect the final results of the calculations. It is not uncommon for two chemometric methods to interpret the same data to produce significantly different results. In such cases, the choice between the results obtained must again be subjective.

Thus, statistical and chemometric approaches to assessing the accuracy of results in qualitative analysis should always be regarded as only auxiliary tools, supporting the knowledge, experience, and research intuition of the analyst.

1.7 Calibration in Quantitative Analysis

The purpose of quantitative analysis is **to establish the amount (content, concentration) of an analyte** (one or more) in the sample being analyzed, that is, the determination of the analyte. Quantitative analysis is therefore formally related to qualitative analysis in the sense that knowledge of the amount of an analyte in a sample is information that naturally supplements the analyst's knowledge of that constituent. On the other hand, the determination of an analyte in a sample in an amount greater than the limit of quantification is at the same time evidence of its presence in that sample. Most often, however, quantitative analysis is undertaken without prior identification of the analyte, predetermining the type of analyte under study and usually knowing the location of the corresponding analytical signal.

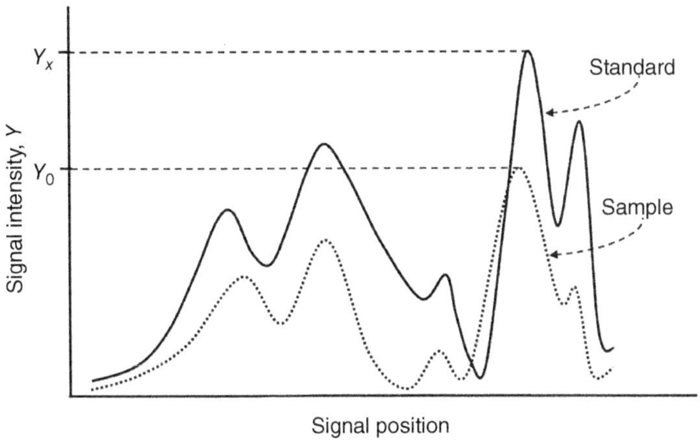

Figure 1.9 Measurement images of the sample and standard used in quantitative analysis: the analyte is determined from the intensities of the Y_0 and Y_x signals obtained for the unknown amount of analyte in the sample and the known amount of analyte in the standard, respectively.

In quantitative analysis, the primary measure of analyte quantity is the intensity[3] **of the analytical signal**, as can be seen in Figure 1.9. When proceeding to the determination of a specific analyte in a sample, among the possible signals generated by the analyte with a given measuring instrument, the signal with the position at which it shows the highest intensity is selected. If the calibration is empirical, the intensity of the signal measured for a standard (one or more) with a known amount of the analyte is measured under the same conditions (possibly changed only within random error).[4]

In quantitative analysis, the real function is a continuous function because the signal intensity is a continuous quantity. Over a range of analyte amounts, it can take a linear or nonlinear form. Furthermore, in some calibration methods, it is transformed to a decreasing or increasing and decreasing function in different analyte concentration ranges. In such a situation, one cannot count on the values of the intensities of the signals measured for the sample and the single standard being equal (just as the values of the positions of the sample and standard signals are equal in qualitative analysis), i.e. the model function developed using the single standard accurately approximates the true function. If the signal intensities of the sample and standard are significantly different (as in Figure 1.9), the determination of the analyte, although theoretically possible, is risky from the point of view of the accuracy of the analytical result obtained.

Empirical calibration in quantitative analysis, therefore, consists of constructing a model function, $F = G(c)$ in mathematical form from measurements usually made for two or more chemical standards containing the analyte in quantities bounding

3 In some analyses, particularly those detected by separation methods, the area after the signal (peak) is alternatively taken as a measure of analyte quantity.

4 As before, this refers to a specific comparative calibration method, most commonly used in quantitative analysis.

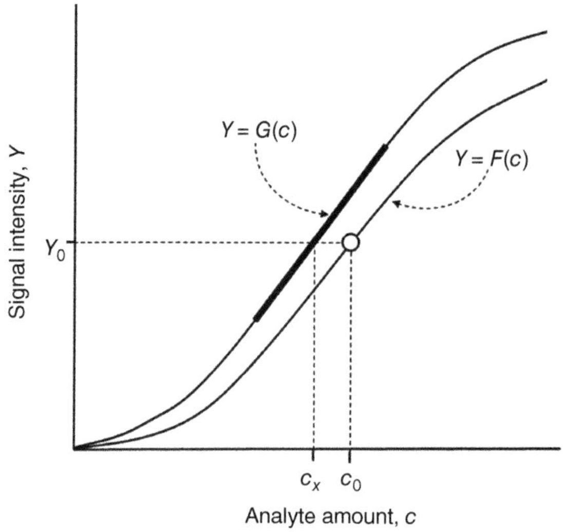

Figure 1.10 Analytical calibration in qualitative analysis: analyte in amount c_0 in the sample (empty point) is determined in amount c_x based on the signal intensity, Y_0, measured for the sample and on the empirical or theoretical model function (bolded line) formulated using chemical or mathematical standards, respectively.

the required range. Usually, a range is chosen in which the model function is most likely to be an exact fit to the linear part of the real function. The amount of analyte in the sample, c_x, is determined from the signal intensity value Y_0 measured for the sample and using the evaluation function: $c_x = G^{-1}(Y_0)$. This procedure is shown in Figure 1.7. As can be seen, the analytical result, c_x, is as close to the true result, c_0, as the model function is to the true function at the point defined by the signal Y_0.

In theoretical calibration, the model $Y = G(c)$ shown in Figure 1.10 is formulated using one or more mathematical formulas. The transformation of the measured signal for the sample, Y_0, to the analytical result, c_x, follows in analogous way as in the empirical calibration.

In quantitative analysis, analytical uncontrolled effects are even more of a problem than in qualitative analysis because the analytical signal is more prone to change its intensity than its position under the influence of various factors. Therefore, the occurrence of any type of effect must be expected during analyte determination. The effect that is particularly problematic in quantitative analysis, but of little significance in qualitative analysis, is the so-called multiplicative interference effect, manifesting itself as a linear or nonlinear change in the intensity of the analytical signal, the greater the concentration of interferents in the sample. Ways of eliminating and compensating for this effect will often be discussed in later chapters.

A separate problem is that uncontrolled effects, regardless of the factors that cause them, are usually manifested by a decrease rather than an increase in the intensity of the analytical signal. Sample processing involves much more loss than gain of analyte (unless the analyst deliberately increases the concentration of analyte in the sample, but then this is a controlled action). The detection effect is generally manifested by a gradual reduction in signal intensity as the analyte concentration in the sample increases. Interferents causing a multiplicative interference effect also tend to cause a gradual reduction in signal intensity. As a consequence, the measurement sensitivity of the analyte is reduced, which is associated with the possibility of larger random errors in quantitative analysis.

Calibration in quantitative analysis is also much more difficult than calibration in qualitative analysis because it requires the formulation of a model function that approximates a continuous real function of unknown shape and position. In addition, it should approximate it not in single points, but in a certain range of analyte amounts, since the amount of analyte in the sample is unknown or can be known only to some approximation. In this situation, the question is justified: how can the **accuracy** of an analytical result obtained by a given analytical method be evaluated in quantitative analysis and how reliable is this evaluation? There are two ways to recommend and use in analytical practice: the application of a reference method or the use of reference material (preferably certified).

The analytical **reference method** should be a well-developed and verified (validated) method to document its high analytical quality. In particular, it should be characterized by a high accuracy of the determination of a given analyte in a given sample. As in qualitative analysis, the method should also be based on different physicochemical principles than the method undergoing accuracy testing. In such a situation, comparison of the analytical result obtained by the reference method with the result obtained for the same analyte in the same sample under analogous experimental conditions by the test method may be a good way to assess the accuracy of the latter. The problem may, of course, be to find a reference method of adequate quality and suitably adapted to the test method.

Another possibility to assess the accuracy of an analytical result is to use a certified reference material [9]. In chemical analysis, there is a substance, usually multicomponent, sufficiently homogeneous and stable, whose chemical composition is determined (at least in part) by interlaboratory analyses and is confirmed by a certificate. Accuracy is tested by selecting a reference material so that it is as similar as possible to the samples analyzed by the method in terms of properties and chemical composition. A sample of the reference material is then analyzed by the method under specified experimental conditions and the result obtained is compared with the certified amount of that analyte. This difference may indicate the accuracy of the method being tested. The problem, of course, is the availability of certified material sufficiently similar to the sample assayed.

In recent years there has been a tendency to use simple chemical standards added to a sample to determine the so-called **analyte recovery** to evaluate the accuracy of analytical results. This approach, although much simpler and less demanding than the methods described above, is nevertheless fallible and can only be used under strictly defined conditions. This will be demonstrated in a later chapter of this book devoted entirely to this subject.

In quantitative analysis, particularly important parameters that testify to the quality of an analytical method (so-called validation parameters) are, in addition to accuracy, precision, and uncertainty of the analytical result.

Precision is assessed by the random scatter of the analytical result, i.e. the values of that result determined several times by a given analytical method under the same or only slightly changed experimental conditions. This is expressed in the form of **repeatability** and **reproducibility.** The former is the precision established under conditions in which the analytical procedure is performed according to the specified analytical method by the same analyst, using the same equipment, and in the shortest possible time (at most one day). The latter is the precision established under conditions in which one of the above factors (analyst, equipment, day) has been deliberately changed.

It should be noted that a precision determined – as is quite often done – solely on the basis of the scatter of only the measurement results determined with the sample cannot be regarded as a measure of the precision of the analytical result, much less as a measure of the quality of the analytical method. Such a way of proceeding ignores the contribution that the calibration procedure, that is, the preparation and measurement of standards and the transformation of the measurement signal to the analytical result, makes to the general precision value.

Uncertainty is defined as the interval within which the value of an analytical result can be located with satisfactory probability [10]. The overall value of uncertainty consists of the component uncertainties with which the various steps and actions that make up the analytical procedure are performed. Some of these component values can be calculated as experimental standard deviations from the statistical distribution of the results of a series of measurements. Other values, which may also be characterized by standard deviations, are evaluated from assumed probability distributions based on the analyst's experience or the information available to the analyst. It is obvious that all steps of the calibration procedure should be included in the evaluation of the overall uncertainty of the analytical result.

1.8 General Rules for Correct Calibration

Based on the above considerations, one may be tempted to define some general rules of conduct that will allow the calibration process to be carried out primarily so that the analytical result is subject to the lowest possible random and systematic errors. These rules should also be as consistent as possible with the rules of green analytical chemistry [11]. This means that when determining the correct calibration procedure, the analyst should take into account the minimization of factors that pose a threat to our environment.

The first very important factor is the **selection of an appropriate analytical method**. If the analyst has several qualitatively equal methods at his disposal leading to the identification or determination of an analyte in a given sample or is starting to develop a new analytical method, he should take care that the chosen or developed method is **as simple as possible** chemically and instrumentally. Several factors support this. A simple analytical method contains relatively few sources of uncontrolled effects, which promotes their effective elimination. It is also important that the simpler the method, the easier and more accurately the standard's handling can be made to resemble that of the sample, and, as a result, uncontrolled effects can be effectively compensated for. The simpler the analytical method, the greater the chance of using fewer and less reagents and producing little waste, i.e. of following the basic rules of green analytical chemistry.

Irrespective of the analytical method chosen, it is essential that all the steps required by the analytical method are carried out with the **greatest possible**

correctness and care – i.e. in accordance with all the rules of the analytical art. The human factor plays an important role in both empirical calibration and theoretical calibration. The sources of random and systematic errors associated with the construction of the empirical model function are all incorrectly or carelessly performed laboratory operations, and in theoretical calibration – incorrect theoretical assumptions, erroneous or inaccurate calculations, and all approximations with which the mathematical description of phenomena and processes is made. Note that personal errors can dominate the error of an analytical result regardless of other steps taken to eliminate or compensate for uncontrolled effects.

During the implementation of an analytical method, a very important issue is the skillful, balanced use of both ways: **elimination and compensation of uncontrolled effects**. In particular, this applies to empirical calibration with chemical standards. This is because it is supported by theoretical and practical considerations outlined above. Both ways should be used in such a way that they complement each other and are as effective as possible. It is best to be guided by specific, proven, and customary principles as well as by one's own analytical experience.

Thus, for example,

- during the analysis measurements for sample and standard should be made under identical experimental conditions set at optimum levels and with the same measuring instrument,
- do not use too many reagents to eliminate uncontrolled (preparative, interference) effects (especially those endangering health and life), but rather try to compensate for these effects,
- try as far as possible to make the chemical composition of the standards similar to that of the sample (e.g. by means of reference materials) and treat the standards in the same way as the sample,
- pay close attention to the compatibility of the chemical form of the analyte in the sample and in the standard and in case of doubt take appropriate instrumental or chemical steps to ensure this compatibility.

The same guidelines also apply to theoretical calibration, though of course taking into account the appropriate chemical (relative to the sample) and mathematical (relative to the standard) ways.

It is important to remember that the calibration process is inherently linked to the overall analytical process. It is therefore important that all calibration activities are performed in a **correct and careful manner**, just like other non-calibration operations. This seems to be a trivial and obvious statement, but reality often does not bear this out. As shown in scientific publications, analysts directing their efforts to create new and ingenious analytical procedures often neglect the calibration aspects. It is not uncommon, therefore, that the lack of a proposal for a clearly defined, suitably adapted calibration procedure within the analytical procedure developed is, in effect, the cause of unsatisfactory results in terms of their accuracy and precision.

Most random and systematic errors are made at the stage of sample and standard preparation for measurements. Minimization of these errors is facilitated by **mechanization and automation** of this stage of the analytical procedure. One way of doing this in quantitative analysis is to process the sample and standard in flow mode, examples of which will be presented in later chapters of this book. The automation of analysis not only allows, as will be shown, for an improvement in the quality parameters of the analytical method but also provides greater operational safety and supports lower reagent consumption and waste reduction. It is therefore very important to implement calibration procedures into analytical methods performed in such a mode.

Finally, one more matter of importance in the context of the subject of this book should be mentioned: **the proper choice of the calibration method**. This is, in fact, one of the most important factors determining the correctness of the whole analytical procedure, since it affects not only the precision and uncertainty of the analytical result but also its accuracy. However, in order for this choice to be appropriate, adequate to the different circumstances in which the analysis is carried out, it is necessary to have a good knowledge of the different calibration methods, sometimes very rare, but applicable in qualitative and quantitative analysis. The acquisition of this knowledge is precisely the main purpose of this book.

References

1 (2013). *Cambridge Advanced Learner's Dictionary*, 4e. Cambridge: Cambridge University Press.

2 https://www.vocabulary.com/dictionary/calibration (accessed 12 September 2022).

3 Skoog, D.A., Holler, F.J., and Crouch, S.R. (2007). *Principles of Instrumental Analysis*, 6e. Belmont, CA: Thomson Brooks-Cole.

4 JCGM 200 (2012), *International Vocabulary of Metrology – Basic and General Concepts and Associated Terms (VIM)*, 3rd edition. JCGM 200.

5 Guilbault, G.G. and Helm, M. (1989). Nomenclature for automated and mechanised analysis. *Pure and Applied Chemistry* 61 (9): 1657–1664.

6 Danzer, K. and Curie, L.A. (1998). Guidelines for calibration in analytical chemistry. Part 1. Fundamentals and single component calibration. *Pure and Applied Chemistry* 70 (4): 993–114.

7 Kościelniak, P. (2022). Unified principles of univariate analytical calibration. *TRAC Trends in Analytical Chemistry* 149: 116547.

8 Liland, K.H., Almøy, T., and Mevik, B.H. (2010). Optimal choice of baseline correction for multivariate calibration of spectra. *Applied Spectroscopy* 64 (9): 1007–1016.

9 ISO Guide 35:2006 2006. Reference Materials. General and Statistical Principles for Certification.

10 ISO/IEC Guide 98:1993 1993. Guide to the Expression of Uncertainty in Measurement.

11 Anastas, P.T. (1999). Green chemistry and the role of analytical methodology development. *Critical Reviews in Analytical Chemistry* 29 (3): 167–175.

2

"Calibration-Free" Analysis

2.1 Novel Approach

Chapter 1 posed the question: can chemical analysis be performed without analytical calibration? In general, belief based on scientific sources, the answer to these questions is positive. In metrology, the term **primary method** is known, defined as "... a method ... whose operation is completely described and understood, for which a complete uncertainty statement can be written down in terms of SI units, and whose results are, therefore, **accepted without reference to a standard of the quantity being measured**." [1]. Such methods from the field of chemical analysis include gravimetry, titrimetry, coulometry, and isotopic dilution mass spectrometry (IDMS).

The approach represented by IUPAC is similar, although in this case, it uses the concepts of calibration-free methods or absolute methods [2]. In the work by Hulanicki [3] one can read: "**When the functional relationship can be completely described on the basis of physical constants and universal quantities, the method can be considered to be absolute**", and further: "Gravimetric, volumetric and coulometric methods ... can be regarded as absolute methods, under the condition that experimental conditions are chosen such that their efficiency is theoretically predictable (preferably 100% efficiency)."

What is characteristic of the definitions mentioned above is their idealistic approach to analytical methods. After all, it is difficult to find in analytical chemistry a method whose "operation is completely described and understood" or "the functional relationship can be completely described on the basis of physical constants and universal quantities." Nor can it be required that the method being characterized by 100% efficiency in practice. These rigorous conditions in themselves call into question the actual existence of absolute methods, including even those that are commonly regarded as such. It is also difficult to agree that methods considered as absolute methods or primary methods need not require a "standard of the quantity being measured."

Both IUPAC papers also note the term standard-free [2] or standardless [3] methods. These names should only be applied to methods in which the analytical signal is relatively robust to changes in instrumental conditions and once calibrated under recommended conditions can be referred to many subsequent analyses (this type

Calibration in Analytical Science: Methods and Procedures, First Edition. Paweł Kościelniak.

Table 2.1 Empirical and theoretical analytical calibration based on different standard and models.

Standard type	Model type	Calibration type
Chemical, Biological, Physical	Empirical	Empirical
Mathematical	Mathematical, fixed and flexible	Theoretical

of procedure is characteristic, e.g. for X-ray fluorescence spectrometry). It should be emphasized, however, that in the analytical literature the terms calibration-free, absolute methods, standard-free, and standardless methods are generally treated as synonyms.

The position represented in this work is that **every analytical method requires calibration** [4]. The calibration nature of a method is demonstrated by the following features:

- the type or amount of analyte cannot be determined accurately and directly from the analytical signal, i.e. from a real function with a fixed form that does not rely on uncontrolled effects,
- the method requires the use of a standard (not necessarily in chemical form) needed for the formulation of an empirical or theoretical model function that approximates the real function.

Analysis of the above features will provide a basis for verifying the calibration nature of some analytical methods where such nature is overlooked or questioned. In particular, this applies to gravimetry and coulometry. Titrimetry and IDMS are described later in this book not as analytical methods but even calibration methods, and their calibration nature is explained in Chapter 6. Special attention will be given to theoretical calibration (considered absolute [3]) and its application in qualitative and quantitative analysis.

Based on these considerations, a general division of analytical calibration can be proposed taking into account different types of calibration standards and model functions. This classification is presented in Table 2.1.

2.2 Empirical Calibration

The most direct method of substance detection and identification is **sensory analysis**, i.e. the determination of the type or amount of analyte in a sample using the senses. This takes advantage of the natural human ability to detect, distinguish and evaluate the intensity of sensory impressions, in particular smell, color, and taste. It is possible, for example, to visually detect iron compounds in a clod of soil, to taste the presence of sodium chloride in a food sample, or to determine the type of psychoactive compound using the sense of smell.

Because of the direct, as it seems, possibility of determining the type of analyte on the basis of the registered analytical signal (color, taste, and smell), can we not consider sensory analysis as an example of a calibration-free analysis? No, because every human being has a well-developed sensory memory, which means that once a sensory impression (color, taste, and smell) is assigned to a specific, appropriate receptor. Sense memory, therefore, acts as a kind of **biological standard**, which we acquire in life through empirical information, and the impression we get with our senses is compared with the memorized relationship between the analytic signal and the type of analyte can, in this case, be called an **empirical sensory model**.

Sensory analysis makes it possible to identify both a component of a sample and the entire sample in terms of both type and commonly understood quality. Comparison of two samples with the unaided eye or with the aid of a microscope inherently involves their general characteristics such as color, shape, or morphology. Very often, this external observation of the samples is already sufficient to determine their similarity or dissimilarity (with a certain probability). Sensory ability is also used to compare whether the analyte under study (sample component, sample) produces an impression of greater or lesser intensity than another analyte (reference). In this way, the analysis performed is semiquantitative. For a more accurate, but still only estimated determination of the analyte content in the sample, various types of graphical or point scales or even chemical compounds providing sensory impressions of different, "controlled" intensities are used. These should be regarded as additional standards supporting the biological model in the correct evaluation of human sensations.

Sensory memory plays a larger role in analytical practice than we are prepared to admit. After all, in many cases, the analyst makes a decision about the presence or absence of an analyte in a sample without using a chemical standard, but only on the basis of a visual evaluation of the measurement result provided by the instrument. He relies then again essentially on sensory (visual) memory, which in such cases we are inclined to call rather the analyst's intuition or experience.

As in any other chemical analysis, errors of both random and systematic nature are made in sensory analysis, resulting from various types of uncontrolled effects. Errors result, for example, from disturbances of perceived sensory impressions by components other than the analyte, accidentally present in the sample or in its environment. The color, smell, or taste characteristic of the analyte may then be altered by these substances. If these changes are caused by substances introduced into the sample during the analysis, we are dealing with a typical interference effect, whereas if they are caused by components in the sample environment, we can rather speak of a preparative effect. Another source of preparative effect are factors determining the environmental conditions, such as temperature, humidity, color, or illumination.

The results of sensory analysis are particularly vulnerable to the human factor. Sensory impressions perceived by humans are naturally subjective and the accuracy of the analysis depends to a large extent on both the individual minimum sensory sensitivity of the analyst (the so-called sensory minimum) and his ability to recognize and remember impressions (the so-called sensory memory). Also of great importance is the phenomenon of so-called sensory fatigue, i.e. a gradual decline in the ability of this perception over time. The memory and sensory model may even be completely lost or indicate a completely erroneous relationship between our sensory impression and the type or amount of analyte. As the pandemic period beginning in 2020 teaches us, these types of cases can be caused by COVID-19 disease.

Calibration-free empirical methods in quantitative analysis commonly include **gravimetric method** in particular. In this case, the analyst uses a balance, which is the measuring instrument. The calibration standard in gravimetry is **a physical standard** in the form of an analytical weight of a well-defined and known mass. The relationship between the response of the balance and the mass of the weight is an **empirical model function**, shown in Figure 2.1 as a graph of a linear function. Based on this model and the measurement result obtained for the sample, the analytical result, i.e. the mass of the sample, is determined.

The uniqueness of gravimetry, revealed by the specific perception of this method compared to other analytical methods, is due to several reasons. One of them is that the response of the measuring instrument (measurement signal) is in this case measured in mass units, i.e. in the same units in which the analytical result is obtained. This gives the misleading impression that the result is determined directly from the weight reading, without the use of any standard or model function, i.e. without analytical calibration. In addition, unlike in other analytical methods, the result is obtained immediately, without the need to perform calculations or use any graph. Moreover, the analytical result is generally the same as the weight reading, because the weight automatically matches its mass to the mass of the sample being analyzed (as schematically indicated in Figure 2.1).

The gravimetric method is one of the most precise and accurate analytical methods. This is due, on the one hand, to the naturally high accuracy with which the

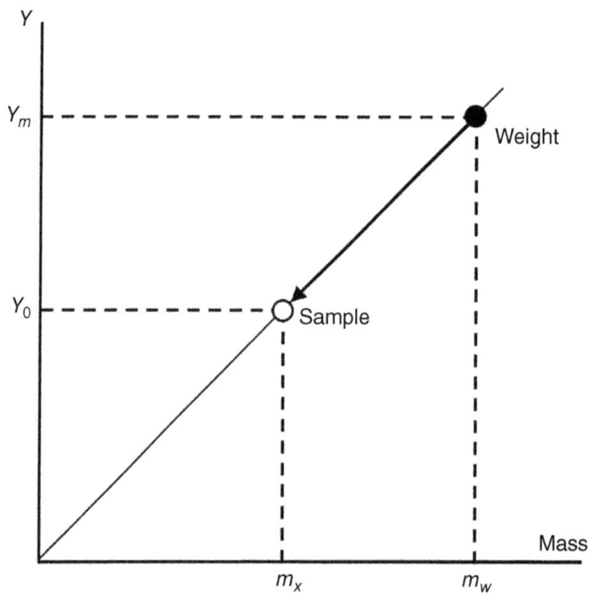

Figure 2.1 Analytical calibration in gravimetry: weight of sample (analytical result), m_x, is obtained based on weight of weight (physical standard), m_w, and on the corresponding responses of a balance (measurement instrument), Y_0 and Y_m. Source: Adapted from Kościelniak [4].

mass of the analytical weight can be determined, and, on the other, to the simple, very well-developed, and reliable construction of the analytical balance. Nevertheless, the indications of the balance are always subject to some random fluctuations, which affects the precision of the results. Furthermore, both elements – the mass of the weight and the construction of the balance – are subject to changes over time, which then become a source of uncontrolled effects and, consequently, a source of analytical errors. Gravimetric results can also be systematically erroneous as a result of temperature and environmental conditions. It is a well-known rule that the temperature of the substance to be weighed should be brought to ambient temperature (to compensate for its effect on the analytical signal), and that disturbance of the position of the analytical balance may affect the weight reading of the substance to be weighed.

The methods of quantitative analysis that do not require calibration also sometimes include **potentiometry and absorption spectrophotometry**. This is due to the relatively accurate mathematical description of the relationships between the measurement signal and the concentration of the analyte, based on the nature of the phenomena occurring in both methods. In potentiometry, this relationship is described by the Nernst equation:

$$E = E^0 + \frac{R \cdot T}{n \cdot F} \cdot \ln[\mathrm{Me}^{n+}] \tag{2.1}$$

where E^0 is the normal potential of the electrode, R is the gas constant, T is the absolute temperature of the solution, F is Faraday's constant, n is the number of electrons involved in the reaction, and $[\mathrm{Me}^{n+}]$ is the molar concentration of the ion being determined. In absorption spectrophotometry Lambert–Beer's law applies:

$$A = k \cdot l \cdot c \tag{2.2}$$

where A is the absorbance, l is the length of the environment in which the absorption of radiation takes place, and k is the so-called absorption coefficient.

Equations (2.1) and (2.2), however, cannot be treated as real functions, but only **empirical model functions** because the described relations are not completely strict and depend on the current experimental conditions. Although the values of electrode normal potentials can be found in tables, they are determined for pure systems. In practice, electrode reactions always proceed in more complex solutions, often in the presence of ions causing interference effects. In Eq. (2.2), the value of the absorption coefficient depends on the instrumental conditions (e.g. on the monochromaticity of the radiation). Deviations from Lambert–Beer's law also occur when reactions involving absorbing ions occur in the solution under study (e.g. polymerization, condensation, or complex formation reactions). As a result, in both cases calibration is performed using **chemical standards**, and Eqs. (2.1) by (2.2) provide only some theoretical assistance in constructing calibration graphs.

From a calibration point of view, the law of absorbance additivity in absorption spectrophotometry is very interesting. It applies to multicomponent solutions and says that the absorbance of a sample is equal to the sum of absorbances of its individual components. This allows for unique consideration of additive interference effects by multicomponent analysis. However, for the reasons mentioned above, this type of analysis is also performed based on empirical calibration with chemical standards.

Thus, none of the analytical methods presented above can be considered as a calibration-free method, since each of them requires the use of an empirical calibration standard. A standard – biological, physical, or chemical – is necessary to construct a model function and to compensate for uncontrolled effects with it to obtain an analytical result of sufficiently high accuracy.

2.3 Theoretical Calibration

Theoretical calibration is performed, as already mentioned, on the basis of a theoretical description of phenomena and processes specific to the measurement used for sample analysis. With regard to their structure, all theoretical models used in chemical analysis can be broadly divided into two types:

- fixed models, resulting directly from the laws and rules governing a particular method, not subject to modification by the analyst,
- flexible models, formulated by the analyst and, consequently, possible to modify, and even giving the possibility to describe phenomena and processes characteristic of a given method in different mathematical forms.

In both cases, the formulated models can be used directly to determine the type or amount of analyte without using an empirical standard, but only on the basis of the measurement signal obtained for the analyte present in the analyzed sample. The models are formulated using mathematical standards that play the same role as the empirical standards – they serve to make the models similar to the real function by compensating for uncontrolled effects.

It should be noted, however, that in many cases theoretical modeling fails the test in the sense that it provides results that are very far from the real ones. This is particularly true when analyzing samples with complex compositions causing strong interference effects. In many such situations, it is necessary to use data describing the actual experimental conditions of the analysis as empirical standards. In extremely difficult cases, theoretical models are replaced by empirical model functions formulated using chemical standards.

2.3.1 Fixed Models

Several chemical analysis methods can be used to identify substances based on theoretical or theoretical–empirical calibration. These include, for example, **infrared (IR) spectrometry**. Theoretical studies, supported by experimental analysis of infrared spectra, have shown that certain functional groups in the molecules of organic compounds absorb infrared radiation in a characteristic range of frequencies corresponding to the stretching and deforming vibrations of these

Table 2.2 Functional groups, characteristic frequencies of their stretching vibrations, and organic compounds identified.

Bonding	Position (cm^{-1})	Compound
O—H	3500–3550	Carboxylic acids
N=H	3200–3600	Amines
C—H	3300	Alkynes
C—H	3010–3100	Aromatic compounds, olefines
C—H	2970–2850	Aliphatic compounds
C≡C	2100–2270	Alkines
C=O	1650–1780	Aldehydes, ketones, esters, carboxylic acids
C=C	1600–1680	Alkenes
C=C	1450–1610	Aromatic compounds

molecules. These frequencies are independent of the structure of the rest of the molecule, so they act as calibration theoretical (mathematical) standards. From these data, a determination of the structure and type of analyte sought can be formulated. The arrangement of bands on the infrared spectrum in the frequency range below $1500\,cm^{-1}$ is so characteristic of a given organic molecule that this range is even called the "fingerprint region." Table 2.2 collects the frequencies of stretching vibrations characteristic of the functional groups of various organic compounds.

A certain problem with this type of analysis is that functional groups of different organic compounds can absorb infrared radiation in partially overlapping ranges (as seen in Table 2.2). For samples of complex composition, identification of the analyte from characteristic frequencies is therefore quite difficult and often requires support in the form of empirical calibration performed with chemical standards.

In qualitative analysis, it is possible to distinguish such methods in which the theoretical models are based on strict laws and rules, but, thanks to additional, auxiliary calculations, leave the analyst some room for his own interpretation to varying degrees. These methods may include the X-ray diffraction (XRD) method, the nuclear magnetic resonance (NMR) method, and the mass spectrometry (MS) method.

At the basis of the **XRD** method lies Bragg's law. It relates the wavelength λ of the radiation to the angle of reflection θ, which is the angle between the incident or reflected radiation and the reflecting plane in the crystal:

$$n\lambda = 2d\ \sin\theta \tag{2.3}$$

where n is the order of the spectrum and d is the distance between the crystal planes. Knowing the values of the angles θ in the diffraction image of the radiation passing through the studied crystal at different angles, a three-dimensional map of the electron density in the elementary cell of the crystal is determined based on Bragg's equation. Further mathematical analysis of this map makes it possible to determine the position and distance of molecules to each other in the crystal lattice, which characterizes the chemical compound under study.

An auxiliary mathematical relationship used in X-ray analysis for analyte identification is Moseley's law. It describes the relationship between the wavelength λ of the characteristic X-ray radiation of an element and its atomic number Z:

$$\sqrt{\frac{1}{\lambda}} = k(Z - a) \tag{2.4}$$

where a and k are experimental constants. As can be seen, Moseley's law does not allow a strict assignment of the analyte type (Z) to the measurement quantity (λ), but it does indicate a certain regularity occurring between these quantities. Thus, although it does not lead directly to the identification of the analyte, it facilitates this identification to a large extent and may contribute to increasing the accuracy of the analytical result.

Both Eqs. (2.3) and (2.4), are the mathematical standards that, considered separately from the sample, attempt to describe the properties of the sample and the processes occurring with it during analysis.

In XRD analysis, the prerequisite for correct identification of a chemical compound is the ability to obtain a pure monocrystal of that compound. The presence of other crystalline phases in the sample greatly complicates correct mathematical interpretation of measurement results. Similar difficulties are also caused by any defects in the crystal lattice of the analyte. Finally, some identification problems may be caused by the phenomenon of polymorphism, i.e. the occurrence of different crystallographic variations of the same chemical substance.

In the **NMR** method, the intramolecular magnetic field around an atom in a molecule changes its resonance frequency in the presence of an external, constant magnetic field, thus giving information about the details of the electron structure of the molecule and its individual functional groups. One of the basic identification parameters is the so-called chemical shift, which is attributed to the screening effect of the electron cloud in the vicinity of the nucleus, which depends on the electron density and thus on the nature of the bonds with neighboring atoms and the type of these atoms. This parameter can be calculated for protons in different chemical surroundings based on formulas derived from the theoretical description of phenomena underlying the NMR method. Table 2.3 shows examples

Table 2.3 Values of calculated chemical shifts σ in ^1H NMR spectrum for A–CH_2–B type molecule with different functional groups A, B.

A lub B	σ	A lub B	σ	A lub B	σ
–H	0.34	–CF_2	1.12	–I	2.19
–CH_3	0.68	–CF_3	1.14	–OH	2.56
–C=C	1.32	–F	3.30	–OR	2.36
–C≡C	1.44	–Cl	2.53	–OPh	2.94
–Ph	1.83	–Br	2.33	–N_3	1.97

of theoretical values of chemical shifts that play, from a calibration point of view, the role of mathematical standards.

In the case of samples with rich structure, there may occur the interference effect manifesting itself as a change in the value of chemical shifts characteristic for a given molecule. It is assumed, therefore, that in such a situation this parameter is helpful in correct assignment of the order of appearance of signals of appropriate atoms on the spectrum, while it cannot be used as the exclusive criterion for correctness of the performed spectrum interpretation. To increase the identification ability of the method, empirical parameters characteristic of the NMR method, such as number, multiplicity, and height of the peaks as well as the distance between peaks resulting from spin–spin coupling in the molecule, are taken into account.

Mass spectrometry has a special place in identification analysis. The mass spectrum depicts the disintegration of molecules of the analyzed substance into smaller, charged fragments formed under the influence of various ionizing agents. The spectrometer differentiates the mixture of incoming ions according to their mass-to-charge ratio (m/z) and the spectrum is specific for the chemical. As a result, the m/z values act as a standard to identify the test compound. Figure 2.2 shows the mass spectrum of a selected organic compound with signals from different fragments of the analyte molecule highlighted.

From the m/z value of the signal corresponding to the molecular ion of that compound, the m_c mass of the original chemical compound under test can be calculated from the formula:

$$m_c = (m/z) \cdot (z - m_p) \tag{2.5}$$

where m_p is the sum of the masses of particles or ions that have given a charge by attaching to the initial particle. A more in-depth interpretation of the spectrum is

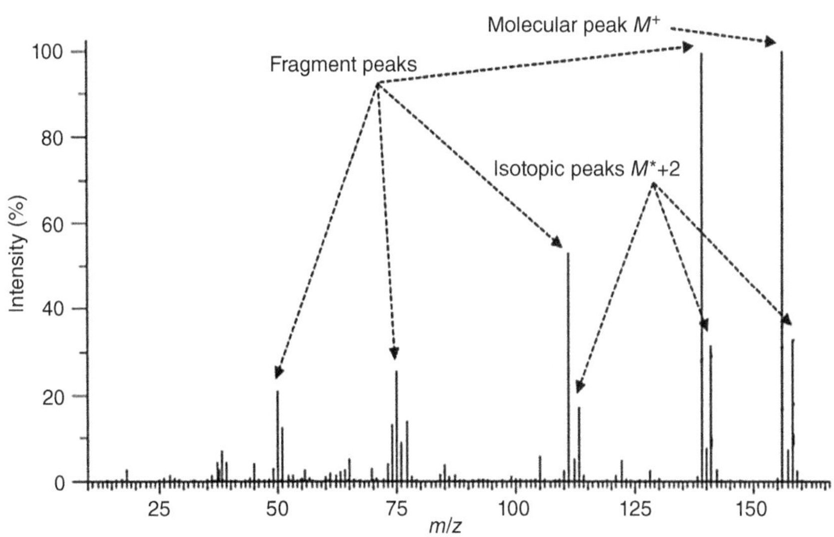

Figure 2.2 Mass spectra obtained for *m*-chlorobenzoic acid.

possible by identifying (in terms of location and relative intensity) the peaks corresponding to ions that have formed as a result of fragmentation of the molecule and the peaks corresponding to ions composed of isotopes.

A modern instrumental solution is tandem mass spectrometry (MS/MS), where two or more mass analyzers are coupled together. Particles of a given sample are ionized, the first spectrometer separates these ions according to their mass-to-charge ratio, and then ions with a specific *m/z* ratio are broken down into smaller fragment ions (e.g. by collision-induced dissociation, ion–molecule reaction, or photodissociation). These fragments are then fed into a second mass spectrometer, which in turn separates the fragments according to their *m/z* ratio and detects them.

Tandem mass spectrometry allows the identification and separation of ions that have very similar *m/z* ratios in a single mass spectrometer. As an example, Figure 2.3 shows the spectra of two compounds with very similar chemical structures, but with significantly different positions of the individual analytical signals. It is obvious that the measurement image obtained in this way significantly facilitates theoretical interpretation and correct identification of the analyte.

An interesting additional possibility for the identification of a chemical compound in mass spectrometry is also the analysis of the fragmentation pathways of

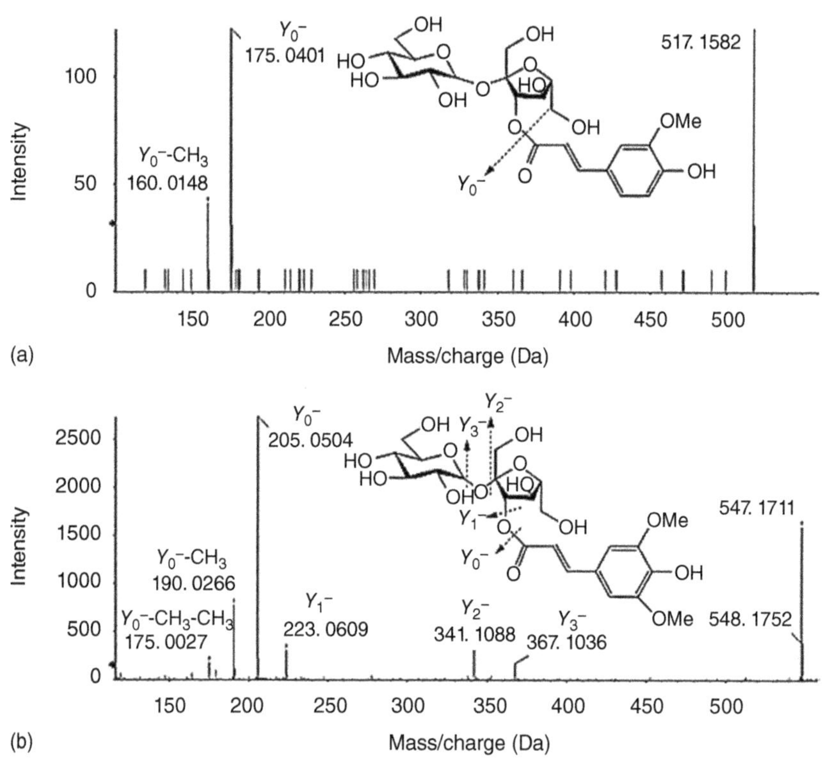

Figure 2.3 MS/MS spectra of sibiricose A5 (a), sibiricose A6 (b). Source: Adapted from Shi et al. [5].

this compound based on the characteristic m/z values (mainly corresponding to fragmentation peaks) revealed on its spectrum. The end result of this conceptual process is the determination of the structure of the compound under study. In the formulation of fragmentation pathways in the role of specific calibration standards, certain rules are helpful, the main one of which states that fragmentation proceeds in such a way that its products are the most stable chemical individuals (ions, radicals) possible.

In quantitative analysis, an example of a method based on theoretical calibration is **coulometry**. It is an electrochemical method based on two Faraday's laws of electrolysis, which state that (i) the amount of chemical change produced by current at an electrode–electrolyte boundary is proportional to the quantity of electricity used and (ii) the amounts of chemical changes produced by the same quantity of electricity in different substances are proportional to their equivalent weights. Both laws expressed mathematically lead to a formula describing the fixed relationship between the electric charge, Q, required for an electrochemical reaction involving a substance of mass m:

$$Q = \frac{n \cdot F}{M} \cdot m \tag{2.6}$$

where n is the number of electrons involved in the reaction, M is the molecular weight of the substance under test, and F is Faraday's constant. On the basis of this relation, it is possible to determine the unknown quantity (mass) of a substance subjected to a specific electrochemical reaction by measuring the corresponding value of electric charge as an analytical signal.

Equation (2.6) meets the requirements for the absolute method posed in the work [3] since the functional relationship is completely described on the basis of physical constants and universal quantities. Nevertheless, it should be taken into account that Faraday's laws of electrolysis are nevertheless valid only when well-defined experimental conditions are met, above all: the stoichiometry of electron reaction is known, and the electrode reaction proceeds with 100% efficiency.

In practice, these conditions are not possible to fulfill. Side reactions may occur with, e.g. the solvent, the electrode material, the components of supporting electrode, or the products of the electrolysis undergoing secondary electrode reaction. Other components of the sample may also react, producing the typical interference effect. A separate problem in coulometry is the accuracy of determining the amount of electricity required for an electrochemical reaction. In addition, when the electrode potential is held constant, accurate and reproducible results can be obtained by minimizing the residual current, which is virtually impossible to eliminate completely.

Due to the fact that, the function (2.6) is therefore subject to numerous uncontrolled effects, it cannot be treated as a real function and the coulometric method as an absolute method. Function (4) should therefore be considered as a mathematical model function. It is not an exact representation of the actual state (the real function), but only an approximation of the real function to varying degrees depending on the extent to which the experimental conditions deviate from those required theoretically. The model function is established by two Faraday's laws, so they act as mathematical standards (in place of chemical standards usually used in electrochemical methods).

It is worth noting that a mathematical standard derived from such an unquestionable source as Faraday's law cannot be subject to any modifications that take into account the current state of the sample and any uncontrolled effects that the analyte in the sample undergoes. It, therefore, provides no way to compensate for these effects. The only way to increase the reliability of determinations is therefore to eliminate these effects.

In spite of the reservations against coulometry as a calibration-free method, it must be clearly stated that this method finds a unique place among analytical methods. This is due to the extremely strict definition, not found in other methods, of the dependence of the analytical signal on the quantity of a substance. Owing to this, under specific, well-recognized conditions and with the use of modern measuring instruments, it is possible to obtain results of exceptional precision and accuracy.

2.3.2 Flexible Models

The formulation of "own" mathematical models in quantitative analysis that enable the determination of analytes without the use of chemical standards has long been in the field of interest of analysts as an extremely important and attractive analytical goal. However, studies of this type, carried out over the years for various measurement methods, have revealed many problems and difficulties. These come down to the fact that the form of the real function usually depends on too many chemical and instrumental parameters to be accurately included in the theoretical model. In many cases, these parameters depend on unknown physicochemical quantities, and mathematical approximation of them in various ways does not always give satisfactory results. The difficulties are enhanced by the fact that, as a rule, models must be formulated individually with respect to particular analytes or even types of samples analyzed. Separate problems arise when it is necessary to take into account in the theoretical model the occurrence of uncontrolled effects (e.g. interference) to which the sample is subjected.

The first attempts to create flexible models concerned spectrometric methods, in particular, **atomic absorption spectrometry (AAS)**. Originator and developer of this analytical method, A. Walsh, wrote in 1955 in his landmark paper: "In spite of the remarkable advances in technique ... there has been practically no progress whatsoever in solving the fundamental problem of devising an absolute method, i.e. a method which will provide an analysis without comparison with chemically analyzed standards or synthetic samples of known composition." [6]. Despite his generally critical view of the AAS method in terms of absolute analysis, he held out hope that further research in this area would yield positive results because, from a theoretical point of view, the method " ... is expected to be much less susceptible to interelement effects" (which, unfortunately, subsequent experimental studies have not confirmed). An additional argument was the possibility of determining the value of the absorption coefficient by measuring the ratio of two intensities, which is much simpler to achieve than the measurement of emission intensities in absolute units.

Several years later, V. L'vov basically dashed these hopes with regard to the method of AAS with flame atomization (FAAS), showing in detail many obstacles practically impossible to overcome [7]. He saw the problems not so much in the lack of reliable data on the pertaining atomic constants and some other parameters, but in the fact that "... the present models are not capable of describing the process of formation of a free-atom layer for the nebulizer/slot-burner flame atomizer with the accuracy desired for analytical purposes." Influenced by this opinion, further analytical efforts were focused on bringing absolute analysis to **AAS with graphite furnace atomization (GFAAS)**.

The first significant step forward was the formulation by V. L'vov of a model to describe the vaporization and atomization behavior of the analyte atoms in the graphite furnace [8]. This model represents a mass balance between the supply and loss functions and includes atom generation by zero-order kinetics and atom loss by diffusion. In the following years, the formulation of theoretical models in GFAAS was addressed by many other researchers presenting various approaches (e.g. systematic theoretical description, thermodynamic modeling, kinetic modeling, and Monte Carlo [MC] simulation) with varying success. These efforts basically boiled down to an accurate theoretical determination of the measurement sensitivity of a specific analyte by calculating the value of so-called characteristic mass, m_0, of an element, i.e. the amount of analyte giving 0.0044 integrated absorbance (A_{int}).

In one of the last works in this series [9], the characteristic masses were calculated based on the former convective–diffusive vapor-transport models. The following equation was used:

$$m_0 = 6.36 \cdot 10^{-14} \cdot \frac{A_r \cdot \Delta\nu_D \cdot Z(T) \cdot r^2}{H(\alpha,\omega) \cdot \gamma \cdot f \cdot \delta \cdot g_l \cdot \tau_d \cdot \exp(-E_l/kT)} \tag{2.7}$$

where A_r is the relative atomic mass, $\Delta\nu_D$ – the Doppler broadening of the analytical line, $Z(T)$ – the atomic partition function or state sum at temperature of $T(K)$, r – the inner radius of the graphite tube, $H(\alpha, \omega)$ – the intensity-distribution integral, γ – the factor accounting for the fine and hyperfine splitting and for the Doppler profile of the hollow cathode lamp, f – the oscillator strength of the electronic transition, δ – the coefficient accounting for the presence of adjacent lines in the spectrum of the primary light source, g_l and E_l – the statistical weight and energy of the lower level of the transition of the analytical line, respectively, τ_d is the residence time accounting for the concentration diffusion of analyte atoms, and k is the Boltzmann coefficient. The values of the physicochemical parameters were determined on the basis of other works. The values of characteristic masses calculated from Eq. (2.7) for several elements are compared with the experimental values in Table 2.4.

The presented example shows very well the complexity of the issue of theoretical modeling in quantitative analysis. Due to the still limited knowledge of phenomena and properties characterizing the GFAAS method, the formulated model is undoubtedly simplified despite the fact that it was formulated on the basis of knowledge and achievements of many authors. The model does not take into account potential interference effects or other uncontrolled effects that may exist in the sample. Parameters in this model, playing a role of physicochemical standards, were necessarily

Table 2.4 Theoretical and experimental characteristic mass data obtained for the GFAAS method under gas-stop conditions during atomization.

Element	Wavelength (nm)	Characteristic mass, m_0 (pg)		Element	Wavelength (nm)	Characteristic mass, m_0 (pg)	
		Calculated	Experimental			Calculated	Experimental
Ag	329.1	3.65	3.7	Mn	279.5	5.58	4.4
As	193.7	27.8	35.4	Mo	313.3	9.48	10.1
Cd	228.8	2.40	1.46	Ni	232.0	22.4	21.7
Co	240.7	15.4	16.5	Pb	283.3	41.6	31.6
Cr	357.9	6.18	5.10	Sb	217.6	53.2	47.6
Cu	324.8	8.57	0.1	Se	196.0	33.7	46.2
Fe	248.3	12.5	12.0	Sn	286.3	50.1	58.7
Hg	253.7	360	325	V	319.4	49.8	38.4
Mg	285.2	0.81	0.44	Zn	213.9	1.23	1.01

Source: Adapted from Bencs et al. [9].

calculated with some approximations under strictly defined conditions that may not always be strictly fulfilled. All of this causes many of the theoretical data presented in **Table 2.5**. Table 2.4 significantly deviate from the experimental values, which indicates unsatisfactory representation of the real function by the model function.

It is, therefore, clear that creating more reliable flexible models in the GFAAS method, which will be able to replace empirical models, requires further in-depth theoretical research supported by new concepts and assumptions. This is also true for the vast majority of other analytical methods.

On the other hand, some studies show that full knowledge of the phenomena and processes governing a method is not always necessary, as long as the analyst is able to isolate and take into account those among them that have a dominant influence on the form of the real function. This is proved by the results of theoretical modeling obtained for the method of **flame photometry** [10]. In this case, not only the dependence of signal intensity on the concentration of alkali metal (potassium) was considered, but also the mutual influence of analyte and interferent (sodium) on this signal. The theoretical model has the following form:

$$
\begin{aligned}
p_{\mathrm{K}}{}^{(T)} &= p_{\mathrm{K}} + \frac{K_{\mathrm{K}} \cdot p_{\mathrm{K}} \cdot \left(\sqrt{\left[\left(\frac{p_{\mathrm{e}}}{2}\right)^2 + K_{\mathrm{K}} \cdot p_{\mathrm{K}} + K_{\mathrm{Na}} \cdot p_{\mathrm{Na}}\right]} - \frac{p_{\mathrm{e}}}{2}\right)}{K_{\mathrm{K}} \cdot p_{\mathrm{K}} + K_{\mathrm{Na}} \cdot p_{\mathrm{Na}}} + \gamma_{\mathrm{K}} \cdot p_k \\
p_{\mathrm{Na}}{}^{(T)} &= p_{\mathrm{Na}} + \frac{K_{\mathrm{Na}} \cdot p_{\mathrm{Na}} \cdot \left(\sqrt{\left[\left(\frac{p_{\mathrm{e}}}{2}\right)^2 + K_{\mathrm{K}} \cdot p_{\mathrm{K}} + K_{\mathrm{Na}} \cdot p_{\mathrm{Na}}\right]} - \frac{p_{\mathrm{e}}}{2}\right)}{K_{\mathrm{K}} \cdot p_{\mathrm{K}} + K_{\mathrm{Na}} \cdot p_{\mathrm{Na}}} + \gamma_{\mathrm{Na}} \cdot p_{\mathrm{Na}}
\end{aligned}
\tag{2.8}
$$

where $p_{\mathrm{Me}}{}^{(T)}$ is the total partial pressure of metal Me (K, Na) in the flame, p_{Me} and p_{e} – pressures of atoms Me and electrons, respectively, and K_{Me} – ionization constant for metal Me. Constant coefficient γ_{Me} is defined by $\Sigma_A(p_{MeA}/P_{Me})$, where p_{MeA} is the pressure of all compounds Me*A* occurring in the flame.

To simplify theoretical modeling it was assumed that:

- the effect of self-absorption of K is negligible (this assumption is valid for low atomic concentration),
- chemical equilibria are reached in the flame; this can be expected to hold for the outer cone of the flames usually used in flame spectrometry,
- compounds KC1 and NaCl, in which K and Na were prepared in the liquid samples were completely dissociated in the flame,
- pressure p, does not depend significantly on the concentration of K and Na in the liquid samples.

It was also assumed that pressure $p_{Me}{}^T$ of atoms Me in the flame is proportional to the concentration of Me in the sample, and that pressure p_{Me} is proportional to the signal intensity.

As a result, the results presented in Figure 2.4 were obtained, which show that only the parameters included in formula (2.8) determine the accuracy of fitting the model function to the real function. It is worth emphasizing that satisfactory results were obtained (see Figure 2.4c) despite the nonlinear character of the real function.

Model (2.8) cannot be considered a complete model function because it does not take into account the value of the proportionality factor linking the analytical signal to the concentration of analyte in the sample and requires normalization with experimental values. Nevertheless, it represents a significant step toward the creation of such a function. In its presented form, it also gives a correct picture of the changes in the intensity of the analytical signal occurring under the influence of the interferent and forms a good theoretical basis for the study of various interference effects in flame photometry.

Various attempts to adapt theoretical calibration to quantitative analysis also apply to X-ray spectrometry methods, in particular to **electron probe microanalysis (EPMA)**. In this method, the quantitative analysis based on chemical standards is difficult or even not possible for some reasons. One of them is that the composition of the standards at the microscopic scale is not necessarily identical with the bulk nominal composition. Moreover, certain standards tend to evaporate or oxidize very rapidly in measurement conditions. Therefore, all approaches to avoid empirical calibration in EPMA are especially attractive.

In some cases, the proposals of "standardless" calibration in EPMA include the use of chemical standards in essence (e.g. PuO_2 standard is used to determine by extrapolation the intensities of the Mα and Mβ X-ray lines of such elements as Am and Cm). Sometimes the standards have an unusual form, not seen in other measurement methods (such as a database of characteristic intensities created from mono-elemental standards measured under different excitation conditions). For calibration purposes, however, purely mathematical approaches are also used [11].

For instance, in the method called TWIX, the X-ray path lengths in samples at different take-off angles are used to find the sample composition [12]. To this end, two spectra are measured with the sample tilted at a suitable angle for two different azimuth angles: one of them obtained by a horizontal rotation of 180° with respect

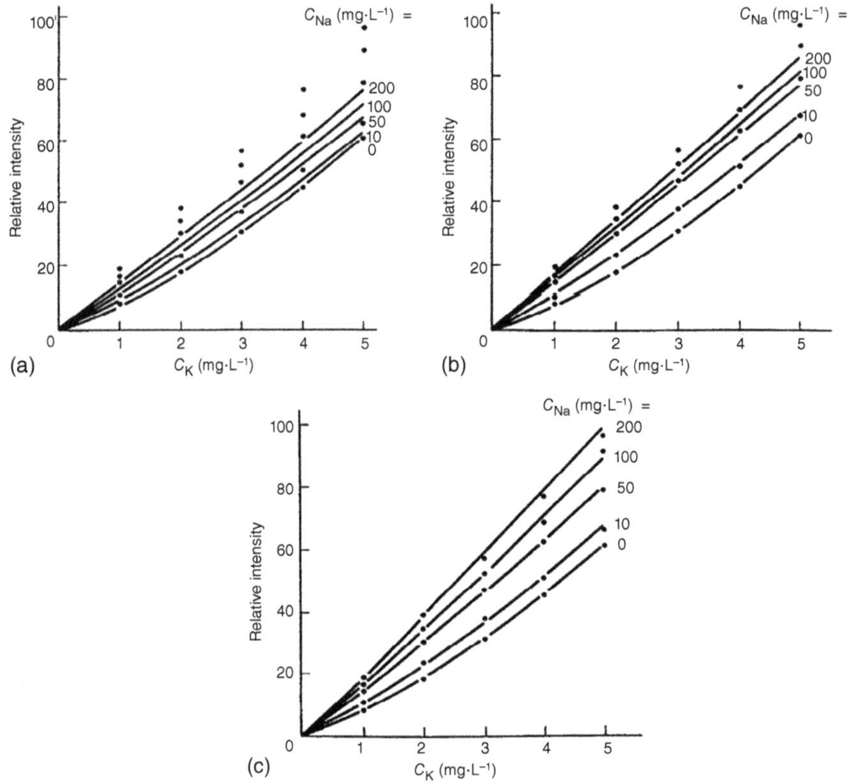

Figure 2.4 The consistency of the theoretical model (Eq. (2.8)) (solid lines) with experimental data (points) after taking into account: ionization of free atoms of potassium and sodium (a), and additionally formation of compounds *KA* (b) and ionization of the flame gas (c). Source: Kościelniak and Parczewski [10], fig 1 (p. 885)/with permission of Elsevier.

to the other one. The calculated ratio of X-ray intensities corresponding to the two mentioned configurations plays a role of an analytical signal. The advantage of this relative approach is that the evaluation of many atomic and instrumental parameters not well-known is avoided.

For many years, attempts have also been made to calculate X-ray intensities by means of MC simulations. Due to high flexibility, the MC method can simulate intensities emitted from very complex geometries. However, for calibration purposes the method requires the knowledge of physical parameters, which are not always well-known and, therefore, they must be verified by comparison with experimental measurements. In work [13] the results obtained in this way for Pb and U were found to be in good agreement with those obtained using empirical calibration when both analytes were main components of the samples analyzed. The examples are shown in Figure 2.5.

There are some several other mathematical approaches to theoretical calibration in EPMA that give results with sufficient level of accuracy for many practical

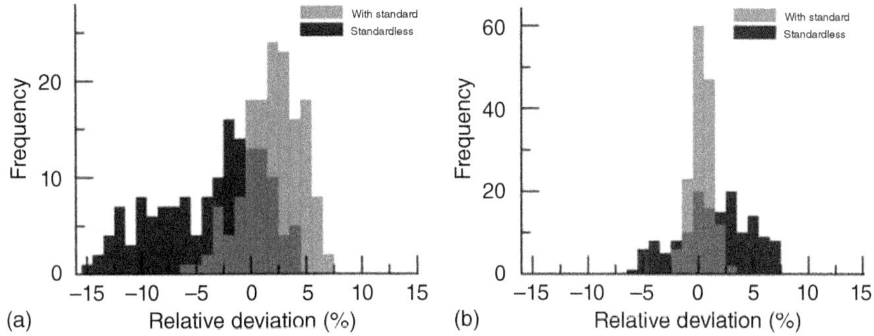

Figure 2.5 Relative deviations (%) of the determination of Pb in PbTe (a) and U in UO_2 (b) obtained when calibration without and with chemical standards was performed. Source: Moy et al. [13], fig 5 (p. 7)/with permission of American Chemical Society.

applications. However, as pointed out in [11], to achieve better results it is necessary to improve even more the description of the spectrometer efficiency, and of certain atomic parameters like ionization cross sections and fluorescence yields.

The formulation of theoretical models does of course not only apply to spectrometric methods. Very good results can be obtained, for example, in **anodic stripping voltammetry (ASV)**, in particular with thin mercury film microelectrodes. Due to their small surface nonplanar diffusion does occur and the steady-state currents are provided, which forms ideal deposition conditions for ASV. Additionally, the small currents passed by microelectrodes result in negligible ohmic losses and this allows electroanalysis to be carried out in poorly conductive media without the need for a supporting electrolyte. Furthermore, the metal is accumulated during the electrodeposition period in a very small volume of electrode, and, therefore, is completely reoxidized during the anodic scan.

These properties of microelectrodes allow for simplification of the conditions under which measurements are made and consequently minimize the number of physicochemical parameters describing these conditions. For instance, it was shown that ASV responses recorded in low ionic strength aqueous solutions are not affected by migration for divalent cations whose analytical concentration is below 0.1 mM [14]. It was also revealed that although relatively long deposition times have been employed no effect due to natural convection is observed [15].

Assuming such simplified conditions it was proved that the relation between the analyte signal (i.e. the charge, Q, passed during the deposition step) and the analyte concentration can be described by the following formula [15]:

$$Q = K \cdot n \cdot F \cdot D \cdot a \cdot \left(t_d + \frac{E_p - E_d}{v} \right) \cdot c \tag{2.9}$$

where n is the number of electrons transferred, F – the Faraday constant, D – the diffusion coefficient of the electroactive species, a – the radius of the microdisc electrode, t_d – the deposition time, v – the scan rate, E_d and E_p are deposition and peak potential, respectively. The parameter K is a geometric factor dependent on the ratio of the height of the sphere cap to the radius of the substrate electrode, a.

Table 2.5 Comparison between the analytical results (μg l^{-1}) obtained theoretically (c_{calc}) and experimentally (c_{exp}) for Cd, Pb, and Cu determined in different rain samples by ASV.

	Cd		Pb		Cu	
Sample	c_{calc}	c_{exp}	c_{calc}	c_{exp}	c_{calc}	c_{exp}
I	0.8	0.9	9.5	9.2	5.0	5.1
II	1.1	1.0	3.1	2.8	4.8	4.8
III	1.1	1.0	7.4	7.3	6.2	5.7
IV	0.7	0.7	6.4	6.5	5.7	5.7
V	0.9	0.8	7.2	7.0	4.4	3.9

Source: Adapted from Abdelsalam et al. [15].

Model (2.9) was applied to the determination of Cd, Pb, and Cu in the rain samples obtaining results with very good agreement with experimental data as shown in Table 2.5.

The most spectacular theoretical modeling results have been achieved for the **laser-induced breakdown spectroscopy (LIBS)** method. In this method a powerful laser pulse is focused on the sample surface, resulting in the ejection of its material, which leads to the formation of a plasma plume. The emitted radiation from the plasma plume is recorded, and the composition of the sample is determined from the measured spectra by identifying the observed spectral lines.

LIBS offers many advantages compared to other elemental analysis techniques, such as a strong potential for analysis in situ and in real time, not requiring the sample preparation step, ability to analyze the samples in any phase (solid, liquid, or gas) in microscale and without destroying them, broad elemental coverage (including lighter elements), and extremely fast measurement time (usually a few seconds for a single spot analysis). Another property of this method that makes it suitable for characterizing a kind of the sample analyzed is the possibility of creating the sample composition through the multicomponent analysis.

On the other hand, the LIBS signal depends not only on the concentration of the analyte but also on the composition and aggregation state of the matrix. For this reason, the determination of trace and minor elements in the samples of complex matrix requires calibration with chemical standards. The empirical calibration process is, therefore, often the most tedious and lengthy stage of analysis. Moreover, matrix-matched standards needed to compensate effectively the matrix effect are simply unavailable in many practical situations. Even so, in many cases, it is very difficult to obtain or buy the standards that allow for creating empirical models accurately approximating the real function. This is because the character of the plasma and the behavior of the atoms depend on many variables related to the origin of the ablated material: composition, crystallinity, optical reflectivity, optical transmissivity, and morphology of the surface.

In this situation, theoretical modeling attempts have been made quite intensively, realizing the difficulties that arise from the complexity of the phenomena of laser plasma properties and the processes occurring in it. The first basic model was developed in 1999 by Cucci et al. [16]. The construction of this model is based on the following assumptions:

- the plasma composition is representative of the actual material composition prior to the ablation,
- in the actual temporal and spatial observation window, the plasma is in local thermal equilibrium (LTE) conditions,
- the radiation source is optically thin,
- self-absorption effect doesn't occur.

Assuming such simplified conditions, the dependence of the analytical signal, I_λ^{ki} (integral line intensity measured at wavelength λ) on the concentration of the emitting atomic species (i.e. element with a determined electric charge), c, is defined by the formula:

$$I_\lambda^{ki} = F \cdot A_{ki} \cdot \frac{g_k \cdot e^{-(E_K/k_B T)}}{U_s(T)} \cdot c \tag{2.10}$$

where N_s is the number density of the emitting atoms for each species, A_{ki} – the probability of transition between the upper (k) and lower (i) energy states, g_k and E_k – the degeneracy and energy of upper-level k, respectively, k_B – the Boltzmann constant, T – the plasma temperature; and $U_s(T)$ – the partition function for the emitting species at the given plasma temperature.

By experimentally measuring the intensity of the plasma emission spectrum and querying the parameters A_{ki}, g_k, and E_k in the spectral database, one can convert the spectral line to a point in a two-dimensional Boltzmann plane, as presented in Figure 2.6. Different points obtained for an analyte lie on straight line, with intercept proportional to the logarithm of the analyte concentration. The concentration values of each species may be obtained by the sum of all species as 100%.

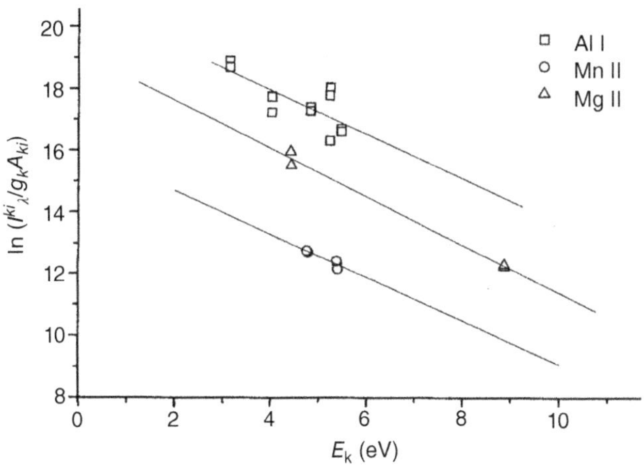

Figure 2.6 Boltzmann plot obtained for Al, Mn, and Mg on the basis of data taken from the LIBS analysis of aluminum alloy. Source: Ciucci et al. [16], fig 1 (p. 961)/With permission of SAGE Publications.

Table 2.6 Results of multicomponent analysis of a soil sample obtained by LIBS using theoretical model and by ICP-OES in the conventional calibration way.

	Contents (%)			Contents (%)	
Analyte	**Calculated**	**Experimental**	**Analyte**	**Calculated**	**Experimental**
SiO_2	34.7 ± 2.7	52.9 ± 0.5	MgO	3.2 ± 0.8	4.7 ± 0.1
Al_2O_3	15.5 ± 1.6	16.9 ± 0.2	Na_2O	1.9 ± 0.2	4.4 ± 0.0
Fe_2O_3	11.9 ± 1.7	0.3 ± 0.2	TiO_2	3.2 ± 0.3	1.7 ± 0.0
CaO	9.7 ± 1.5	8.8 ± 0.1	K_2O	0.7 ± 0.1	0.5 ± 0.2

Source: Adapted from Anzano et al. [18].

The great advantage of the model (2.10) was that it could produce results less affected by matrix effects than conventional empirical calibration. Not surprisingly, it has found increasing use over time in the analysis of many types of samples. In 2010, dozens of papers on its use for the analysis of solid samples were already reported [17]. At the same time, the model has been repeatedly modified and supplemented to increase its reliability and efficiency. This is because its main drawback is the gradual decrease in the accuracy of the calculated results (while maintaining satisfactory precision) as the analyte content in the sample decreases. A good picture of this state of affairs is given by recently obtained results of multicomponent analysis of Antarctic soils, which are partially summarized in Table 2.6 [18]. A significant improvement in this respect is brought by the additional use of internal standards, but then the calibration loses its theoretical value.

The accuracy of the theoretical formula as a model function depends on the accuracy of the calculation of the individual components of this formula, i.e. the mathematical standards. The LIBS method is known to determine increasingly accurate values of parameters such as spectral intensity, the spatial and temporal window of local thermodynamic equilibrium, temperature, electron density, and others [19]. Achieving more satisfactory theoretical modeling results, however, requires further in-depth research into the mechanisms of plasma processes. An alternative is a hybrid approach combining theoretical calibration with empirical calibration. The first studies of this type on LIBS are promising and perhaps, further development of this method in terms of calibration will also go in this direction.

The examples cited above for the development of flexible models do not, of course, exhaust all the research that has been conducted in this area for other analytical methods. This outline is only meant to give a general idea of the varying degrees of difficulty and the different chances for use in analytical practice in the near future. Systematic progress in this field of chemical analysis must be expected, which will certainly come with technological, computational, material, and above all with the development of knowledge of the physical and physicochemical processes underlying the operation of analytical methods. This is a fascinating subject, worthy of a separate, extensive monographic study.

The remainder of this book will be devoted exclusively to empirical calibration methods and procedures implemented with chemical standards.

References

1 Bureau International des Poids et Mesures (BIPM) (1995). Report of the first meeting. *Comité Consultatif pour la Quantité de Matière* 1: Q5.

2 Danzer, K. and Curie, L.A. (1998). Guidelines for calibration in analytical chemistry. Part 1. Fundamentals and single component calibration. *Pure and Applied Chemistry* 70 (4): 993–114.

3 Hulanicki, A. (1995). Absolute methods in analytical chemistry. *Pure and Applied Chemistry* 67 (11): 1905–1911.

4 Kościelniak, P. (2022). Unified principles of univariate analytical calibration. *TRAC Trends in Analytical Chemistry* 149: 116547.

5 Shi, Q., Chen, J., Zhou, Q. et al. (2015). Indirect identification of antioxidants in Polygalae Radix through their reaction with 2,2-diphenyl-1-picrylhydrazyl and subsequent HPLC–ESI-Q-TOF-MS/MS. *Talanta* 144: 830–835.

6 Walsh, A. (1955). The application of atomic absorption spectra to chemical analysis. *Spectrochimica Acta* 7 (2): 108–117.

7 L'vov, B.V., Katskov, D.A., Kruglikova, L.P. et al. (1976). Absolute analysis by flame atomic absorption spectroscopy: present status and some problems. *Spectrochimica Acta, Part B: Atomic Spectroscopy* 31 (2): 49–80.

8 L'vov, B.V. (1984). The investigation of atomic absorption spectra by the complete vaporization of the sample in a graphite cuvette. *Spectrochimica Acta, Part B: Atomic Spectroscopy* 39 (2, 3): 159–166.

9 Bencs, L., Laczai, N., and Ajtony, Z. (2015). Model calculation of the characteristic mass for convective and diffusive vapor transport in graphite furnace atomic absorption spectrometry. *Spectrochimica Acta, Part B: Atomic Spectroscopy* 109: 52–59.

10 Kościelniak, P. and Parczewski, A. (1982). Theoretical model of the alkali metals interferences in flame emission spectrometry. *Spectrochimica Acta, Part B: Atomic Spectroscopy* 37 (10): 881–887.

11 Trincavelli, J., Limandri, S., and Bonetto, R. (2014). Standardless quantification methods in electron probe microanalysis. *Spectrochimica Acta, Part B: Atomic Spectroscopy* 101: 76–85.

12 Völkerer, M., Andrae, M., Röhrbacher, K. et al. (1998). A new technique for standardless analysis by EPMA-TWIX. *Mikrochimica Acta* 15: 317–320.

13 Moy, A., Merlet, C., and Dugne, O. (2015). Standardless quantification of heavy elements by electron probe microanalysis. *Analytical Chemistry* 87 (15): 7779–7786.

14 Daniele, S., Bragato, C., and Baldo, M.A. (1997). An approach to the calibrationless determination of copper and lead by anodic stripping voltammetry at thin mercury film microelectrodes. Application to well water and rain. *Analytica Chimica Acta* 346 (2): 145–156.

15 Abdelsalam, M.E., Denuault, G., and Daniele, S. (2002). Calibrationless determination of cadmium, lead and copper in rain samples by stripping voltammetry at mercury microelectrodes. Effect of natural convection on the deposition step. *Analytica Chimica Acta* 452 (1): 65–75.

16 Ciucci, A., Corsi, M., Palleschi, V. et al. (1999). New procedure for quantitative elemental analysis by laser-induced plasma spectroscopy. *Applied Spectroscopy* 53 (8): 960–964.

17 Tognoni, E., Cristoforetti, G., Legnaioli, S. et al. (2010). Calibration-free laser-induced breakdown spectroscopy: state of the art. *Spectrochimica Acta, Part B: Atomic Spectroscopy* 65 (1): 1–14.

18 Anzano, J.M., Cruz-Conesa, A., Lasheras, R.J. et al. (2021). Multielemental analysis of Antarctic soils using calibration free laser-induced breakdown spectroscopy. *Spectrochimica Acta, Part B: Atomic Spectroscopy* 180: 106191.

19 Fu, H., Ni, Z., Wang, H. et al. (2019). Accuracy improvement of calibration-free laser-induced breakdown spectroscopy. *Plasma Science and Technology* 21 (3): 034001.

3

Calibration Methods in Qualitative Analysis

The issue of analytical calibration in qualitative analysis has very rarely been addressed. It is difficult to find anything about these issues in textbooks on analytical chemistry or even in the scientific literature. Moreover, in the analytical community it is commonly believed that the calibration process is not about identification but only about determination of analytes. The same is true of the concept of a calibration method. While it is widely known in quantitative analysis, the possibility of performing identification of analytes by various different methods is very rarely recognized. Therefore, the problem arises: how to present these methods so that their calibration character does not raise any doubts?

3.1 Classification

The starting point for the classification of empirical calibration methods in qualitative analysis is the aforementioned general division of methods into comparative, i.e. when the sample and standard are treated separately, and additive, i.e. when the standard is added to the sample. This division refers not only to laboratory aspects but also has a deeper substantive sense. This is because in the first case the model function is formulated independently of the real function and in the second case on the basis of this function. The advantages and disadvantages of both ways of proceeding are a separate issue, which are discussed later in this chapter.

Empirical calibration has an advantage over theoretical calibration: the nature of chemical samples and standards is similar. This creates the possibility not only of combining a sample with a standard but also of adding another substance to the sample and standard that is inert to or reacts with the analyte. If these procedures do not limit the ability of the calibration procedure to accurately and precisely determine the type or amount of analyte in the sample being analyzed, but provide certain additional analytical advantages, they then form separate calibration methods. Based on this reasoning, the empirical methods of qualitative analysis have been classified as shown in Figure 3.1 [1].

An overview of the methods presented in Figure 3.2 will be the focus of this chapter. Their nomenclature has been adapted to that nomenclature that is

Calibration in Analytical Science: Methods and Procedures, First Edition. Paweł Kościelniak.

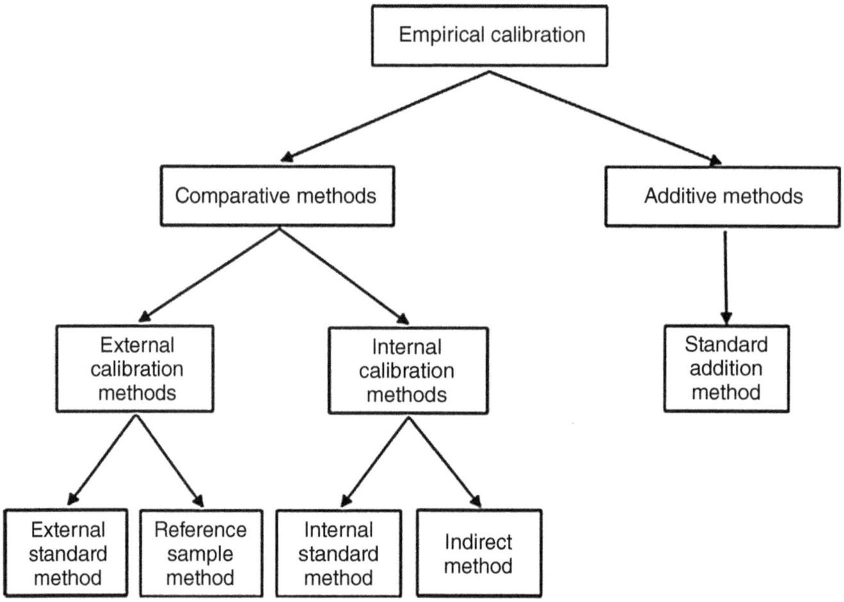

Figure 3.1 Classification of empirical calibration methods based on chemical standards in qualitative analysis.

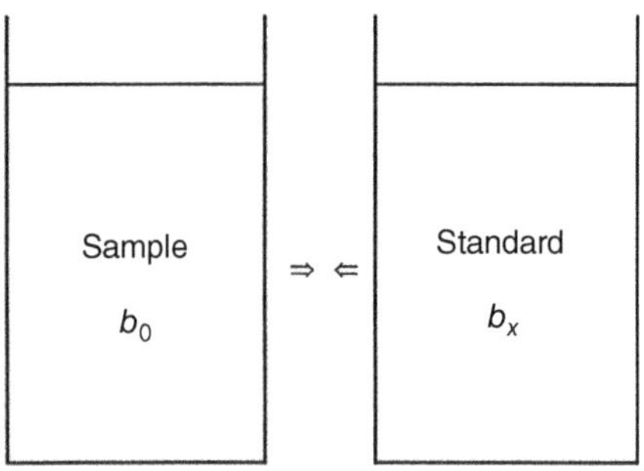

Figure 3.2 Preparation of the sample and standard in qualitative analysis in accordance with the external standard method; b_0 and b_x are expected and known analytes, respectively.

commonly used and accepted in quantitative analysis. This was not difficult because, as it turned out, the role of chemical standards in both analytical areas is similar. The occurrence of uncontrolled effects in qualitative analysis and the robustness of particular calibration methods to these effects will also receive considerable attention. All these issues will be discussed using numerous experimental examples, mainly from forensic toxicology and forensic trace analysis.

Identification analysis occupies an extremely important place in forensic research, especially in the examination of objects (traces) left at the scene of a crime, i.e. acting as evidence materials. Determining the type of such materials, its similarity to other samples, or belonging to a specific group of samples are analytical objectives very often set before a forensic expert. In addition, the analytical results obtained are of great social importance, given that they can decide about the guilt or innocence of a person suspected of a crime.

A variety of empirical calibration approaches are used in forensic research, so they provide a very good picture of a variety of calibration situations and problems in qualitative analysis. However, it should be clearly emphasized that all the calibration methods described here are general in nature and can be used in all other fields and areas requiring the performance of identification analysis.

3.2 External Calibration Methods

As seen in Figure 3.1, the vast majority of empirical calibration methods in qualitative analysis are comparative in nature. Of these, an approach based on the following principles should be considered basic:

- The sample and standard are prepared separately from each other.
- No other substance is added to the sample and standard to act as an additional standard.
- The standard does not react with any constituent in the sample.

This type of calibration strategy can be called the **external calibration method** (in analogy to the name of a similar procedure in quantitative analysis).

In qualitative analysis, the external calibration method is generally implemented in two versions, i.e. when:

- the analyte is an unknown component of the sample under study (e.g. an element, inorganic or organic compound); and the standard is a synthetic or (less frequently) natural material (substance) containing the known component sought in the sample;
- the analyte is the sample as a whole; and the standard – another sample of known or even unknown composition, which may be called the reference sample.

Thus, in the first case we are dealing with the **external standard method** and in the second case with the **reference sample method**.

In practice, the exact implementation of the external calibration method depends on many factors, including the type and number of analytes, as well as the type of sample and the type of measurement image provided by the measurement method used. From a calibration point of view, it is important that these factors are selected so that the similarity of the values of the analyte-identifying parameters in the sample and standard images (i.e. the mapping of the real function with the model function) gives the possibility of identifying the analyte with sufficiently high probability. According to another rule, the sample and standard should be measured under identical conditions that remain constant throughout the analytical procedure.

3.2.1 External Standard Method

Figure 3.2 shows schematically the principle of the **external standard method**: According to the principle of the method, a chemical standard is prepared or selected to contain at least one component of known type, b_x, which is expected to be present, b_0, in the sample.

After the sample and standard have been properly prepared and measurements have been made for them under identical experimental conditions, the analytical result is obtained by comparing the values of the parameters identifying the analyte in the standard with the values obtained for the sample. If the analyte present in the standard, b_x, is with sufficiently high probability the same as the analyte in the sample, b_0, then the analytical result is positive ($b_0 = b_x$); otherwise, it is negative ($b_0 \neq b_x$). The practical side of the external calibration method is shown by the following experimental examples.

The above conditions for calibration by the external standard method are met in qualitative analysis by the sensory method discussed earlier, because the sensory impression of the analyte is compared with the impression received by a separately localized human sensory memory. It is worth noting that in this case, the analyte can be a sample component or the sample as a whole. In fact, this calibration method is also used to identify analytes in all cases where theoretical models and mathematical standards are used. However, the standards most commonly used in qualitative analysis are the chemical standards.

One of the oldest identification approaches, classified as classical analysis, is subjecting an analyte in a sample to a suitably selected characteristic chemical reaction. The reaction results in a visible change detected with the aid of a sense (e.g. appearance of color, precipitate, odor) and indicating the presence of the analyte in the sample. The examples are shown in Table 3.1. Most often the change occurs in solution, but the presence of the analyte in the sample can also be indicated by the characteristic coloring of the flame or the formation of a "pearl" of the appropriate color. By burning the sample, it is possible to distinguish between the organic and inorganic compounds in it by sight and smell. Characteristic reactions are widely used in laboratory tests to detect specific substances quickly and easily under a variety of environmental conditions.

This type of analysis is also largely sensory in nature. However, the relationships between the analytical signals (reaction result) and the type of analyte were, as a rule, originally established from experiments performed with chemical standards of each analyte. The results of these experiments are usually so precise and reproducible that they do not need to be continuously verified. Once collected, e.g. in the form

Table 3.1 Examples of identification analysis of chemical compounds based on characteristic reactions.

Analyte	Reagent	Analytical signal
Fe^{3+}	SCN^-	Appearance of color
Pb^{2+}	J^-	Appearance of a colored precipitate
F^-	$FeSCN^{2+}$	Disappearance of color
CO_3^{2-}	HCl	Appearance of gas bubbles
Amines	$CHCl_3$	Appearance of smell
Phenols	$FeCl_3$	Appearance of color

of tables, they can be used for identification of particular analytes in subsequent analyses if only the parameters of these analyses are maintained as recommended.

In qualitative analysis, analyte-specific physical constants, such as boiling point, melting point, refractive index, and torsion angle of polarized light, play a large identification role. The type of analyte can also be detected from characteristic values of physicochemical constants. For example, a number of organic compounds can be detected from the values of their solubility constants in water, acids, bases, or complexing solutions. In thermogravimetric analysis, measuring the mass of an analyzed sample as the temperature increases determines the temperature interval over which the mass of the sample remains constant. As before, all of these values are determined primarily from chemical standards, although they can largely be assisted by mathematical (thermodynamic) calculations.

Most commonly, however, qualitative analysis involves subjecting a sample to measurement with an instrument capable of providing information about the qualitative composition of the sample in the form of a continuous signal. The intensity of the signal changes with a change in a quantity characteristic of the measurement method used (e.g. wavelength, migration time, current voltage). Each signal maximum value theoretically indicates the presence of some component in the sample, and the nature of this component is determined by the position of this maximum value in the measurement standard.

Such measurement images greatly facilitate calibration by the external standard method using a chemical standard. If the standard contains the desired analyte and the position of the signal corresponding to the analyte is known, the presence of the analyte in the sample can be determined by comparing the sample and standard images obtained under the same measurement conditions. In the example shown in Figure 3.3, the presence of the analyte (zinc) in the sample was determined in this way along with other components that were unknown and not of interest at the time.

In instrumental qualitative analysis, the external standard method can be used to identify the sample as an analyte. In this situation, the measurement image of the

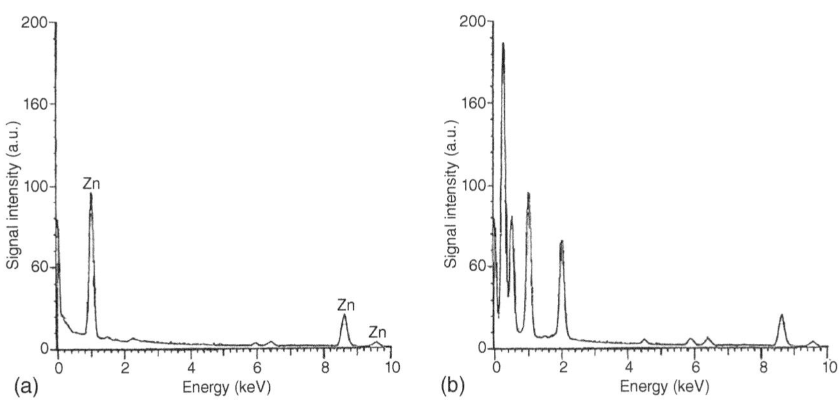

Figure 3.3 Comparison of spectra obtained by X-ray fluorescence (XRF) for the zinc standard (a) and the motor oil sample containing zinc (b): the presence of the analyte in the sample is evidenced by the peaks revealed at 1.0, 8.6, and 9.6 eV [1].

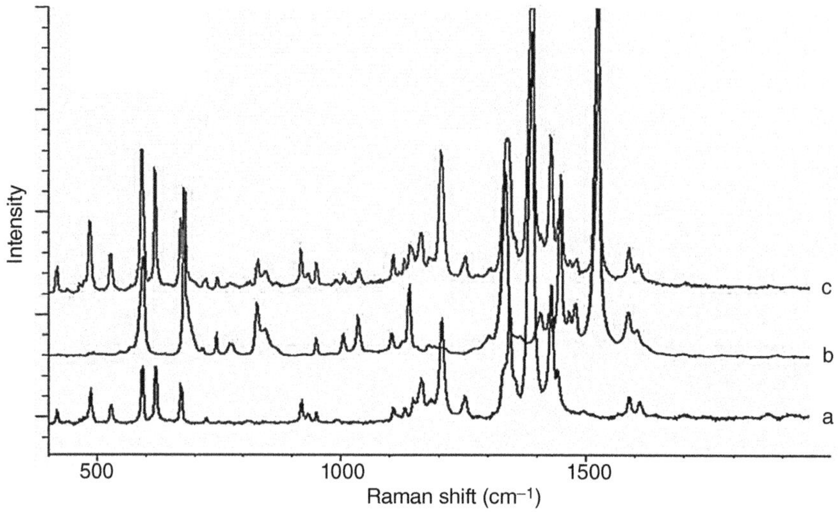

Figure 3.4 Identification of a blue gel pen ink sample (c) on the basis of Raman spectra using standards of two pigments, dioxazine violet (a) and copper phthalocyanine (b) [2].

sample contains information about the various components of the sample, and a standard of each of these components can be helpful in identifying the sample. To be more certain of the accuracy of the analytical result, it is useful to use standards of two or more components. Such an example of qualitative analysis is shown in Figure 3.4.

This type of analysis can be quite problematic, especially for organic samples with complex compositions. It is evident in Figure 3.4, even using standards of two compounds does not give certainty as to which compound is in the sample, because the spectrum of the sample is complex and exact matching of the spectra of the standards is virtually impossible. Making a final decision is complicated by the fact that some small signals may not correspond to specific native components of the sample, but may come from impurities or result from natural, random fluctuations in the measurement signal.

In the identification of organic compounds, databases are a helpful tool in the identification of analytes. These are sets of measurement data, most often in the form of measurement images, which are created for known, pure chemical compounds using a selected measurement method under strictly defined experimental conditions. Such data sets can be created in a given laboratory for the analyses most frequently performed there or more extensive commercial databases can be used. The latter are generally very extensive, and usually the search for the chemical standard corresponding to the tested sample is performed by means of dedicated computer programs. Figure 3.5 shows an example of nicotine identification results obtained using a commercial database from standard spectra measured by mass spectrometry (MS).

Each database can be thought of as a multielement chemical standard. Identification of the analyte in the tested sample is performed by searching in the database for

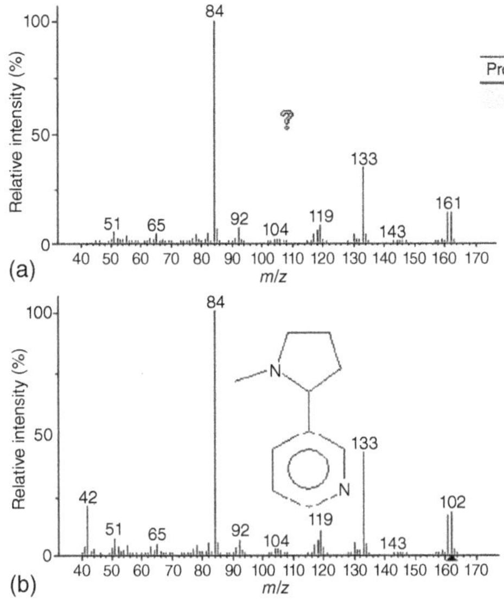

Probability (%)	Name
77.0	Pyridine, 3-(1-methyl-2-pyrrolidinyl)-, (*S*)-
77.0	Pyridine, 3-(1-methyl-2-pyrrolidinyl)-, (*S*)-
77.0	Pyridine, 3-(1-methyl-2-pyrrolidinyl)-, (*S*)-
77.0	Pyridine, 3-(1-methyl-2-pyrrolidinyl)-, (*S*)-
77.0	Pyridine, 3-(1-methyl-2-pyrrolidinyl)-, (*S*)-
21.8	Pyridine, 2-(1-methyl-2-pyrrolidinyl)-
77.0	Pyridine, 3-(1-methyl-2-pyrrolidinyl)-, (*S*)-
0.85	Pyridine, 3-(1-methyl-2-pyrrolidinyl)-, 1-oxide, (*S*)-
0.22	Anabasine
0.05	Pyridine, 3-(2-piperidinyl)-
0.22	Anabasine
0.85	Pyridine, 3-(1-methyl-2-pyrrolidinyl)-, 1-oxide, (*S*)-
0.00	Pyridine, 2-amino-3-(1-methylpyrrolidin-2-yl)-
0.00	Pyridine-1-acetamide, [*N*-(2-carboxy)pheyl]-
0.00	Desmethylcotinine
0.00	3-Pyridinecarbonitrile, 2-methoxy-4,6-dimethyl-
0.00	Pyridine, 2-amino-3-(1-methylpyrrolidin-2-yl)-
0.00	Phenylamine, 3-(pyrrolidin-1-yl)-

Figure 3.5 Comparison of the MS spectra of an analyte (a) and of a compound found in the database (b) leading to the identification result: the analyte is pyridine, 3-[1-methyl-2-pyrrolidinyl] (nicotine) with 77% probability [1].

such an image of the standard, which is to the greatest extent similar (with particular consideration of the position of analytical signals) to the image of the sample. Due to the complexity of measurement images of organic compounds and the occurrence of inevitable, random differences in the position of signals in the two images, one cannot expect a perfect mapping of the sample and standard images and obtain an unambiguous answer about the type of identified analyte. As a rule, a computer program offers, as seen in Figure 3.5, a range of answers with calculated probabilities of similarity of different patterns to the sample. The final decision on the best solution is, of course, up to the analyst.

In some cases, fortunately quite rare, this decision can be wrong even when it is made using a database. For example, a characteristic phenomenon in mass spectrometry (MS) is that spectral lines recorded for compounds with the same molecular formula but different structure appear at the same values of the m/z parameter. An example of this type is shown in Figure 3.6. If one of these compounds is the analyte and the other is the standard, identification of the analyte based only on the position of the spectral lines (without a more in-depth analysis of the spectra taking into account, for example, the intensities of these lines) will be erroneous. Assuming that such two compounds are different chemical forms of the same analyte, the speciation effect is then responsible for the misidentification of one of them.

Identification of analytes by the external standard method may sometimes be difficult due to accidental or even systematic shift of the measurement images obtained for the sample and for the standard despite maintaining constant experimental conditions during the measurements. This phenomenon occurs particularly in analyses performed with chromatographic and electrometric techniques. This uncontrolled

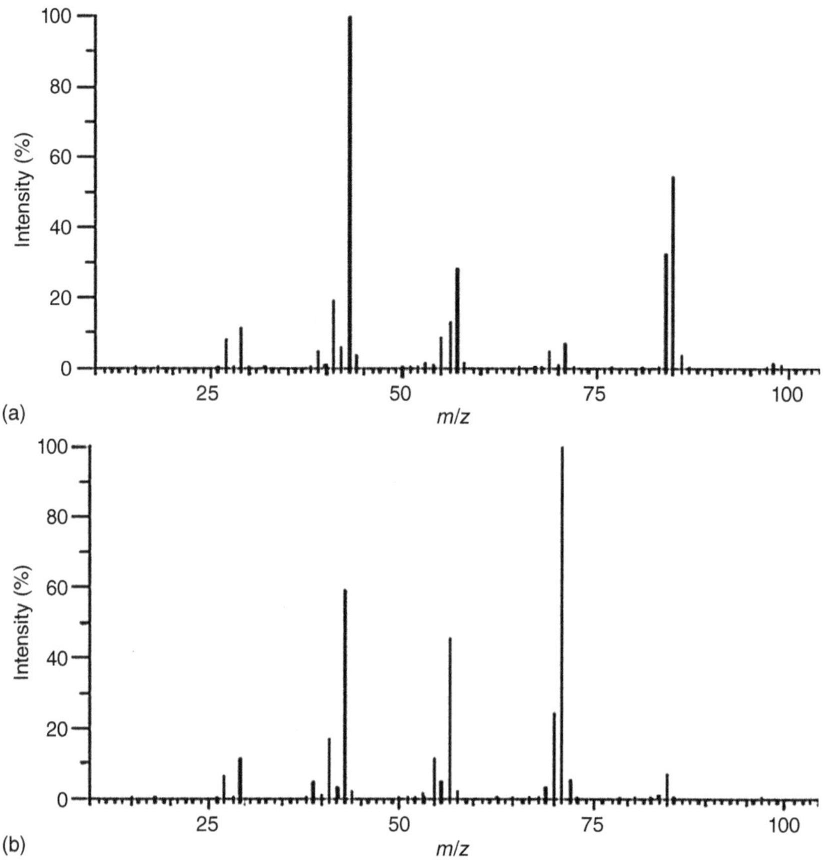

Figure 3.6 Mass spectra obtained for two compounds of the same molecular formulae ($C_{10}H_{22}$): 5-methylnonane (a) and 3,3-dimethyloctane (b) (from own collection).

instrumental effect is because these techniques rely on transport of masses of sample and standard, the speed of which may be slightly different in both cases, especially when the process of separation of components requires a relatively long time. In Figure 3.7, this effect is exemplified by a high-performance liquid chromatography (HPLC) analysis. It can be seen that the shift of the peaks is so great that the position of the peak of the desired analyte (nortriptyline) on the chromatogram of the standard compared to the position of the peaks recorded for the sample can support the conclusion that the analyte is not present in the sample.

Another important problem in qualitative analysis is the overlapping of the analyte signal with the signals of other sample components. This additive interference effect is common in spectrometric methods, as seen for example in Figure 3.3, but it is also often revealed in analysis by separation techniques. In particular, if the object of analysis is a sample containing a number of components with similar properties, their separation may not be complete. This situation obviously makes it difficult or even impossible to identify a given component in the presence of other. The effect

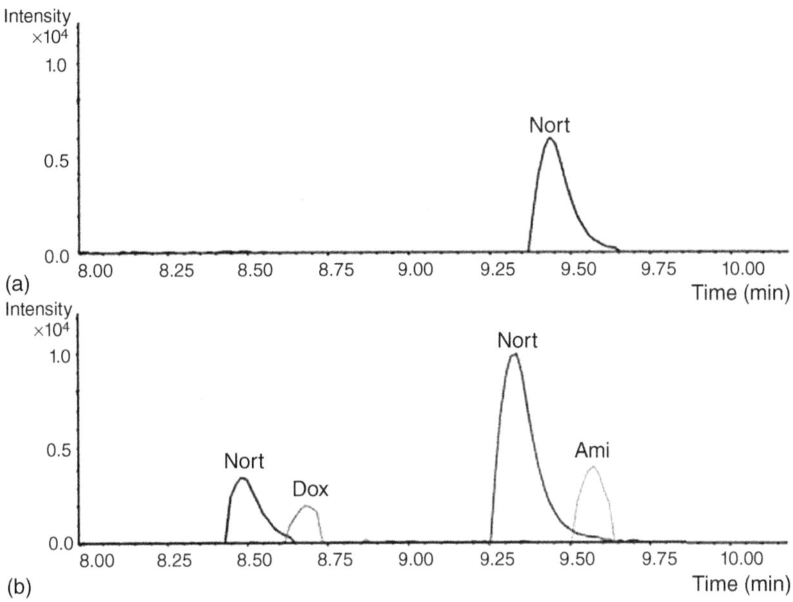

Figure 3.7 HPLC chromatograms obtained for the nortriptyline standard (a) and for the sample (b) containing nordoxepin (Nord), doxepin (Dox), nortriptyline (Nort), and amitriptyline (Ami) (from own collection).

can be minimized by optimizing instrumental conditions to ensure that individual peaks appear after prolonged retention or migration times, or chemically by adding an appropriately selected reagent to the sample and standard. Figure 3.8 shows the extreme efficiency of such activity demonstrated by β-cyclodextrin, which is explained by its ability to easily form inclusion complexes with hydrophobic compounds due to its own hydrophobicity [3].

Measurement images obtained by separation methods often provide too little information for the identification of a sample component to be sufficiently accurate. In such cases, the possibility of combining instruments for component separation with a mass spectrometer as the measuring instrument can be exploited. In this type of system, after component separation, the sample fraction containing the "pure" analyte can be routed directly to the mass spectrometer where the analyte-specific spectrum is recorded. As mentioned earlier, this spectrum gives information not only about the structure of the whole molecule of the compound but also about its individual fragments, which, of course, greatly facilitates its identification. An example of the results obtained by capillary electrophoresis alone (CE) and coupled with mass spectrometry (CE-MS) is shown in Figure 3.9 [4].

Significant identification errors are often caused by preparative effects. During the preparation of the sample and standard for analysis, first of all, a partial loss of analyte may occur and, consequently, the number and intensity of analytical signals may be reduced. Although both parameters have an auxiliary function in identification analysis, they can be very important in special cases. A signal of low intensity obtained for a sample, not "supported" by the presence of other, stronger signals,

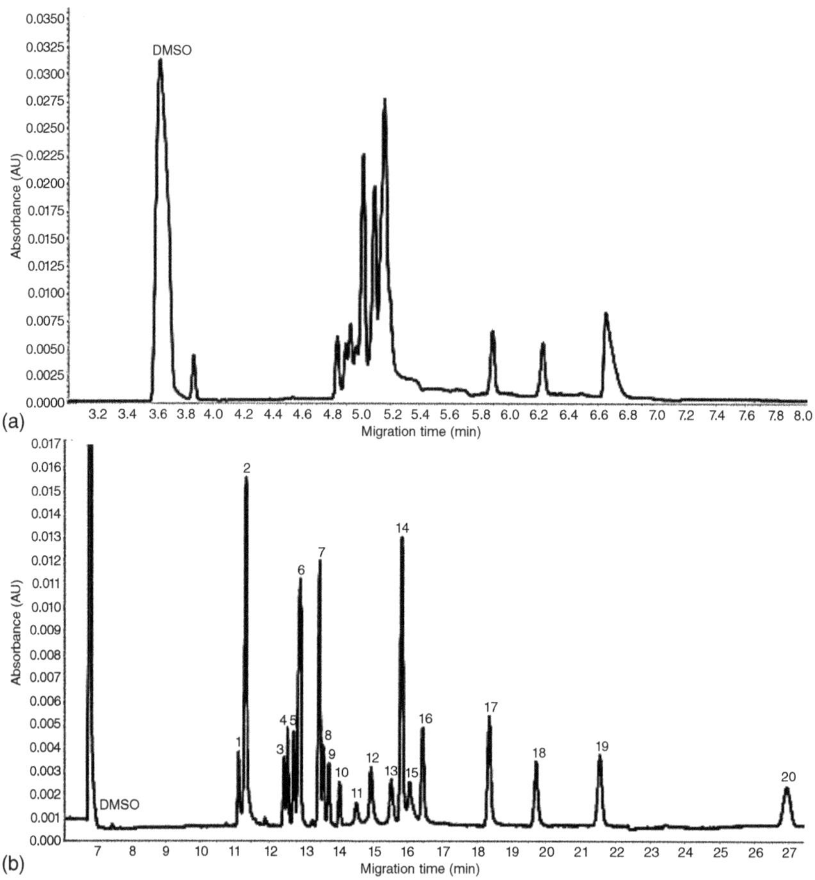

Figure 3.8 Separation of 20 coumarin derivatives using capillary electrophoresis (CE) without (a) and with (b) addition of heptakis(2,3,6-tri-*o*-methyl)-β-cyclodextrin to the background electrolyte. DMSO, dimethyl sulfoxide. Source: Woźniakiewicz et al. [3], fig 3 (p. 5)/with permission of Elsevier.

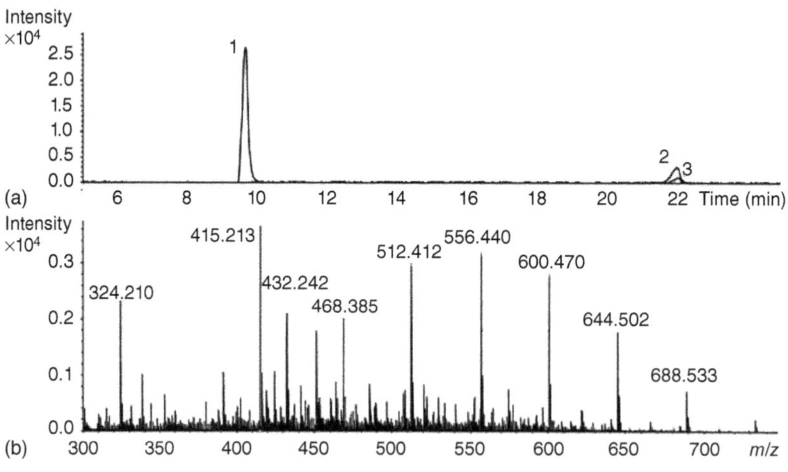

Figure 3.9 Images obtained by CE (a) and CE-MS (b) of an inkjet ink sample extracted from paper. Source: Kula et al. [4], fig 5 (p. 30)/with permission of Elsevier.

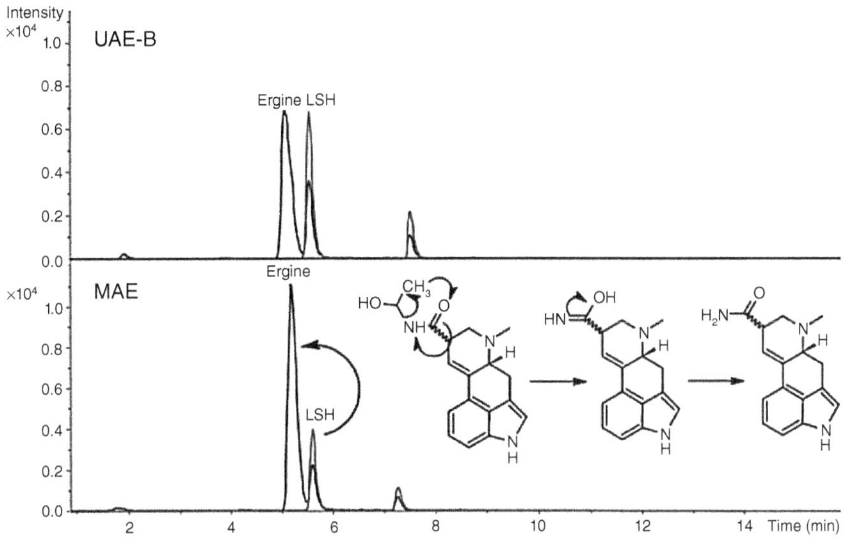

Figure 3.10 Decomposition of α-hydroxyethylamide of lysergic acid (LSH) to ergine in seed samples caused by changing ultrasound-assisted extraction (UAE) to microwave-assisted extraction (MAE) and revealed by liquid chromatography (LC) technique. Source: Nowak et al. [5], fig 3 (p. 6)/Springer Nature/CC BY 4.0.

is always a source of doubt and hinders the correct interpretation of the analytical result even when its location indicates the presence of the component sought.

A change in the peak intensity of an identified compound sometimes occurs at the preparative stage by an unusual and unexpected route. It has been observed, for example, that the extraction conditions of some organic compounds can affect the ratio of their content in the sample [5]. This phenomenon, shown in Figure 3.10, is obviously of major importance in quantitative analysis. Nevertheless, under more drastic extraction conditions and with a lower content of both compounds in the sample, the signal of one of them may decrease to such an extent that it will also cause problems in qualitative analysis during calibration by the external standard method.

A much more dangerous phenomenon is the change in the position of the peak corresponding to a given analyte with a change in its concentration. This effect, fortunately very rare, can of course be the cause of significant systematic errors in calibration by external standard methods when the concentration of the analyte in the standard is different from the concentration of that analyte in the sample. An example is chromatographic separation of some weak organic acids, as shown in Figure 3.11. In this case, the retention time shift is associated with increasing peak asymmetry, which may be due to a change in the ionization ratio of the acidic analytes in non-buffered separation systems such as a mixture of water and acetonitrile. An increase in the concentration likely shifts the equilibrium toward the non-ionized form of the acid, making its stationary phase retention stronger than that of dilute standards. It is worth noting that this explanation of the phenomenon,

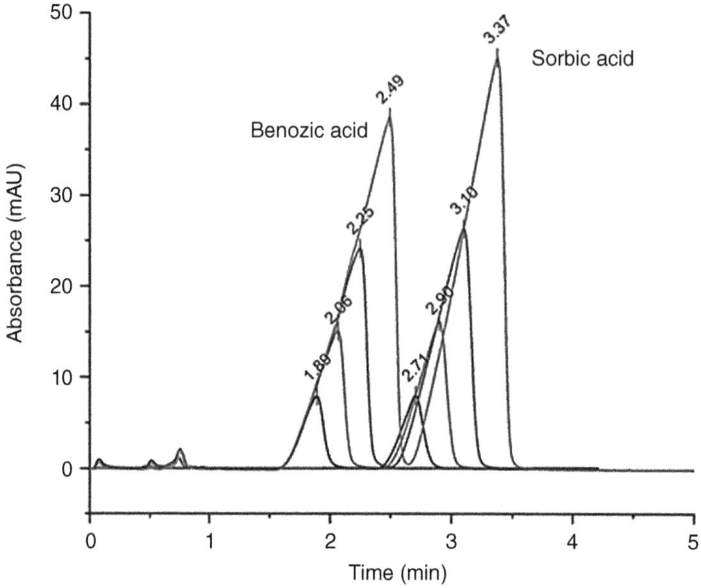

Figure 3.11 The effect of the retention time shift caused by the change in analyte concentration, observed in the HPLC separation of organic acids in the concentration range from 1.0 (lowest peaks) to 10 mg l^{-1} (highest peaks) (from own collection).

which assumes a transition of the analyte from one chemical form to another, gives it the character of a speciation effect.

In the case of complex sample and standard processing, errors are obviously made at every stage of the process. To see the significance of each step in terms of the errors made, statistical or chemometric methods should be used. Table 3.2 shows the

Table 3.2 Examination of measurement errors caused by extraction (b), change of analysis day (c), and change of capillary (d) in comparison with random errors (a) on the basis of the CE signals obtained for dyes contained in a black inkjet printing ink.

	RSD (%)[a)]			
Dye	**a**	**b**	**c**	**d**
Methyl violet	0.44	0.30	2.03	3.34
Victoria blue R	0.58	0.48	3.32	1.56
Victoria blue B	0.48	0.85	3.31	3.98
Rhodamine B base	1.08	1.10	4.04	6.34
Patient blue	2.23	2.02	6.29	11.45

a) Relative standard deviation of measurements conducted:
a – Over one day using one capillary and five sample portions taken from the same extract.
b – Over one day using one capillary and five sample portions taken from five extracts.
c – Over five days using one capillary and five sample portions (a different one each day) taken from the same extract.
d – Over two days using two capillary using two capillaries (a different one each day) and five sample portions taken from the same extract.
Source: Adapted from Kula et al. [4].

results of such a study obtained by one-way analysis of variance (ANOVA) in the case of the identification of dyes present in black inkjet printing inks [4]. The recorded changes in the measurement results obtained with the dye standards showed that the process of extracting the sample from the paper is not, as expected, a critical step in the analytical procedure from the point of view of repeatability of measurement results. It can be seen, moreover, that a more drastic change in conditions (change of analysis day and capillary) resulted in a significant increase in errors to a degree (more than 10%) that gives them a systematic rather than random character.

3.2.2 Reference Sample Method

If in qualitative analysis there is a need to identify the sample as a whole, calibration can be performed not only by the external standard method but also on the basis of comparison of the chemical composition of the test sample with the composition of another appropriately selected reference sample, which acts as a multicomponent chemical standard. Such a calibration procedure can be called the **reference sample method**.

According to the principle of the method, the reference sample is prepared or selected to be as similar as possible to the test sample not only chemically but also physically and structurally (see Figure 3.12). After taking measurements for both samples under identical experimental conditions, their measurement images are compared. If this comparison indicates with sufficiently high probability that the composition of the standard sample, b_x, is the same as that of the test sample, b_0, then the analytical result is positive ($b_0 = b_x$); otherwise, it is negative ($b_0 \neq b_x$).

If the chemical composition of the reference sample is known, its image is compared to that of the test sample based on the signal positions corresponding to known components. While this type of mapping of the real function by a model function is certainly advantageous and promotes accurate identification of the test sample, it is not necessary. A special type of the reference sample method is a procedure involving the use of a reference sample that is known to be similar in chemical composition to the test sample, but whose composition is unknown or not considered in the analytical process. Such a sample may be referred to as a blank chemical standard. The identification of the test sample is then also based on the comparison of its image with the image of the standard sample, where the comparison parameters are usually not only the positions of the signals but also their number and shape.

An example of the application of such a version of the reference sample method is the analysis of measurement images of psychoactive substances (drugs) to determine

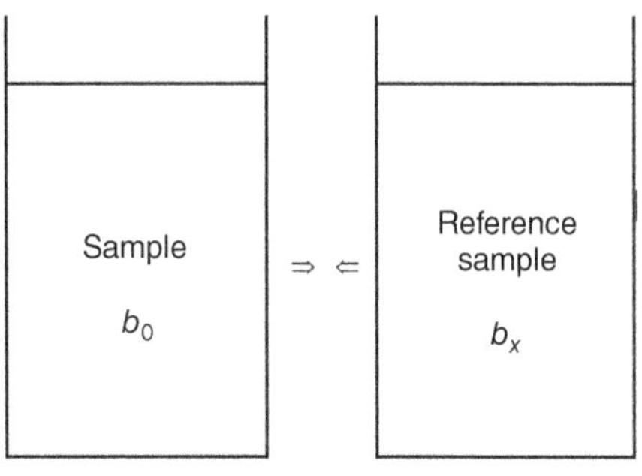

Figure 3.12 Preparation of the sample and standard in qualitative analysis in accordance with the reference sample method; b_0 and b_x are the tested and reference samples, respectively.

Table 3.3 Drug profiling results and their interpretation.

Degree of similarity	Interpretation
Patterns are of very high similarity	Samples are from the same batch of drug obtained in one production run
Patterns are of high similarity	Samples are probably from the same batch of drug, but it is possible that samples come from different batches
Most peaks are similar to each other	Samples have been obtained under similar conditions, i.e. in the same laboratory, but in different batches
Most peaks are different from each other	Samples have been obtained in different laboratories, under different conditions but with the same synthesis method
No similarity of patterns	Samples have been obtained by different methods

their origin. The basis for determining sample similarity, in this case, is the so-called contaminant profile, i.e. measurement signals corresponding not to the main components of the sample but to those that are usually present in the sample in trace amounts. These signals are usually too small to be given a chemical meaning. This unusual analytical procedure is justified by the fact that the trace constituents determine the mode of synthesis of the drug or have been introduced into the test sample as characteristic of that mode. The interpretation of drug profiling results is shown in Table 3.3.

Calibration by the reference sample method is of fundamental importance in forensic analysis. Very often, the measurement image of evidence (test sample) taken from a crime scene is compared with the image of a potentially similar material (reference sample), such as that found on a criminal suspect. The result of the analysis indicating that the analyte image is or is not very likely to be similar to the reference image provides the basis for concluding that the suspect may or may not have committed the crime.

The main difference between the external standard method and the reference sample method at the transformation stage is that the type of sample component sought is usually indicated by only a part of the measurement signals recorded for the sample, while the type of test sample is identified in principle from all the signals measured for the test and reference samples. It seemingly makes identification easier, because both samples contain the same components, then from a theoretical point of view their measurement images should be the same (as long as the amounts of these components in both samples are large enough that the corresponding signals are clearly different from the measurement noise). On the other hand, the absence of any expected component in the test sample causes its measurement image to differ from that of the reference sample, which is the basis for concluding that the two samples are different.

In practice, the interpretation of results in the reference sample method is usually not straightforward, because the measurement images of two of the same samples can differ for many reasons, making it difficult to make a final decision. An example is illustrated by the results shown in Figure 3.13 obtained when testing the

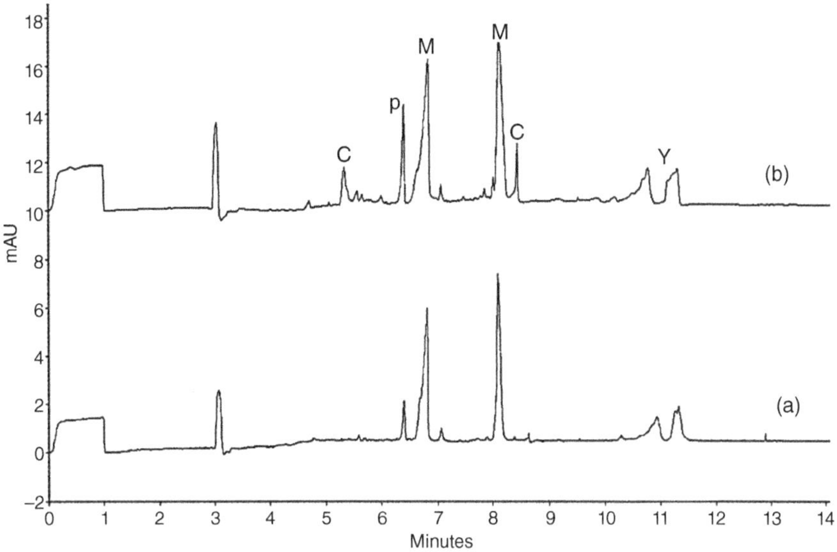

Figure 3.13 Electropherograms of extracts from: (a) postage stamp of unknown origin and (b) ink from Hewlett-Packard Business Inkjet 1200 printer (Y – yellow, M – magenta, C – cyan, p – brighteners from paper). Source: Szafarska et al. [6], fig 2 (p. 82)/with permission of Elsevier.

authenticity of a postage stamp by capillary electrophoresis [6]. The electropherograms obtained for the evidence and reference samples are very similar to each other, the only significant difference being the lack of cyan peaks (at 5.3 and 8.6 minutes) in the image of the evidence sample. Probably, the reddish hue of the postage stamp was obtained in the printing process by mixing of two dyes – magenta and yellow. Given this, it was decided (truthfully) that the challenged postage stamp was printed using HP11 ink (from Business Inkjet 1200) or a similar ink.

An example of the opposite situation is the results of comparative identification of car paint samples [7]. In this case, the spectra obtained by fourier-transform infrared (FTIR) spectroscopy for the test and reference samples were found to be very similar, as shown in Figure 3.14. The main absorption bands, belonging to the main components of the binder, i.e. styrene, acrylic resin, and polyurethane, are almost the same in terms of position, number, and shape. Only intensities of some of them are slightly different.

To dispel doubts about the similarity of the two samples, they were additionally analyzed by pyrolytic gas chromatography (Py-GC) [7]. In the Py-GC system, the sample is introduced into the chromatograph after being brought to a gaseous state in a high-temperature pyrolyzer. This gives the possibility of obtaining signals for much more sample components than IR spectroscopy allows. As can be seen in Figure 3.15, the pyrograms (a) and (b) of the two samples are clearly different from each other at several locations (e.g. at 1.47, 3.91, 6.81, 12.86, 19.80 minutes), which ultimately allowed us to conclude that the two samples are different from each other.

The difficulty of identifying a sample based on its constituent content paradoxically increases when the measuring instrument used is very sensitive and

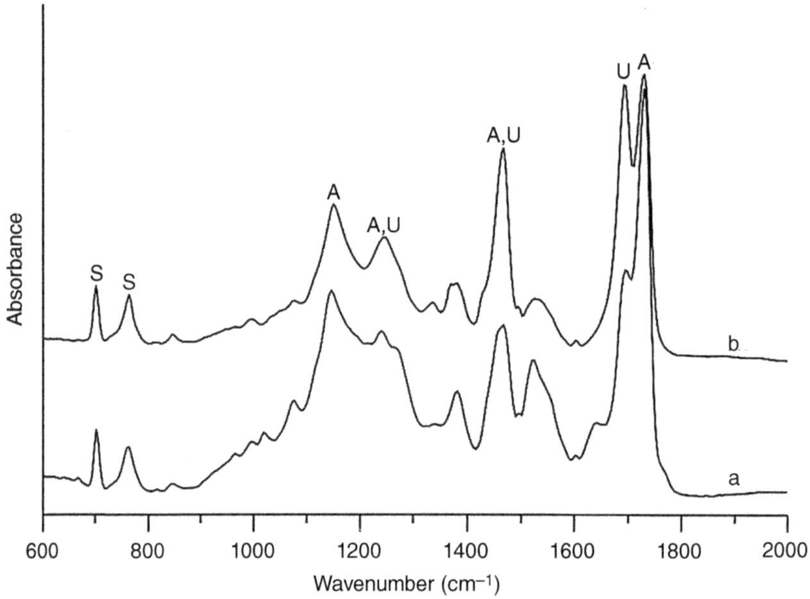

Figure 3.14 FTIR spectra of two paint samples: examined sample (a) and reference sample (b); A – acrylic resin, U – urethane resin, S – styrene. Source: Zięba-Palus et al. [7], fig 1 (p. 112)/with permission of Polish Chemical Society.

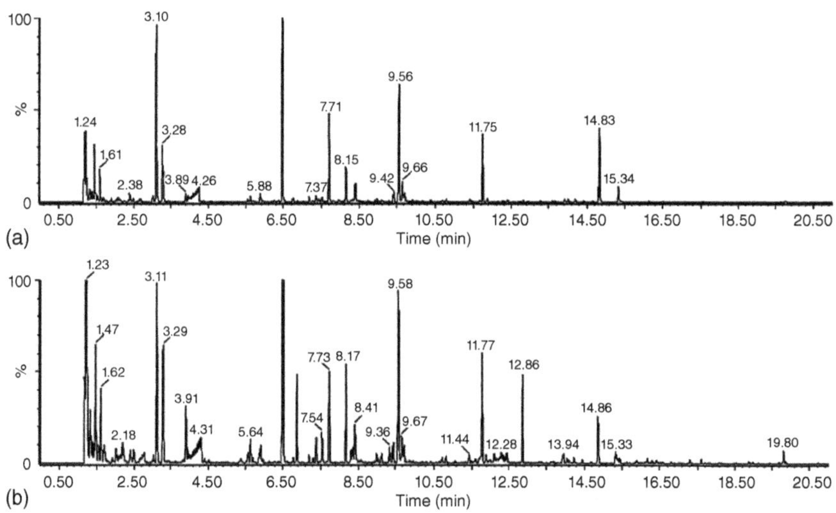

Figure 3.15 Comparison of chromatograms obtained for those samples, (a) and (b), whose infrared spectra are shown in Figure 3.14. Source: Zięba-Palus et al. [7], fig 1 (p. 112)/with permission of Polish Chemical Society.

the measurement image is very complex. The Py-GC method just mentioned is an example of such a method. Although its use potentially increases identification possibilities, too many signals create unfavorable "information noise." In Figure 3.15, the pyrogram of sample B contains several peaks (e.g. at 2.18, 5.64, 11.44, 13.94 minutes) of relatively low intensity, which are absent in the image of sample A. They may come from other components naturally present in the test sample and determining the type of this sample, but it can also be suspected that the pyrolysis process has released components from the sample that are contaminants of the sample or those components that found their way into the sample accidentally during its preparation for measurement.

A different problem can occur when using the reference sample method as a result of accidentally different preparation of the test and reference samples for measurement. This is shown in Figure 3.16 on the example of spectra obtained for a pen paste sample using laser-induced breakdown spectroscopy (LIBS). It can be seen that the spectra are very similar in the position of the peaks, but differ in the intensity of these peaks to such an extent that some of them have become almost invisible against the background of obviously random signal changes. The reason for this phenomenon was the randomly different thickness of the pen paste layer exposed to the radiation of the LIBS spectrometer at two adjacent locations of the writing line. Thus, if the measurement for the test sample is made at a location with too thin a layer compared

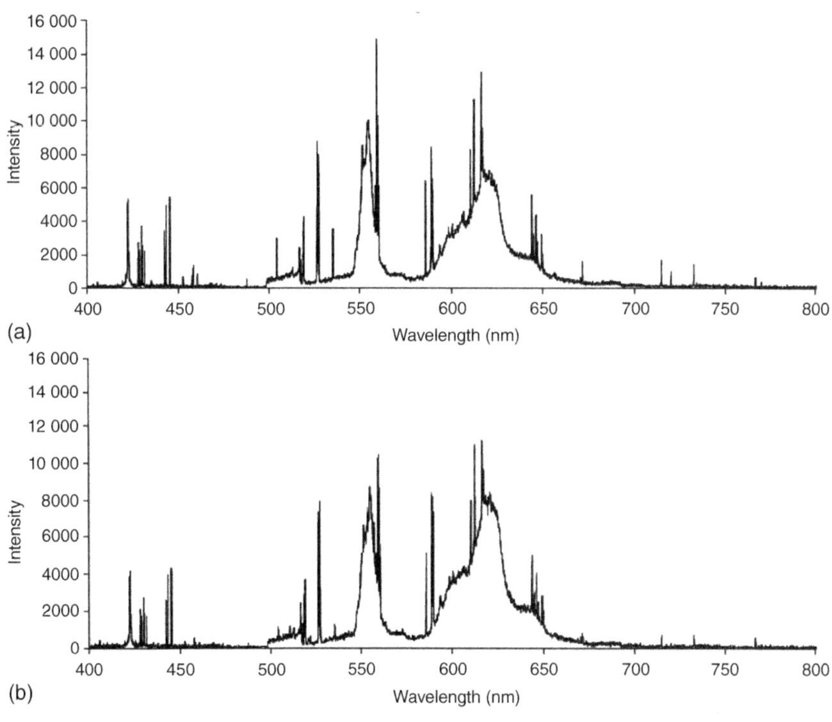

Figure 3.16 LIBS spectra of a ball pen ink sample subjected to measurements in two different places on the writing line (from own collection).

Table 3.4 Number of gunshot residuals (GSR) identified in different places (I–V) of the same gunshot trace using SEM-EDX method; marks ++, +, -, and -- denote the number of residuals more than two times greater, greater, less, and more than two times less than CL (arbitrarily chosen limit), respectively (from own collection).

		Number of particles in comparison with CL				
Kind of particle	**CL**	**I**	**II**	**II**	**IV**	**V**
SbBaPb	100	+	+	--	-	+
SbPb	100	+	+	-	+	+
SbBa	100	+	++	+	+	++
BaPb	100	--	-	--	+	-
Sb	200	+	++	++	+	+
Pb	1000	++	-	+	+	--

to the layer thickness of the reference sample (or vice versa), the omission of small peaks in the identification process may result in a false-negative result.

The problem in comparative analysis of solid samples may also be their natural heterogeneity causing often large differences in measurement images. An extreme example of such a phenomenon is the results presented in Table 3.4, which concern the application of the scanning electron microscopy (SEM) with energy-dispersive X-ray (EDX) method for qualitative determination of the composition of gunshot residue on the basis of the number of characteristic gunshot residuals. It is clear that making a similarity assessment of another trace on this basis to determine whether a shot was fired from this or another weapon is very difficult without additional examinations.

All uncontrolled effects that hinder the identification of analytes by the external standard method are also a problem in calibration by the reference sample method. This is particularly true for preparative effects. It can be particularly difficult when the sample to be analyzed is bound to the base on which it was originally placed. Separation of the sample from the base is usually very difficult and sometimes even impossible or not allowed (e.g. when the sample is evidence and there is a fear of destroying it). The use of chemicals for this purpose is associated with the risk of sample contamination. There are, of course, measurement methods that make it possible to analyze a sample on a base, but then the measurement signals from the components of the sample may be "enhanced" by the signal caused by the components of the base, which obviously creates identification difficulties. The problem is illustrated in Figure 3.17A with the example of identifying automotive paint on a base using the attenuated total reflection (ATR) method. It is clearly seen that the spectrum of the test sample contains many bands and peaks originating from the base, which significantly alter the measurement image from the reference sample.

One way to compensate for this effect is to apply a reference sample to the same base as the test sample and make an identification from the spectra of both samples with the base. However, if it is possible to make a spectrum of the substrate itself

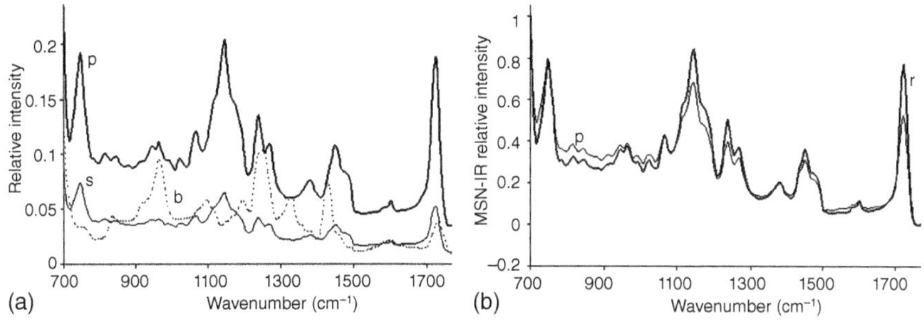

Figure 3.17 ATR spectra of: (a) pure Tutto paint (p), the base (plastic foil) (b), Tutto paint on the plastic foil (s), and (B) pure Tutto paint normalized (p), obtained by the MSN-IR method (r). Source: Szafarska et al. [8], fig 2 (p. 507)/with permission of Elsevier.

under the same conditions, the difference between the spectra of the test and reference samples can be compensated for using mathematical approaches. Figure 3.17B shows the effect of one version of such methods [8]. The procedure involves normalizing both spectra and the base spectrum, and then subtracting the base spectrum from the test sample spectrum. As can be seen, the spectrum of the test sample can be obtained in this way to a large extent similar to the spectrum of the reference sample.

A specific preparative effect occurs in DNA (deoxyribonucleic acid) analysis, which is known to be of extremely importance in forensic research. Forensic science is DNA analysis. The result of a DNA test in a sample taken from a person is basically unique with respect to other people and hence the very high evidentiary value of this result. The shape of the measurement image in DNA analysis does not depend on such factors as the type of biological sample taken (e.g. saliva, sperm, hair), the way it was taken, and the procedure of preparing it for measurement. The only serious phenomenon that reduces the identification potential of DNA analysis is the gradual degradation of the sample under the influence of chemical compounds or storage time [9]. This phenomenon manifests itself by a gradual decrease in peak intensity, sometimes to their complete disappearance, as seen in Figure 3.18.

In general, in analysis by the reference sample method, it is much easier to conclude that two samples are distinctly different from each other than that they are similar enough to be treated as the same. Therefore, it is a good idea to start the identification process with simple measurement methods that do not require much time or cost. If the test and reference samples are different, the difference can often be detected, for example, by microscopic examination, i.e. observation of their physical and morphological properties. However, if at this stage there is doubt about the difference between the samples, other methods should be used, providing progressively more information about their chemical composition. Such a procedure is particularly valid in the analysis of forensic traces. Looking at it from the calibration side, one can say that it contributes to the enrichment of the real and model functions to best approximate the real function with the model function.

An example of this research strategy is presented in Table 3.5 [10]. In this case, progressively different microscopic observation techniques and several

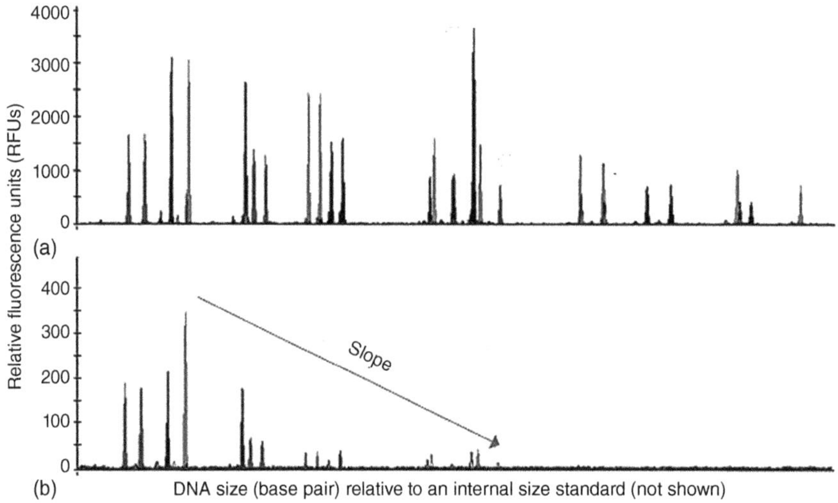

Figure 3.18 DNA profiles obtained for a sample of good-quality (a) and poor-quality (b) due to its degradation [9].

physicochemical methods with different identification capabilities were used to investigate the similarity of the two paint samples. In addition, measurements by these methods were made for several layers of the proof and reference samples, taking advantage of the fact that the different paint layers in the two samples were gradually superimposed on each other. Only the information gathered in this way made it possible to conclude that the two paint samples are different from each other (*nota bene* consistent with reality).

The reference sample method in both versions is used not only for individual sample identification but also to determine, on the basis of chemical composition, similarities or differences between several or many samples with similar visual characteristics and physical properties. The samples are then sequentially measured under identical experimental conditions in such a way that, during a given analysis, one sample is treated as a standard and all others as tested samples. On the basis of the obtained results, a classification of samples is made, assigning to the same group samples of the same or similar chemical composition. This is called **discriminant analysis**.

This type of identification analysis is of fundamental importance in the work of a forensic expert. Having the classification of samples of a given type at his or her disposal, he or she can subject the evidential sample to measurements and on the basis of the obtained image classify it to a specific group of samples. The classification serves as a kind of database. Assigning an evidentiary sample to a specific group of samples makes it much easier at a later stage of proceedings to decide whether or not a crime has been committed.

For example, Table 3.6 shows the results of discriminant analysis of black printing inks obtained by CE-MS [4]. After the initial electrophoretic separation of the components, the classification of the samples was performed by MS spectra based

Table 3.5 Results of qualitative analysis (layer by layer) of two car paints (evidence and reference ones) using microscopic techniques: stereoscopic in reflected light (MS/O), polarizing in transmitted light (MP/P), polarizing in polarized light (MP/S), and fluorescence with UV filter (MF/UV), as well as using FTIR, Raman and SEM-EDX spectrometric methods.

	Color				Chemical composition		
Layer	**MS/O**	**MP/P**	**MP/S**	**MF/UV**	**FTIR**	**Raman**	**SEM-EDX**
Evidence sample							
1	Red	Bordeaux	Red	Red	Alkyd, acryl, melamine	—	C, O, Cl, Ti, Fe
2	White	Brown	Brown	Blue	Alkyd	TiO_2, $CaCO_3$	C, O, Mg, Al, Si, Ca, Ti
3	Grey	Brown	Brown	Yellow	Epoxide	TiO_2, aluminosilicates	C, O, Al, Si, Ca, Ti
4	—	—	—	Grey	—	—	C, O, Zn, Al, Si, P, S, Ti, Mn, Fe, Ni
Reference sample							
1	Red	Bordeaux	Red	Red	Alkyd, acryl, melamine	—	C, O, Al, Si, Cl, Ti, Fe
2	White	Brown	Brown	Blue	Alkyd	TiO_2, $CaCO_3$	C, O, Mg, Al, Si, Ca, Ti
3	Ash–grey	Brown	Brown	Yellow	Epoxide	TiO_2, aluminosilicates	C, O, Al, Si, Pb, S, Ca, Ti, Ba
4	—	—	—	Grey	—	—	C, O, Al, Si, P, P, Ti, Mn, Fe, Ni, Zn

Source: Nieznańska et al. [10], table VI (p. 10)/Institute of Forensic Sciences.

Table 3.6 The results of CE-MS analyses of black inkjet printing inks.

Sample	Group	Manufacturer	Model	Black cartridge number	Color cartridge number	*m/z* value
1	I	Hewlett–Packard	Business Inkjet 1200	HP10	HP11	250.169; 295.144; 312.168; 317.120; 340.200; 350.186; 408.227; 431.192; 474.298; 531.130; 548.347; 559.158; 581.139; 604.217
2	II	Hewlett–Packard	Deskjet 4280F	HP300	HP300	360.125; 450.637; 486.154; 548.346
3	III	Hewlett–Packard	Deskjet 5550	HP56	HP57	506.956; 508.953; 531.128; 548.344; 559.155; 581.135; 604.213
4	IV	Hewlett–Packard	Deskjet 5652	HP56	HP57	320.288; 506.957; 508.956; 531.127; 548.347; 559.158; 581.138; 604.213
5	V	Hewlett–Packard	Photosmart B109a	HP364	HP364	312.165; 548.344
6	VI	Brother	DCP 135C	LC970B	LC970	548.350; 562.367
7	VII	Brother	MFC 5440	LC900B	LC900	562.368; 592.381
8	VIII	Canon	I 965	BCI6	BCI6	477.107; 552.411; 656.142
9	IX	Canon	MP 240	PG510	CL511	556.440
10	X	Canon	IP 1900	PG37	CL38	441.198; 524.294; 552.409; 568.323
11		Canon	IP 1900	MP210	CL38	441.198; 524.294; 552.409; 568.323
12	XI	Lexmark	2530X	34	35	476.308
13	XII	Epson	92D	T0711	T07	552.412

Source: Kula et al. [4], table 1 (p. 21)/Elsevier.

on the characteristic values of the m/z parameter. As can be seen, the applied research method allowed to distinguish 12 out of 13 samples coming from different printer models regardless of whether the printers were manufactured by the same or different manufacturers. Interestingly, this was possible even when the inks were taken from the same two cartridges installed in two different printer models (samples 3 and 4).

Discriminant analysis is often supported by mathematical and chemometric methods that ensure the use of objective classification criteria. Thus, in the work [11], samples taken from automotive paint coatings of red color with different shades of red were compared. The color information of the samples was obtained by visible light microspectrophotometry (MSP). To enhance the identification capability of the method, each sample was measured in reflection mode using a light beam falling perpendicular to the top surface of the paint sample (RS) and on a cross section of the paint chip (RCS). An objective mathematical approach in the form of the CIELAB color model was used to characterize each spectrum. The differences between the color shades, E, of each pair of samples were calculated from the following formula:

$$\Delta E = \sqrt{(L_1 - L_2)^2 + (a_1 - a_2)^2 + (b_1 - b_2)^2} \tag{3.1}$$

where L_i, a_1, and b_i ($i = 1, 2$) are the coordinates representing the lightness of the color, position between red and green colors, and position between yellow and blue colors, respectively. Based on the results of previous studies, it was assumed that the two samples characterized by $E < 5.0$ values obtained by both techniques, RS and RCS, could be classified as equal shade samples.

The samples were compared in pairs (each with each other). In each case, two samples played the role of test and reference samples with respect to each other. The results obtained are shown in Table 3.7. Despite the visual similarity of all the samples tested, only three pairs of them (samples 1 and 3, 2 and 4, and 3 and 4) turned out to be samples with the same shade of red color. This allowed us to suppose that all four samples (1–4) could come from the paint of the same car.

In another work [12], there have been attempts to discriminate red lipstick samples for forensic purposes. The problem was especially difficult because sample of lipsticks of the same color have a very similar chemical composition. The study was performed with the use of 38 lipsticks produced by 20 different manufacturers on the basis of ATR spectra. The discrimination process was performed using chemometric methods. Preliminary results obtained according to principal component analysis (PCA) are shown in Figure 3.19. At this stage, six groups of samples were able to

Table 3.7 ΔE values obtained by MSP method and both RS and RCS techniques for pairs of bright red solid car paints (1–7).

Sample	RS	RCS	RS	RCS	RS	RCS	RS	RCS	RS	RCS	RS	RCS
	1		2		3		4		5		6	
2	4.2	5.7										
3	3.7	3.5										
4	4.2	5.6	1.3	4.7								
5	8.9	2.9	7.7	6.8	10.2	3.0	6.6	4.3				
6	6.1	8.6	3.3	12.6	6.8	8.8	4.5	8.8	10.2	6.3		
7	22.0	9.3	22.0	14.6	19.5	9.9	22.9	12.3	29.3	8.2	21.2	7.0

Source: Adapted from Trzcińska et al. [11].

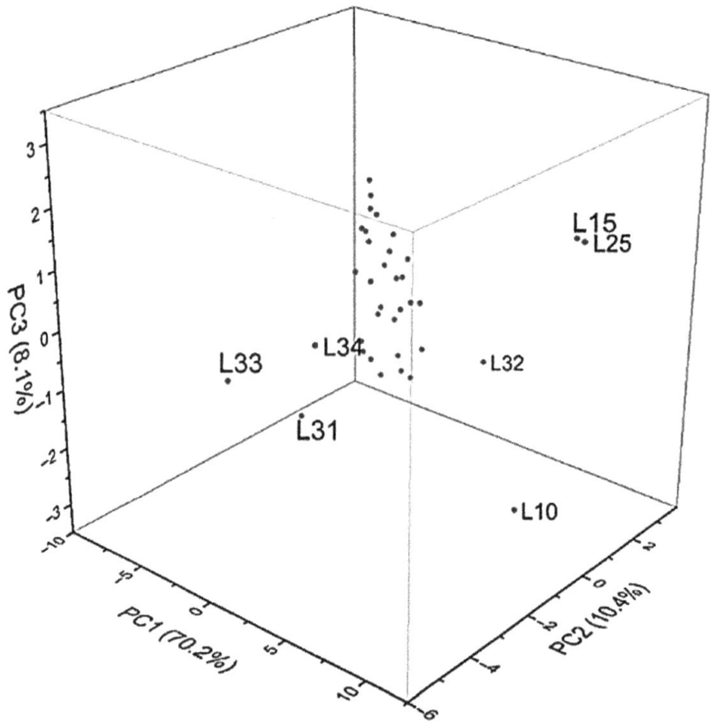

Figure 3.19 The PCA score plot of the first three components for all investigated lipsticks, taking into account ATR spectra registered in the range from 4000 to 650 cm^{-1}. Source: Gładysz et al. [12], fig 3 (p. 133)/with permission of Elsevier.

be distinguished: {L10}, {L15, L25}, {L31, L33}, {L32}, {L34}, and the group, which includes all the remaining samples. The use of cluster analysis identified an even larger number of dissimilar groups.

The reference sample method is also used when there is a need to study changes in the chemical composition of samples under the influence of various factors. The tests are conducted in a controlled manner, establishing the values of a particular factor at fixed levels, and at each stage a measurement image of the sample is taken, keeping the measuring conditions at the same level. It is often used to study the effect of environmental conditions on the chemical composition of a sample or changes in the interference effect that occur under different sample preparation conditions.

Figure 3.20 presents the results of analyses performed for forensic purposes, showing the change in the chemical composition of car oil as it is used up (after the car has traveled a certain distance) [13, 14]. In this case, the spectrum of fresh oil (0 km) was treated as the spectrum of a reference sample, used to compare subsequent spectra of the test samples. The chemometric analysis showed that after another 1000 km, the oil spectrum differs significantly from the previous one, which is the basis for estimating the mileage of the car involved in the forensic event [14].

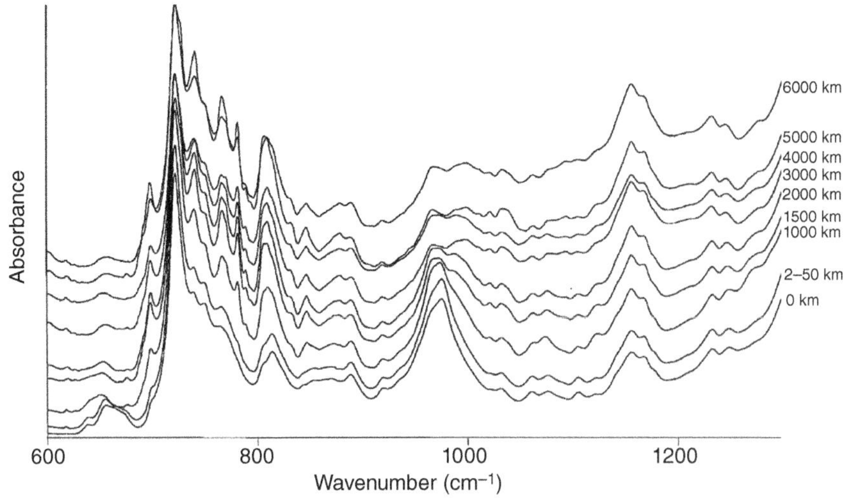

Figure 3.20 FTIR spectra of the oil sample (Elf Sport Super) received after a car has covered a certain distance. Source: Zięba-Palus et al. [13], fig 1 (p. 36)/with permission of Elsevier.

3.3 Internal Calibration Methods

As seen in Figure 3.1, **internal calibration methods** also belong to comparative methods. They satisfy the following conditions:

- The sample and standard are prepared separately from each other.
- Another substance of known type and absent in the sample can be added to the sample and to the standard in usually equal and known amounts.

The added substance can either:

- be chemically inert with respect to the analyte and other sample components; or
- react with the analyte, but not with other sample components.

If both substances meet additional, specific conditions (discussed in what follows), then a comparative calibration involving them takes on the characteristics of the **internal standard method** and the **indirect method**, respectively.

Internal calibration methods require specific sample and standard preparation for measurement, and, consequently, different interpretation of measurement results from that used in external calibration methods. They also have their own specific analytical objectives. Therefore, despite their similarity to external calibration methods, both approaches are separate calibration methods.

3.3.1 Internal Standard Method

In the **internal standard method**, the added substance is called an internal standard (IS). Care should therefore be taken not to confuse the name and role of this

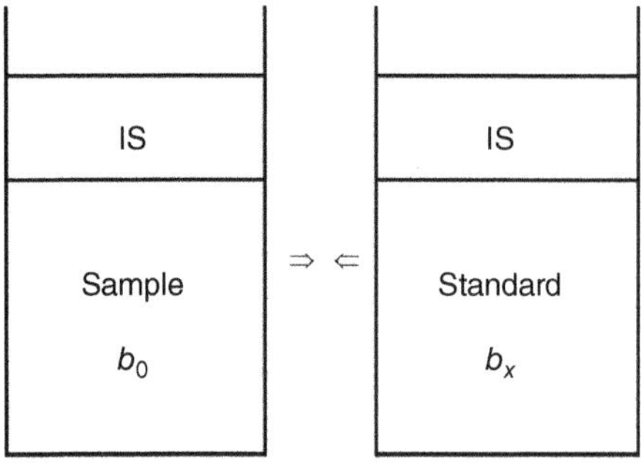

Figure 3.21 Preparation of the sample and standard in qualitative analysis in accordance with the internal standard method; b_0 and b_x is expected and known analyte, respectively, IS – internal standard.

standard with an "external" standard that is prepared separately from the sample and contains a known type of analyte. A scheme for the preparation of the sample and the external standard in the internal standard method is shown in Figure 3.21.

The sample and the external standard are measured under identical conditions and the positions of the signals for both the analyte (a) and internal standard (IS) in the sample (S), $t_{a,S}$, $t_{IS,S}$, and in the external standard (St), $t_{a,St}$, $t_{IS,St}$, are determined. After the measurements are made, the analyte is identified in the sample by comparison of not the absolute (as in the external calibration methods), but the relative values, $r_{a,S}$ and $r_{a,St}$, of the positions of the peaks, using the formulas:

$$r_{a,S} = \frac{t_{a,S} - t_0}{t_{IS,S} - t_0} \tag{3.2}$$

$$r_{a,St} = \frac{t_{a,St} - t_0}{t_{IS,St} - t_0} \tag{3.3}$$

where t_0 is the initial moment (particularly taking the value 0) of measurement of images for both the sample and external standard.

As mentioned earlier, in the analysis of various analytical techniques, especially separation techniques, there is often a phenomenon consisting in a shift of the measurement images obtained for the sample and for the external standard in spite of keeping constant the experimental conditions during the measurements. If the shift of the peak recorded for the internal standard with respect to the analyte peak is proportional to the time of appearance of both peaks, this effect can be compensated to a large extent by such a calibration procedure. Compensation is all the more likely and more complete (at least in theory) if the internal standard meets the following additional conditions:

- It is similar to the analyte in terms of physicochemical properties and structure.
- The corresponding signal is located in the sample and standard images relatively close to, but does not overlap with, peaks from the analyte.

Thus, it can be said that the internal standard acts as a source of additional information to support the accurate representation of the real function by the model function. It is clear that in such a role, the internal standard can be helpful not only in identifying compounds treated as an analyte, but also when the compounds are the basis for identifying the sample as an analyte. It also does not matter then whether the components of the model sample are known or not.

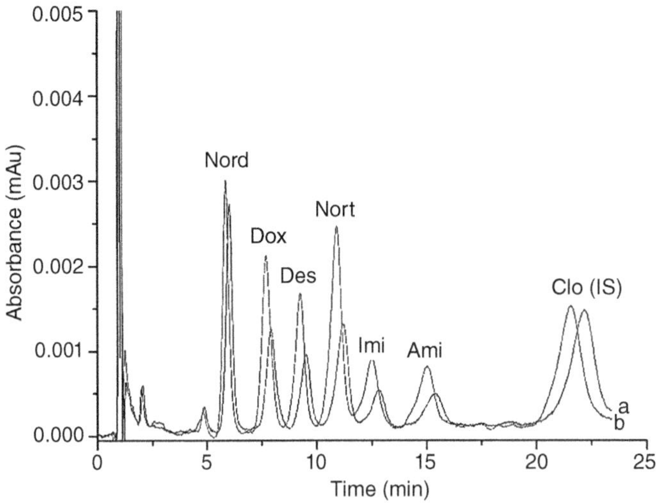

Figure 3.22 HPLC chromatograms obtained for a sample (a) and standard (b), both containing nordoxepin (Nord), doxepin (Dox), desipramine (Des), nortriptyline (Nort), imipramine (Imi), amitriptyline (Ami), and clomipramine as the internal standard (IS). Source: Adapted from Woźniakiewicz et al. [15].

An example of the use of an internal standard in a multicomponent qualitative analysis is shown in Figure 3.22. In this case, the internal standard was selected in such a way that, like the analytes, it is a drug belonging to the group of tricyclic antidepressants. It is evident that the shifts of the peaks recorded for the analytes and the peak of the internal standard gradually increase as the retention times corresponding to these peaks increase, which justifies performing the calibration based on the values of the retention times calculated from Eq. (3.2). Using absolute retention times could cause some difficulties in identifying individual analytes.

Choosing an internal standard belonging to the same group of compounds to which the analyte(s) belongs involves the risk that it is naturally present in the sample. If you are analyzing a series of samples, the amount of the standard can vary from sample to sample. The addition of a standard from outside will therefore also cause the intensity of its signals to vary in the measurement images of successive samples. Nevertheless, these signals will have the same position relative to the other peaks, and differences in their intensities should not be a source of analytical error. It is certainly better, however, if the internal standard is absent from the samples, because then any doubt in interpretation is avoided.

The problem of analytical signal position stability seen in chromatographic methods is even more pronounced in the capillary electrophoresis method. Here, the effect of signal shifts is often caused by fluctuations in electroosmotic flow (EOF), which occurs when sample or buffer components interact and adsorb on the inner surface of the capillary, changing their physicochemical properties. The resulting shifts in migration times, especially during long separation sequences, often reach several minutes or even more. In some cases very good results can therefore be obtained by treating the EOF peak as an internal standard peak. The calibration process remains the same as in the classical internal standard method (it is based on formulas (3.2) and (3.3)), but does not require the use of a specific chemical. The data in Table 3.8 show the effects of this calibration approach.

It has been shown in [16] that despite relative migration time is more stable than the absolute values upon alteration in the flow rate, some shift should always be expected. This effect, determined for a selected pair of compounds acting as the analyte and internal standard, is presented in Figure 3.23. The magnitude of this shift depends on the position distance of the peaks of the two compounds and on their mutual location relative to the flow direction. It has been shown that if the molecular mobility of these compounds is known, this relationship can be predicted theoretically by a mathematical model, as seen in Figure 3.23.

Since in the internal standard method the sample and the external standard are treated separately, as in the external standard method, the susceptibility of both

Table 3.8 The absolute and relative values (with EOF as the internal standard) of migration time corresponding to the signals of the yellow (Y), magenta (M), and cyan (C) dyes in standards and in the ink sample taken from Hewlett–Packard Business Inkjet 1200 printer (from own collection).

Measurement mode	Material examined	Peaks					
		EOF	M	M	C	Y	Y
Absolute migration time (min)	Standard	3.05	6.73	7.73	8.50	10.69	11.05
	Sample	3.07	6.80	8.09	8.63	10.93	11.32
Relative migration time	Standard	1.00	2.21	2.53	2.79	3.50	3.62
	Sample	1.00	2.21	2.64	2.81	3.56	3.69

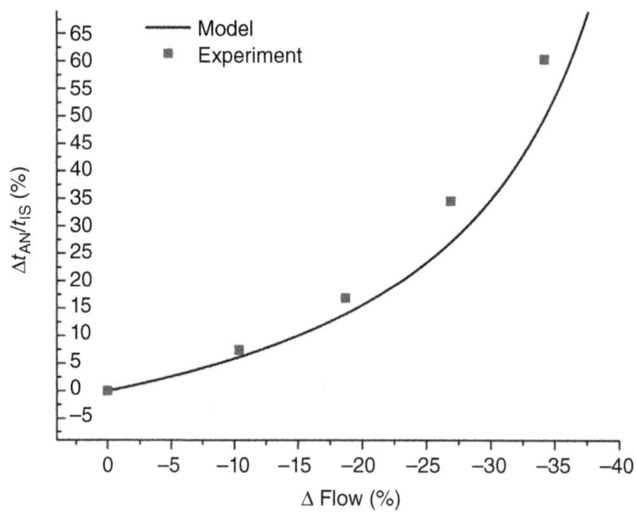

Figure 3.23 The average relative shifts in migration time ratios (%) as a function of the actual flow rate decrease. Source: Nowak et al. [16], fig 2 (p. 4)/with permission of Elsevier.

methods to various kinds of previously described uncontrolled effects – apart from the effect of changes in the position of the signals in time – is similar. In particular, identification difficulties should be expected in cases of overlapping signals, signals of very low intensity, or measurement images with a very large number of peaks. In such situations, one should take the steps previously mentioned, i.e. perform an additional analysis of the sample using a different measurement method.

The specific problem of the internal standard method is the proper selection of the standard, ensuring that its "behavior" is similar to that exhibited by the analyte under the given measurement conditions. This requires special knowledge and experience of the analyst (although this problem, as will be shown, is of much greater importance in the version of this calibration method adapted to quantitative analysis). Furthermore, the addition of a "foreign" substance to the sample can produce additional, undesirable uncontrolled effects. External interference with sample composition may even be unacceptable in certain situations. This is, for example, often the case in forensic testing, where such interference may compromise the evidentiary character of the sample. Nevertheless, the internal standard method is a widely accepted calibration method and is often used in qualitative analysis.

3.3.2 Indirect Method

In the **indirect method**, a substance of known type is added to the sample and the external standard to react with the analyte in the standard, b_0, and the same, the desired analyte in the sample b_x (see Figure 3.24). Both solutions are then measured under conditions that are optimal for the reaction product of the reagent with the standard analyte. This product thus takes over the role of the analyte in the sense that the position of its signal, and not that of the analyte, indirectly indicates the presence of the analyte in the tested sample.

When an analyte is present in the sample, the product obtained after reaction with this component should be the same as the product obtained after reaction with the analyte contained in the standard. The signals obtained for this product should therefore be located at the same positions in the measurement images of the standard and the sample and thus indicate the presence of the desired analyte in the sample. Otherwise, the reaction of the reagent with the sample should not take place or should lead to a different product. Both situations are then revealed by the absence

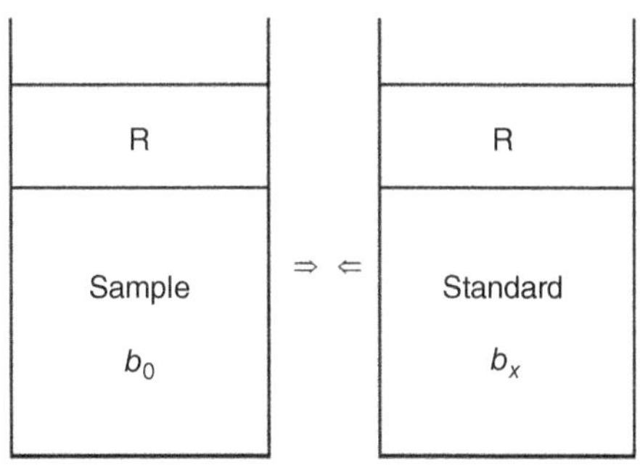

Figure 3.24 Preparation of the sample and standard in qualitative analysis in accordance with the indirect method; b_0 and b_x are expected and known analytes, respectively; R – reagent.

of a signal at the expected position for the analyte, which is evidence of the absence of the analyte in the sample.

The reagent added to the sample and standard should meet the following basic conditions:

- Not be a natural component of the sample being analyzed
- React selectively with the analyte
- Lead to a product that does not react with other sample components and whose signal is clearly separated from the signals of these components.

In addition, the entire calibration procedure should provide better conditions for identifying the analyte than handling it without a reagent (i.e. calibration by the external standard method). This is a necessary condition for it to be considered a separate calibration method.

The most common indirect calibration uses the derivatization technique, i.e. conversion of a chemical compound into a product of a similar chemical structure, called a derivative. The technique was used to identify e.g. fluoxetine in blood samples by HPLC [17]. The method is based on a selective reaction between 7,7,8,8-tetracyanoquinodimethane (TCNQ) and molecules with a secondary amine moiety. In this reaction, one or two (if secondary amine is in excess) cyano groups in TCNQ molecules are replaced by an amine group from a drug molecule and the obtained products (quinodimethanes) exhibit an intense purple color. The reaction mechanism is shown in Figure 3.25.

A blood sample was spiked with fluoxetine and nortriptyline as external and internal standards, respectively, and then subjected to the microwave-assisted extraction (MAE). After appropriate preparative stage, dried residues were dissolved in TCNQ solution and analyzed by HPLC with diode-array detection at 567 nm. The signal measured for the reaction product was compared in terms of position with the signal obtained for the sample prepared in the described way but not spiked with fluoxetine.

The signal obtained by UV–VIS spectrophotometry for fluoxetine derivative is shown in Figure 3.26. As seen, it is very well separated from both fluoxetine and TCNQ signals. Moreover, only a limited number of substances extracted from blood may absorb light at 567 nm. Selectivity of derivative reaction reduces the risk of interaction of the analyte (or IS) with other endogenous or exogenous compounds. Finally, the calibration method used is more reliable than direct calibration as the

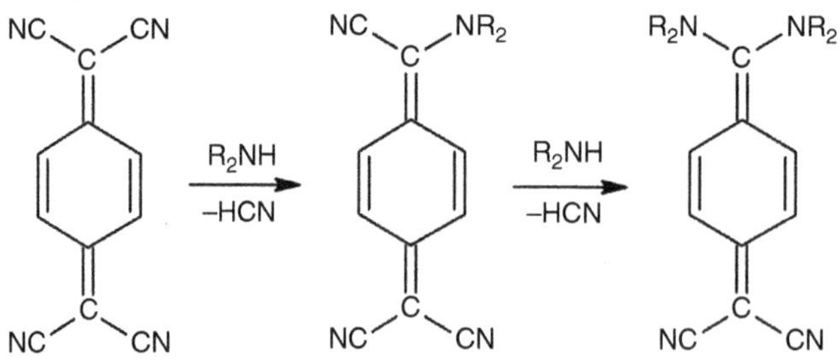

Figure 3.25 Mechanism of formation of derivatives of secondary aliphatic amines (R_2NH) with TCNQ.

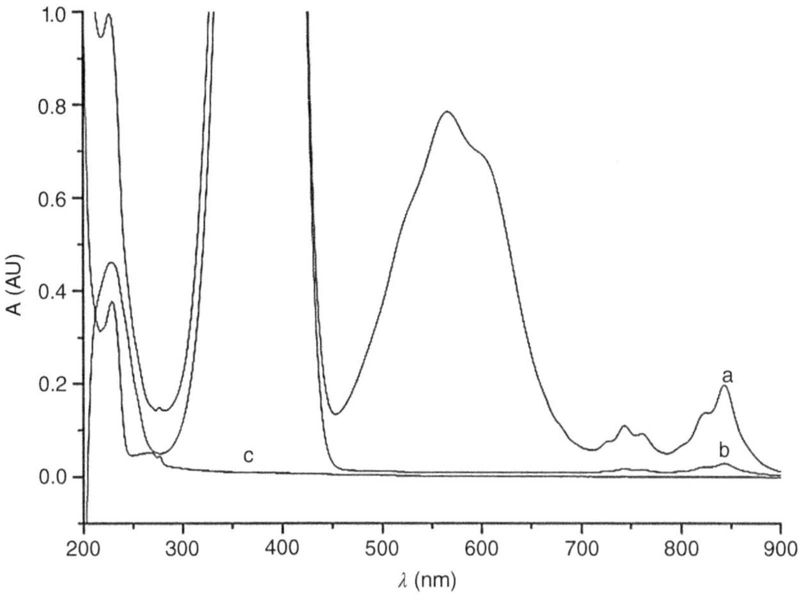

Figure 3.26 UV–VIS absorption spectra of (a) TCNQ-fluoxetine derivative (5 μg ml^{-1}), (b) TCNQ solution, (c) fluoxetine solution (5 μg ml^{-1}). Source: Woźniakiewicz et al. [17], fig 3 (p. 673)/with permission of Polish Chemical Society.

fluoxetine is rapidly metabolized in blood and its content quickly falls to less than therapeutic level.

The above studies also show the possibility of combining the indirect method with the internal standard method in a single calibration procedure [17]. Such a procedure usually offers the advantages of both calibration methods alone. However, in the choice of internal standard and reagent, care must be taken to consider the presence of both substances together in a sample. In particular, it is important that they do not react chemically with each other and that their signals are sufficiently well separated.

In forensic analysis, the derivatization process can help to improve the efficiency of group discrimination of samples. It has been found that rubber samples from different tires can be effectively discriminated by pyrolytic gas chromatography–mass spectrometry (Py-GC-MS) [18]. If both main polymers used in the manufacturing process and trace substances that improve rubber properties are included in the analytical process, it is possible to discriminate between rubber samples from different manufacturers and from different tire models produced by the same manufacturer. However, some samples could not be definitely distinguished. In such cases, an online derivatization technique (using tetramethylammonium hydroxide) was applied and, consequently, the presence of fatty acids (as their methyl esters) was additionally detected on the pyrograms. This improved the discrimination ability of the samples, as shown by the data collected in Table 3.9.

A specific possibility of performing indirect identification is offered again by capillary electrophoresis. Wide applicability of this method is limited by the fact that

Table 3.9 Results of the discrimination Py-GC-MS analysis of 42 rubber samples.

	Sample group			
Experimental data	**I**	**IIa**	**IIb**	**III**
No. of the samples in groups	11	5	18	8
No. of the sample pairs differentiated without derivatization	51	9	147	28
No. of the sample pairs differentiated after derivatization	55	10	151	28
Discrimination power without derivatization (%)	93	90	96	100
Discrimination power after derivatization (%)	100	100	99	100

Source: Adapted from Lachowicz et al. [18].

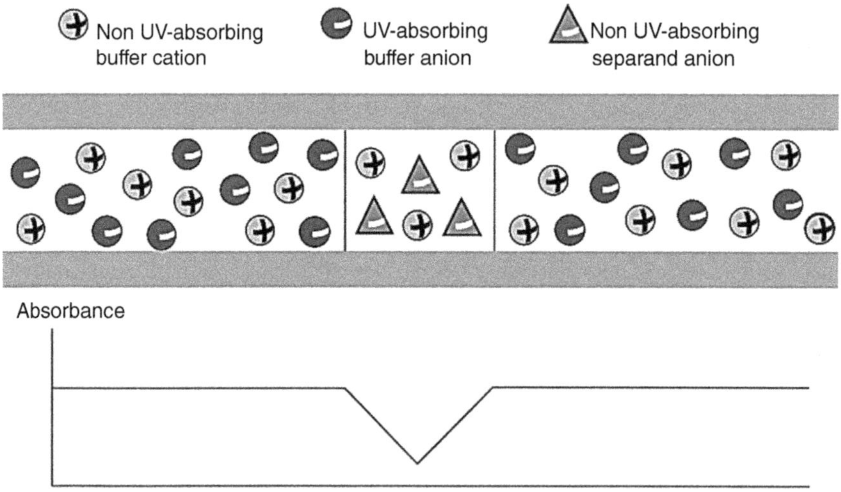

Figure 3.27 The principle of indirect photometric detection in capillary electrophoresis (for details see the text) [19].

most small ions do not absorb in the UV/VIS regions of the spectrum. This problem could be solved in a way named indirect photometric detection (IPD) [19], which is schematically shown in Figure 3.27. In this technique, the CE background electrolyte containing analyte is spiked with a UV-absorbing buffer ion of the same charge as the analyte ion. This additive, known as a visualizing reagent, elevates the baseline. Analyte displaces the visualizing reagent (in accordance with the principle of electroneutrality) and its presence is detected as a negative peak relative to the higher baseline.

IPD (absorbance and fluorescence) of anions by capillary electrophoresis is reviewed in [20] together with discussion about factors that influence the displacement process of analyte ions in background electrolytes containing one or more co-ions.

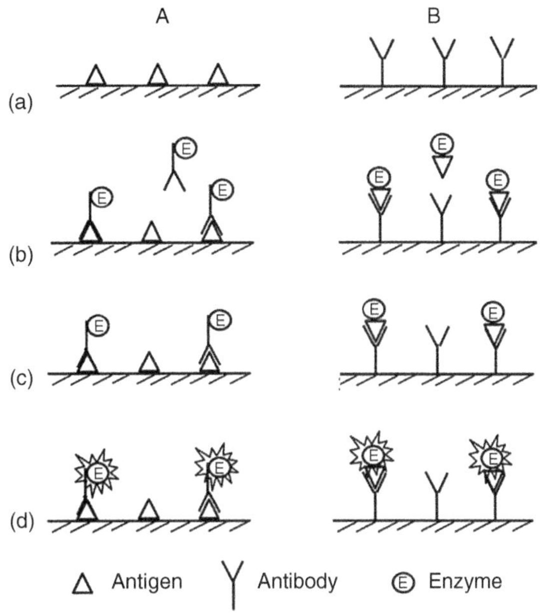

Figure 3.28 Direct ELISA to detect antigen (A) and antibody (B) (steps a–d are explained in the text).

A special type of indirect calibration is used in an analytical procedure named the enzyme-linked immunosorbent assay (ELISA). The assay uses an enzyme to detect or to determine a protein (analyte) in a sample using antibodies directed against the protein to be detected.

The most simple ELISA procedure (called "direct ELISA") is shown schematically in Figure 3.28. When an antigen is to be identified (see Figure 3.28a), it is immobilized on a solid support (usually a polystyrene microtiter plate) (a) and then a matching antibody is applied over the surface so it can bind the antigen (b). This antibody is linked to an enzyme, and then any unbound antibodies are removed (c). In the final step, a substance containing the enzyme's substrate is added (d). The subsequent reaction produces a signal usually registered by UV/VIS spectrophotometry. The standard (control) sample containing no test antigen should be treated in the same way. Analogous procedure relates to identification of antibodies (see Figure 3.28b).

Another type of the ELISA procedure (direct ELISA) involves a two-binding process between a primary antibody and an enzyme-labeled secondary antibody. The primary antibody binding to the immobilized antigen can react with the enzyme-labeled secondary antibody, followed by color development. The procedure gets its name from the fact that the antigen is indirectly detected by the enzyme-labeled secondary antibody. Thus, from a calibration point of view, this procedure can be called doubly indirect.

In analytical practice, the ELIZA is used in other variants (direct and indirect competitive ELIZA, sandwich and open-sandwich ELIZA) adapted to different analytical circumstances (among others, to the size of the protein molecule to be detected). In all cases, the analysis is performed on the basis of an indirect calibration. The advantages of this technique are the high specificity resulting from the antigenic reaction and the relatively simple, rapid, and eco-friendly preparative step, which does not require the use of organic solvents and radioactive substances. Its disadvantages include relatively high cost of analysis (expensive culture cell media are required to obtain a specific antibody), instability of antibody, and high possibility of false-positive or -negative results because of insufficient blocking of the surface of microtiter plate immobilized with antigen. Nevertheless, ELISA test is widely used as a diagnostic tool in medicine, plant pathology, and biotechnology, as well as a quality control check in various industries.

3.4 Standard Addition Method

The only additive calibration method in qualitative analysis is the **standard addition method**. Following the procedure of this method, a standard containing a known analyte, b_x, is added to a test sample in which that analyte, b_0, is expected to be present (see Figure 3.29). The sample is measured before and after the addition of the standard.

The addition of a standard to a sample causes a signal to appear in the measurement image of the sample with the standard at a position corresponding to the analyte contained in the standard. If the intensity of this signal increases compared to the intensity of the peak at the same position in the sample image, the analyte is said to be present in the sample with high probability. The absence of this effect indicates that there is likely no analyte in the sample. In the standard addition method, signal intensity is therefore an additional – besides signal position – parameter taken into account in analyte identification and supports the accuracy of this identification. This ability to operate with two parameters regardless of whether and what other identification parameters are available to the analyst in a given analysis is an added value of the standard addition method.

From a calibration point of view, the additive method of modeling involves formulating a model function on the basis of a real function, which creates the possibility of its better representation. This, as will be seen later, is of great importance in

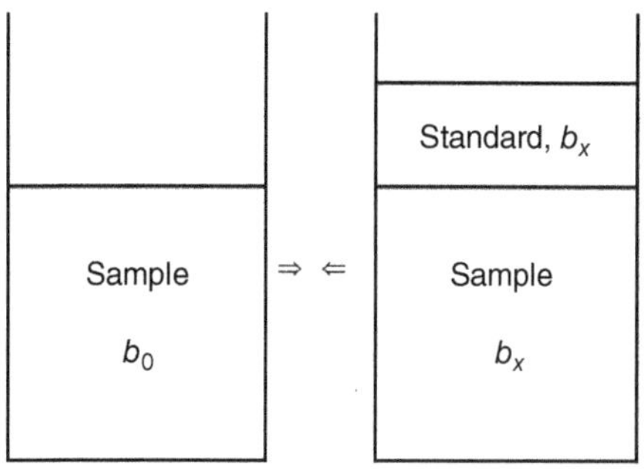

Figure 3.29 Preparation of the sample and standard in qualitative analysis in accordance with the standard addition method; b_0 and b_x are unknown and known analytes, respectively.

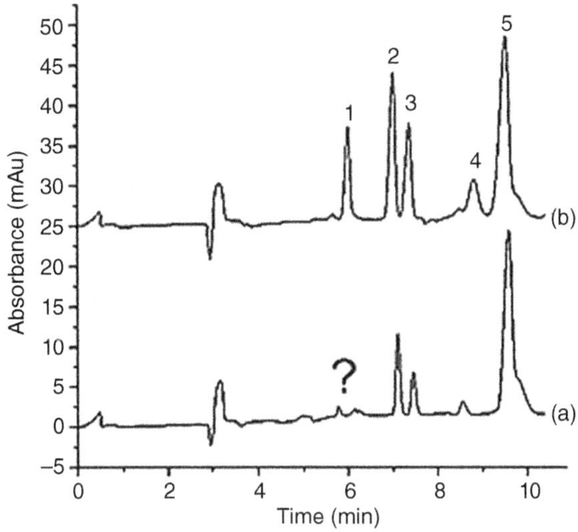

Figure 3.30 Qualitative analysis of the nutmeg extract by capillary electrophoresis with the use of the standard addition method: signals of the sample (a) and of the sample spiked with the standard (b) containing eugenol (1), elemicin (2), methyleugenol (3), thymol (4), and myristicin (5) (from own collection).

quantitative analysis. In qualitative analysis, the method is used rather rarely and rather to verify results obtained by other calibration methods. In this role, however, it can bring significant practical advantages.

An example of the use of the standard addition method in capillary electrophoresis analysis is shown in Figure 3.30. The changes that occurred in this case in the sample measurement image after standard addition allow the presence of the three compounds elemicin, methyleugenol, and myristicin in the sample to be confirmed without question and the presence of thymol to be excluded. One of the two small peaks visible in the sample image may indicate that eugenol is also present in the sample, but both peaks are slightly offset from the eugenol peak in the standard image and this conclusion is very questionable.

To more accurately determine the position of the peaks on the electropherograms of the sample and the sample with template addition, the template addition method can be combined with the internal template method. This combination of methods is shown in Figure 3.31 for an example of protein identification using the CE technique. When the standard was added to the protein mixture sample, no iron-saturated form of transferrin (h-Tf) was found in the sample, while the presumed presence of monoferric forms having iron bound at C- and N-terminal binding site was confirmed. This demonstrates that the sample initially contained small amounts of proteins bound to one Fe atom, i.e. in fact it was contaminated with iron. Such a categorical conclusion could be drawn on the basis of analytical results with very good accuracy, which was achieved by the additional use of an internal standard (albumin).

In a particular version, the standard addition method is applied in gas chromatography for the identification of organic compounds (in particular *n*-alkanes). The analyte is identified in such a way that the sample is measured with the addition of a standard containing selected *n*- and *N*-alkanes having, respectively, fewer and

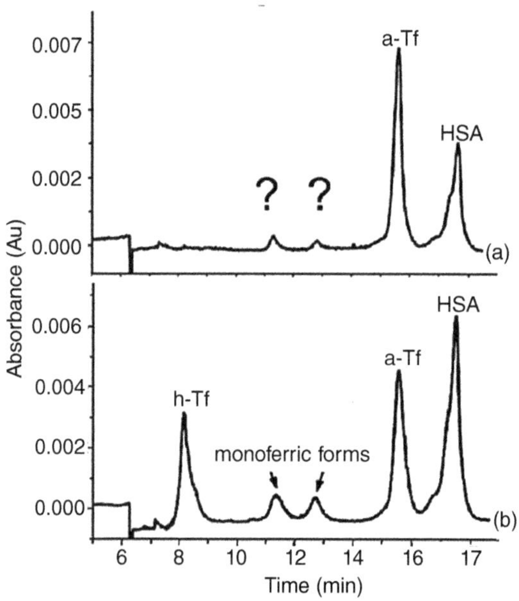

Figure 3.31 Qualitative analysis of the protein mixture based on the standard addition method coupled with the internal standard method: electropherograms of the sample (a) containing iron-free form (a-Tf) and unknown forms of transferrin (?), and of the sample spiked with the standard (b) containing transferrin as a mixture of holo-transferrin (h-Tf) and monoferric forms of transferrin; albumin from human serum (HSA) was used as the internal standard (from own collection).

more carbon atoms than in the analyte molecule. From the revealed retention times for these alkanes, t_n and t_N, and for the analyte, t_a, the so-called Kováts retention index, I_a, expressed by the formula, is calculated:

$$I_a = 100 \cdot \left[n + (N - n) \frac{\log(t_a - t_0) - \log(t_n - t_0)}{\log(t_N - t_0) - \log(t_n - t_0)} \right] \tag{3.4}$$

where t_0 is the zero retention time. This quantity is largely independent of the conditions of analysis and is thus a parameter specific to a given compound at a given temperature and in the presence of a particular liquid phase. The type of analyte can therefore be found from the data in the literature. On the formal side, this approach is an example of an empirical calibration supported by some charge of theoretical factor.

As can be seen from the examples above, the standard addition method is a simple and effective calibration method. It is distinguished from external and internal calibration methods by the fact that when added to a sample, the analyte present in the standard is in the same environment as the analyte naturally present in the sample, and vice versa. Thus, there is a natural commonality and mutual similarity between the physical and chemical conditions of the sample and the standard. Thus, if the standard, when added to the sample, changes even to some extent the properties of the sample, their commonalities are more similar to each other than the properties of the sample itself and the standard itself. This is obviously very important from a calibration point of view.

As shown in the experimental examples, the advantages of the standard addition method become apparent especially when the analyst expects that the peak appearing in the measurement image is from the analyte, but has such a low intensity that there is reasonable doubt about it. However, it must then be ensured that the amount of analyte in the added standard is also sufficiently small to cause a comparably small increase in signal. Otherwise, the signal coming from the added analyte will be too large to tell whether its change is due to the presence of the analyte in the sample or is random.

The method of standard additions can also be used to verify that a peak with a position and intensity clearly indicating the presence of a particular component in a sample is in fact from that component, even though its presence is unexpected or even undesirable. In the case of overlapping peaks or spectral bands, a standard can also be added to the sample to amplify one part of that signal part of the band and thus prove the presence (or absence) of the added component in the sample.

The standard addition method is procedurally very simple and rapid, requiring no introduction of foreign substances into the sample and free from the conditions that must be met for the internal standard method or the indirect method. Considering these advantages, one can be surprised that it is so rarely used in qualitative analysis.

Thus, the analyst has at his or her disposal in qualitative analysis a number of calibration methods that can be used depending on the current situation and the additional objectives that one wishes to achieve. As shown, some of these methods can be combined with each other, achieving the combined benefits characteristic of the component methods. In the following chapters, it will be shown that the possibility of operating a variety of single and combined calibration methods in quantitative analysis is even greater. Most of these quantitative methods are based on similar principles to the qualitative methods, once again demonstrating the consistent and uniform nature of analytical chemistry.

References

1 Kościelniak, P. (2022). Calibration methods in qualitative analysis. *TrAC Trends in Analytical Chemistry* 150: 116587.

2 Kunicki, M., Fabiańska, E., and Parczewski, A. (2013). Raman spectroscopy supported by optical methods of examination for the purpose of differentiating blue gel pen inks. *Problems of Forensic Sciences* 95: 627–641.

3 Woźniakiewicz, M., Gładysz, M., Nowak, P.M. et al. (2017). Separation of 20 coumarin derivatives using the capillary electrophoresis method optimized by a series of Doehlert experimental designs. *Talanta* 167: 714–724.

4 Kula, A., Król, M., Wietecha-Posłuszny, R. et al. (2014). Application of CE-MS to examination of black inkjet printing inks for forensic purposes. *Talanta* 128: 92–101.

5 Nowak, J., Woźniakiewicz, M., Klepacki, P. et al. (2016). Identification and determination of ergot alkaloids in morning glory cultivars. *Analytical and Bioanalytical Chemistry* 408 (12): 3093–3102.

6 Szafarska, M., Wietecha-Posłuszny, R., Woźniakiewicz, M. et al. (2011). Application of capillary electrophoresis to examination of colour inkjet printing inks for forensic purposes. *Forensic Science International* 212 (1–3): 78–85.

7 Zięba-Palus, J., Milczarek, J.M., and Kościelniak, P. (2008). Application of infrared spectroscopy and pyrolytic gas chromatography in examination of automobile paint samples. *Chemia Analityczna (Warsaw)* 53 (1): 109–121.

8 Szafarska, M., Woźniakiewicz, M., Pilch, M. et al. (2009). Computer analysis of ATR-FTIR spectra of paint samples for forensic purposes. *Journal of Molecular Structure* 924: 504–513.

9 Butler, J., Chalmers, J., McVean, G. et al. (2017). *Forensic DNA Analysis: a Primer for Courts*. London: The Royal Society.

10 Nieznańska, J., Zięba-Palus, J., and Kościelniak, P. (1999). Physico-chemical study of car paints coats. *Problems of Forensic Sciences* 39: 77–89.

11 Trzcińska, B., Zięba-Palus, J., and Kościelniak, P. (2013). Examination of car paint samples using visible microspectrometry for forensic purposes. *Analytical Letters* 46 (8): 1267–1277.

12 Gładysz, M., Król, M., and Kościelniak, P. (2017). Differentiation of red lipsticks using the attenuated total reflection technique supported by two chemometric methods. *Forensic Science International* 280: 130–138.

13 Zięba-Palus, J., Kościelniak, P., and Łącki, M. (2001). Differentiation between used motor oils on the basis of their IR spectra with application of the correlation method. *Forensic Science International* 122 (1): 35–42.

14 Zięba-Palus, J., Kościelniak, P., and Łącki, M. (2001). Differentiation of used motor oils on the basis of their IR spectra with application of cluster analysis. *Journal of Molecular Structure* 596: 221–228.

15 Woźniakiewicz, M., Wietecha-Posłuszny, R., Garbacik, A. et al. (2008). Microwave-assisted extraction of tricyclic antidepressants from human serum followed by high performance liquid chromatography determination. *Journal of Chromatography A* 1190 (1, 2): 52–56.

16 Nowak, P.M., Woźniakiewicz, M., and Kościelniak, P. (2018). Flow variation as a factor determining repeatability of the internal standard-based qualitative and quantitative analyses by capillary electrophoresis. *Journal of Chromatography A* 1548: 92–99.

17 Woźniakiewicz, M., Kuczara, J., and Kościelniak, P. (2009). Determination of fluoxetine in blood samples by high-performance liquid chromatography using a derivatization reagent. *Chemia Analityczna (Warsaw)* 54 (4): 667–677.

18 Lachowicz, T., Zięba-Palus, J., and Kościelniak, P. (2011). Application of pyrolysis gas chromatography to analysis of rubber samples. *Problems of Forensic Sciences* 85: 11–24.

19 Heiger, D. and Weinberger, R. (1994). Determination of small ions by capillary zone electrophoresis with indirect photometric detection. Application Note. Agilent Technologies.

20 Doble, P. and Haddad, P.R. (1999). Indirect photometric detection of anions in capillary electrophoresis. *Journal of Chromatography A* 824 (1, 2): 189–212.

4

Introduction to Empirical Calibration in Quantitative Analysis

As shown earlier, the empirical calibration process in quantitative analysis is in some respects different from that in qualitative analysis. This is primarily because the measurement parameter is the signal intensity, which makes the actual function continuous rather than discrete. A further consequence is the need to formulate the model function also in continuous form and to accurately approximate the real function over as wide a range of analyte concentrations as possible. This generally requires the use of at least two standards with different analyte concentrations.

In qualitative analysis, the problem of accuracy of the analytical result is also viewed differently. The finding of the presence of an identified analyte in a sample with a probability of less than 50% when it is actually present in the sample can be considered as an inaccurate result. In quantitative analysis, such a concept does not actually exist. The ability to estimate accuracy in terms of a specific numerical value means that an analytical result is always more or less accurate (or inaccurate), and it can only be accepted as inaccurate (or accurate) on the basis of an arbitrarily determined threshold. Furthermore, the absence of an analyte in a sample can often be correctly estimated with a probability so close to 100% that it is virtually impossible to achieve analyte determination accuracy at this level.

4.1 Classification

Despite these differences, the principles of calibration methods for quantitative analysis are similar to those for qualitative analysis. This is reflected in their classification and nomenclature, as can be seen in Figure 4.1.

All methods can also be broadly divided into comparative methods, in which the sample and analyte standards are prepared for measurement and measured separately from each other, and additive methods, in which the sample is combined with the standard. In quantitative analysis, however, the number of the latter far exceeds the additive method of the standard known from identification analysis. The determination of an analyte also makes more frequent use of the sample dilution process and several different calibration procedures and methods (e.g. the dilution method) not generally used in qualitative analysis are based on this.

Calibration in Analytical Science: Methods and Procedures, First Edition. Paweł Kościelniak.

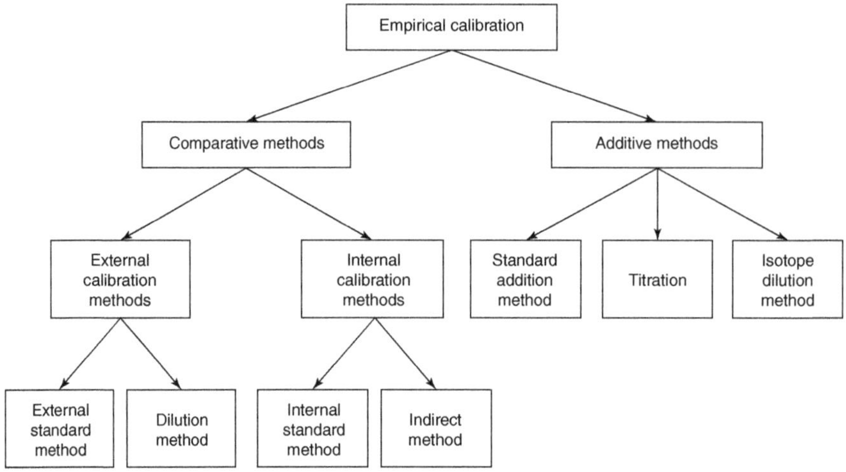

Figure 4.1 Classification of empirical calibration methods based on chemical standards in quantitative analysis.

The more common use of the sample dilution process is a result of the fact that quantitative analysis is more often performed with liquid samples. Furthermore, due to the ease of handling and homogeneity of the liquid sample, the analyte can generally be determined with the highest accuracy in this type of sample. Also of importance is the fact that most of the measurement methods for quantitative analysis are adapted to measurements in this aggregate state. It should be emphasized, however, that the basic calibration methods used in quantitative analysis can also be applied to the determination of analytes in solid and gaseous samples.

What is also characteristic of calibration methods in quantitative analysis are the various ways in which they can be combined and integrated. This type of possibility has already been demonstrated in qualitative analysis, but the range of such operations performed for the determination of analytes is much greater. As will be shown, skillful combinations of basic calibration methods can increase their individual analytical usefulness in various respects, and in particular, allow results to be obtained with increased reliability.

In general, therefore, quantitative analysis is much richer in a variety of empirical calibration approaches than qualitative analysis and is therefore given much more space in the following sections of this chapter. Because of the continuous, linear, or nonlinear nature of the real and model functions, a simple mathematical apparatus was used to describe them, which also allowed us to assess the robustness of some calibration approaches to uncontrolled effects. The analytical capabilities of many calibration approaches are illustrated and compared using numerous experimental examples.

4.2 Formulation of Model Functions

As stated, empirical calibration in quantitative analysis requires the formulation of a continuous model function, $Y = G(c)$. This is accomplished by using chemical standards to produce the model function over a specified range of analyte concentrations in the form of a calibration graph. **A properly created calibration graph is a prerequisite for obtaining a precise and accurate analytical result**.

The chance of properly creating a calibration plot depends largely on the current shape of the real function, $Y = F(c)$. In different calibration methods, this function can take different shapes and different positions in the calibration coordinate system. However, it is most often expected to be an increasing, nonlinear function, taking zero value for zero analyte concentration (i.e. based on the origin of the coordinate system) and increasing monotonically as the analyte concentration increases. It generally does not happen that the real function changes from linear to nonlinear more than twice in the considered analyte concentration range. A graph of such a function is shown in Figure 4.2.

A calibration graph is generally formulated from several (at least two) signal intensity values obtained for chemical standards of different, known concentrations of the analyte. The graph is created by fitting a function (generally nonlinear) to the measurement points thus created. As can be seen in Figure 4.2, the graph may have a shape that generally corresponds to that of the real function, but it may be different degrees and portions away from that function. **A properly created calibration graph should satisfy two essential conditions**:

- **its position should be as little random as possible**, i.e. the model function should be matched as closely as possible to the appropriate number of measurement points,
- **its position should be as close as possible to that of the real function**, i.e. it should take into account the current type, strength, and direction of uncontrolled effects occurring during sample analysis.

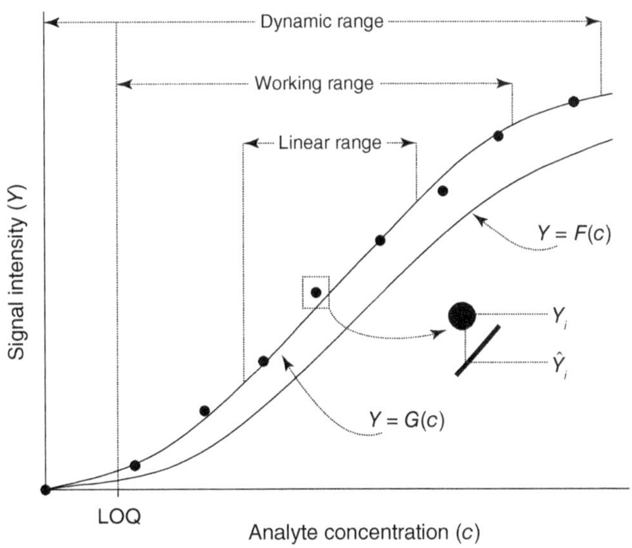

Figure 4.2 Model function (calibration graph), $Y = G(c)$, compared with real function, $Y = F(c)$, and fitted to the experimental points (black points) in dynamic range of the model function, encompassing linear and working ranges; LOQ – limit of qualification; Y_i and $\hat{Y}_i$ – values of intensities measured and calculated from the model function, respectively, both corresponding to *i*th standard.

In practice, there are some natural and arbitrary limits to the range of analyte concentrations that can be considered when creating a calibration plot (see Figure 4.2). The maximum analyte concentration range is defined by the maximum ability of the measuring instrument to record signals for a given analyte (dynamic range). The initial part of this range is limited by the limit of quantification (LOQ), i.e. the smallest concentration of an analyte that can be determined with a given measurement method from the point of view of random fluctuations of the measurement signal. In the final part of the dynamic range, the intensity of the analytical signal usually changes only slightly with a change in analyte concentration, which practically also excludes this fragment from analytical applicability. Between this part and the LOQ value, the so-called working range can be determined, i.e. the range that can practically be used for the determination of the analyte in the analyzed sample.

If the position of the experimental points in the working range indicates a nonlinear form of the model function, its exact formulation requires, as seen in Figure 4.2, the use of up to a dozen standard solutions. In practice, this is tedious and time-consuming. Therefore, **the analyst generally aims to produce a calibration graph in a linear range of analyte concentrations**. If he is absolutely sure that the experimental point system is linear in a certain range, he can use only two standard solutions with appropriate analyte concentrations to formulate a model function in that range.

However, working in the linear range of the calibration graph is not a prerequisite for obtaining an accurate analytical result. If the calibration graph produced with standard solutions containing the analyte itself becomes nonlinear, this effect is probably a detection effect. It is therefore reasonable to suppose that the real function also becomes nonlinear in the same range as the sample and standards are measured with the same detector. In other words, the transition from a linear to a nonlinear range in this situation does not risk the model function not fitting the real function.

Thus, intentionally or accidentally going outside the linear range is acceptable as long as the nonlinearity of the graph does not become too large (e.g. such as that appears outside the working range in Figure 4.2). A progressively increasing slope of the graph to the concentration axis otherwise means that the measurement sensitivity of the analyte in this area decreases, resulting in a reduced precision of the analytical results obtained using this portion of the calibration graph.

It is evident that the linearity of a model function, particularly one formulated with standard solutions containing the analyte itself, does not necessarily mean that within a given range the real function will also be linear and well reproduced by the model function. In other words, **the generally accepted recommendation of using a linear form of the model function does not give confidence in obtaining an accurate analytical result**.

Additional conventional guidelines to assist in the proper construction of a calibration graph are as follows [1]:

- there should be six or more calibration standards,
- the standards should be evenly spaced over the concentration range of interest,
- the range should encompass 0–150% or 50–150% of the concentration likely to be encountered, depending on which of these is the more suitable; and

- standard solutions should be measured under uniform, optimal experimental conditions,
- each standard should be measured at least in duplicate (preferably more), in a random order.

There are some exceptions to all these rules, which depend both on the specific calibration method and on the individual approach of each analyst.

The greater the number of points of standard solutions measured, the more clearly the experimental points obtained approximate the form of the model function. However, this is always some approximation (see Figure 4.2), which is due to the natural random fluctuations in the intensity of the measurement signals. The question then arises: how to mathematically fit a linear or nonlinear function to the measurement points so that it best reflects the arrangement of these points.

The commonly accepted mathematical method used in such cases is the **least squares method**. It leads to such a fitting of the function to the experimental points that the sum of squares of residuals, i.e. the difference of intensities of the measured signal and the signal intensity resulting from the fitted function, $|Y_i - \hat{Y}_i|$ (see Figure 4.1), is minimal:

$$\sum_i (Y_i - \hat{Y}_i)^2 = \min \tag{4.1}$$

Under the assumption that the random error in the preparation of the standard solutions can be neglected, and that the random error (in terms of variance) in the intensity of the measured signal for all standard solutions is equal, then according to this approach, the intercept, B_0, and slope, B_1, of the linear model function

$$G(c) = B_0 + B_1 \cdot c \tag{4.2}$$

are calculated from the formulas:

$$B_1 = \frac{\sum_i (c_i - \bar{c}) \cdot (Y_i - \bar{Y})}{\sum_i (c_i - \bar{c})^2} \tag{4.3}$$

$$B_0 = \bar{Y} - B_1 \cdot \bar{c} \tag{4.4}$$

The method of least squares can of course also be used to formulate nonlinear model functions [2].

The condition of equal variances (i.e. homoscedasticity) is not always met for measurement data. For instance, when the working range of the calibration graph is large, it might be expected that the variance of each data point is quite different. Larger deviations present at larger concentrations tend to influence (weight) the regression line more than smaller deviations associated with smaller concentrations. A simple and effective way to counteract this situation is to use weighted last squares method [3].

Another phenomenon that occasionally occurs when constructing a calibration graph is the appearance of the points that deviate from the clearly linear or nonlinear positions of the other points. This effect can be caused by a variety of reasons, but most often it is simply due to an error made in the preparation of a calibration solution. If there is no doubt that the position of the outliers is due to systematic error, measurements should be repeated or the point omitted when using the least squares method. However, when there is doubt about whether the error is random or systematic, it is risky to make a subjective decision to reject or not to reject a point.

In such cases, an objective mathematical approach called the **single median method** can be used [4]. According to this simple method, the coefficient B_1 of the linear function (4.2) is determined as the median of the values of the coefficients $(B_1)_{ij}$ of the linear functions connecting each two measurement points, in all possible combinations (without repetition):

$$B_1 = \underset{1 \leq i < j \leq n}{\text{med}} (B_1)_{ij} \tag{4.5}$$

and the coefficient B_0 is the median of the values of the coefficients $(B_0)_i$ of the linear functions with coefficient B_1 passing through all the measurement points:

$$B_0 = \underset{1 \leq i \leq n}{\text{med}} (B_0)_i \tag{4.6}$$

The effectiveness of the single median method is shown in Figure 4.3. This method can also be used when modeling nonlinear real functions [5].

Linearity of the calibration graph can be tested initially and informally by examination of a plot of residuals, which is a graph showing the residual values obtained for each of the analyte concentration considered (see Figure 4.4). If the points in a residual plot are randomly dispersed around the horizontal axis (as in Figure 4.4a), a linear regression model is appropriate for the data; otherwise, a nonlinear model is more appropriate (as in Figure 4.4b). The evaluation of both situations is, of course, subject to statistical analysis.

Another issue is the assessment of the quality of fit of the model function to the measurement points. The most commonly used for this purpose is Pearson's correlation coefficient, r, which in its linear version is expressed by the formula:

$$r = \frac{\sum_i (c_i - \bar{c}) \cdot (Y_i - \bar{Y})}{\sqrt{\sum_i (c_i - \bar{c})^2 \sum_i (Y_i - \bar{Y})^2}} \tag{4.7}$$

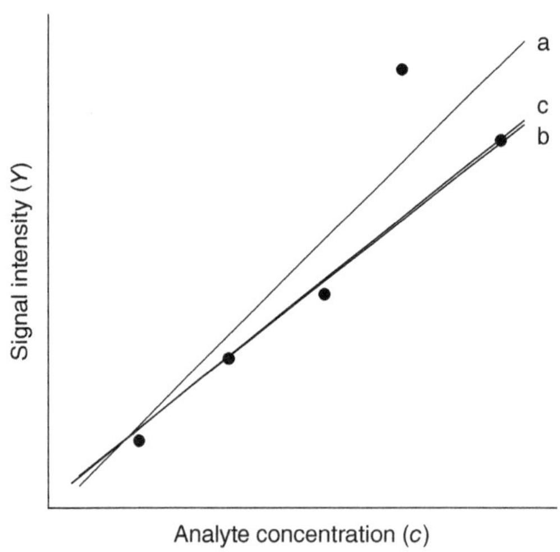

Figure 4.3 Calibration graphs fitted to the experimental points (black points) by the least square method (a and b) and the single median method (c) taking (a and c) and not taking (b) the outlier into account.

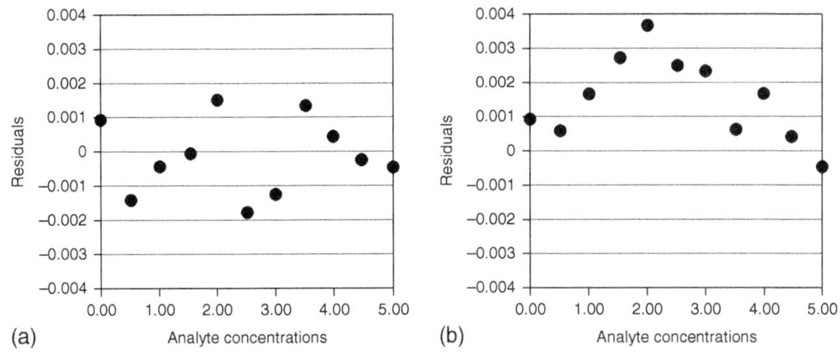

Figure 4.4 Plot of residual characteristics for linear (a) and nonlinear (b) calibration graph fitted to the same experimental points (the mean residual values from the measurements repeated for each concentration are taken into account).

This coefficient takes values from −1 to +1, corresponding to full correlation (perfect fit) of the respectively decreasing and increasing linear function to the measurement points, and value 0 – lack of any correlation between signal intensity and analyte concentration.

However, contrary to popular recognition and common usage, the correlation coefficient is not an appropriate measure when used for calibration purposes. The experimental points of a calibration plot generally express a very good correlation between signal intensity and analyte concentration, because, after all, both quantities are related to each other by a particular form of the model function. As a result, calibration graphs are always characterized by r values very close to +1, differing by the number of decimal nines, and the number of these digits does not clearly indicate the quality of the fit. Moreover, the same value of r can correspond to a linear and a nonlinear set of experimental points. The situation is not improved by the common substitution of the r coefficient for the determination coefficient, R, where $R = r^2$.

In addition to its measurable limitations, the use of a correlation or determination coefficient has negative consequences of psychological nature. Any value greater than 0.995 seems so close to 1 that it gives the impression of a value that confirms a very good fit of the linear function. Often analysts assume that values of $r > 0.995$ accept the linear nature of the model function and do not examine linearity more closely, although other mathematical tools exist for this purpose.

One of them is the quality coefficient, QC [6], which is calculated from the formula:

$$\mathrm{QC} = \sqrt{\frac{\sum_i \left(\frac{Y_i - \hat{Y}_i}{\overline{Y}}\right)^2}{n-1}} \cdot 100\% \tag{4.8}$$

It takes values ranging with equal probability from 0% (for perfect fit) to 100%, with the QC < 5% limit usually taken as the criterion for linear fit. Examination of the form of the model function and evaluation of the fit of this function to experimental points can also be done by statistical tests, for instance, by Fisher–Scedecor test, Mandel's test or Lack-of-fit test [7].

Table 4.1 Values of the different estimators of the goodness of fit of the linear calibration curves to the measurement points.

r	QC (%)	Lack-of-fit test ($F_{crit,0.05}$ = 4.53)	Mandel's test ($F_{crit,0.05}$ = 5.12)
0.9991	2.92	7.49	33.50
0.9987	3.37	13.16	35.74
0.9983	3.95	9.99	55.19
0.9978	4.23	19.42	56.84
0.9975	4.70	10.71	28.65

Source: Adapted from Van Loco et al. [8].

Interestingly, different methods used to evaluate the fit of, e.g. linear functions to a given set of measurement data can yield different results. Table 4.1 compares *r* and QC values and statistical test results obtained for cadmium in flame atomic absorption spectrometry (FAAS) [8]. All the investigated curves were characterized by high correlation coefficient ($r > 0.997$) and low-quality coefficient (QC $<5\%$), but the straight-line model was systematically rejected at the 95% confidence level on the basis of the Lack-of-fit and Mandel's fitting tests. This demonstrates the different sensitivity of these computational tools when applied under identical conditions. Thus, it is one more proof that they can be treated as the auxiliary support only for the analyst's intuition and experience.

How careful observation of the measurement results and the analyst's experience helps in making the right decision is shown in Figure 4.5. The measurement results obtained in this case allowed a very good approximation of their position using a linear function, but further research showed that a nonlinear function formulated using the single median method was an even better fit to the same points. At the same time, it turned out that one of the measurement points should be treated as an outlier.

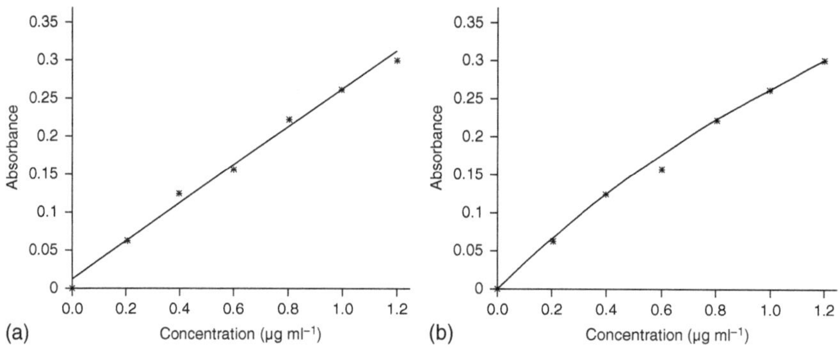

Figure 4.5 Calibration graphs obtained for cadmium in FAAS using median linear (a) and nonlinear (b) method. Source: Kościelniak [5], fig 7 (p. 185)/with permission of Elsevier.

4.3 Examination of Interference Effect

The shape of the real function and its position relative to the model function are determined by uncontrolled effects. During quantitative analysis, the influence of various factors on the analytical signal manifests itself by a change in the intensity of the analytical signal and this change may be different in different ranges of analyte concentration. For this reason, the problems arising from uncontrolled effects in quantitative analysis are more serious and more difficult to solve than in qualitative analysis.

If before the occurrence of the effects the real function is linear, then under the influence of various personal, natural, instrumental, physical, and chemical factors its angle of inclination may change (multiplicative effect) with the preservation of linearity (linearly multiplicative or proportional effect) or in combination with the change of its character from linear to nonlinear (nonlinearly multiplicative or nonproportional effect). The function may also be shifted from its original form by a constant value (additive or translational effect). These changes may occur separately or combine in various combinations into one complex effect. Effects of the abovementioned different characters are shown in Figure 4.6. Of course, regardless of their character, each uncontrolled effect may manifest itself as an increase (positive effect) or decrease (negative effect) with respect to the original values of the linear function.

Thus, when uncontrolled effects occur, formulating a model function (calibration plot) that exactly matches the real function in every part of it (linear and nonlinear) is a kind of analytical art. It is all the more important to take such steps before performing the analysis so that the effects do not occur and do not become apparent when the analytical signal intensity measurement is performed.

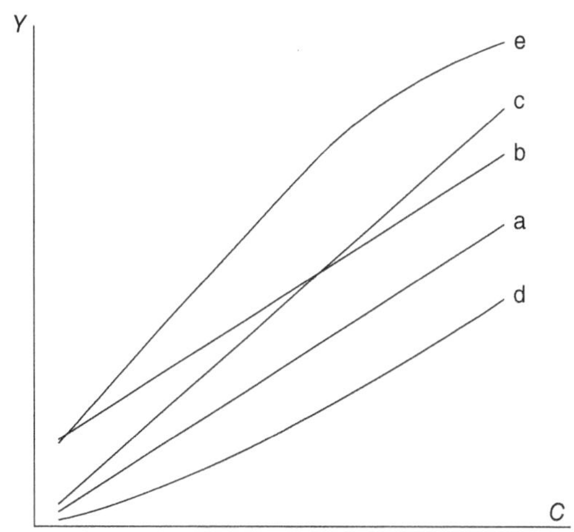

Figure 4.6 Character of uncontrolled effects in quantitative analysis: no effect (a) and additive (b), linearly multiplicative (c), nonlinearly multiplicative (d), complex (e) effects.

Any uncontrolled effect can be responsible for the signal variations shown in Figure 4.6, but with varying probability depending on the ability to eliminate or compensate for it before measurement. If we assume that:

- the human factor is largely eliminated (by correct and careful performance of the analysis) and compensated (as a result of sample and standard preparation by the same analyst),
- instrumental and detection effects are compensated (by keeping the same measurement and detection conditions for the sample and standard)
- the preparative effect is determined by separate experiments and calculations,

then the changes shown in Figure 4.6 can practically be caused by speciation and interference effects.

A speciation effect is usually induced when the analyte in the sample takes on a different chemical form than that of the analyte in the standard. Thus, it can be assumed with high probability that for a given analyte concentration the two forms of the analyte elicit signals of different intensities, and this difference is proportional to the concentration of the analyte in the two forms. The speciation effect is then proportional in nature. Regardless of the form it takes, it can be compensated for if the sample and standard are subjected to appropriate preparative treatments (as will be shown later in the experimental examples).

Due to the unknown composition of the sample, it is most difficult to perform a full elimination and compensation of the interference effect. Special difficulties arise from the following characteristics of this effect:

- it occurs commonly in various measurement methods and its occurrence is difficult to predict,
- the magnitude and nature of the effect depend not only on the type and number of interferents in the sample but also on the ratio of interferents to analyte,
- in a given analysis, a sample component that is an interferent for a particular analyte may not be an interferent for another analyte,
- interferences that occur when applying a particular measurement method to the analysis of a given sample may be caused by other components of that sample and be of a completely different magnitude and nature in a different measurement method.

Interference effects can occur in all of the forms shown in Figure 4.6. Interferences are most often caused by chemical reactions occurring between the interferent and analyte in the sample prior to measurement of the analytical signal or at the time of measurement in the environment in which detection of the measurement signal occurs. They are then multiplicative or complex in nature, although this is difficult to predict because their detailed mechanisms are very different and many are not fully understood. The additive effect is to be expected when the reagents added to the sample before measurement trigger the analytical signal (blank effect) or when the signal coming from the interferent overlaps partially or completely with the signal coming from the analyte. In such a case its intensity is enhanced by a constant value, independent of the analyte concentration in the sample. The interference effect can also occur when the components of the sample have the ability to change their physical properties to such an extent that this change results in a change in the intensity of the analytical signal.

All this means that, of all the uncontrolled effects, the interference effect requires particularly close attention by the analyst in quantitative analysis. In a particular case of analysis, this effect may not occur, of course, but it should always be considered. In other cases, it may also be so large that the systematic analytical errors caused by it may reach tens or even hundreds of percent. It, therefore, requires investigation whenever it is likely to occur, and in particular when:

- a new analytical method is introduced into practice,
- the validated method has been modified chemically or instrumentally,
- the method is to be applied for the analysis of samples (of different chemical compositions) other than the samples analyzed so far,
- the method has been put back into use after a long period of nonuse.

The interference effects are usually detected and examined using standard solutions of known analyte concentrations that act as test samples. Specific amounts of those components of the natural sample that are thought to be interferents are then added sequentially to the simulated sample, and measurements are made for the analyte before and after the addition of the potential interferent. The value of the difference in signals (subjected to possible statistical analysis) indicates the absence or existence of interference of a certain magnitude and direction. If experiments are performed for different concentrations of analyte and interferent, the results can be represented as simulated real functions with variable shapes and positions depending on the type and concentration of interferent. This interpretation of the interference results is shown in Figure 4.7.

The presented method of studying interference effects, although widely used, is rather tedious and relatively inefficient. It requires knowledge (at least approximately) of the type and concentration of interferents in a natural sample, which is of course rarely possible. It is also difficult to study the so-called interference, i.e. the resultant effect of two or more interferents in a sample on the analytical signal, by this route. Increasing the efficiency of the study of interference with the simultaneous possibility of determining interaction values offers the theory of scheduling experiments, as will be discussed in Chapter 11.

In Chapter 9, we will also demonstrate another simple and fast way to test for interference effects recommended by IUPAC under the name "test for general matrix effect" [1]: It involves the appropriate reciprocal treatment of only the natural sample and the analyte standard, without the need for additional reagents, and allows one to determine whether and how much the sample matrix affects the analytical signal without getting into which natural sample components are responsible for the effect. This method is therefore a very good complement to the "simulation method" and both can be used together or each separately.

The detection of statistically significant interference effects obliges one to take appropriate steps to eliminate or compensate for them. There are many possibilities here, some of which have already been outlined.

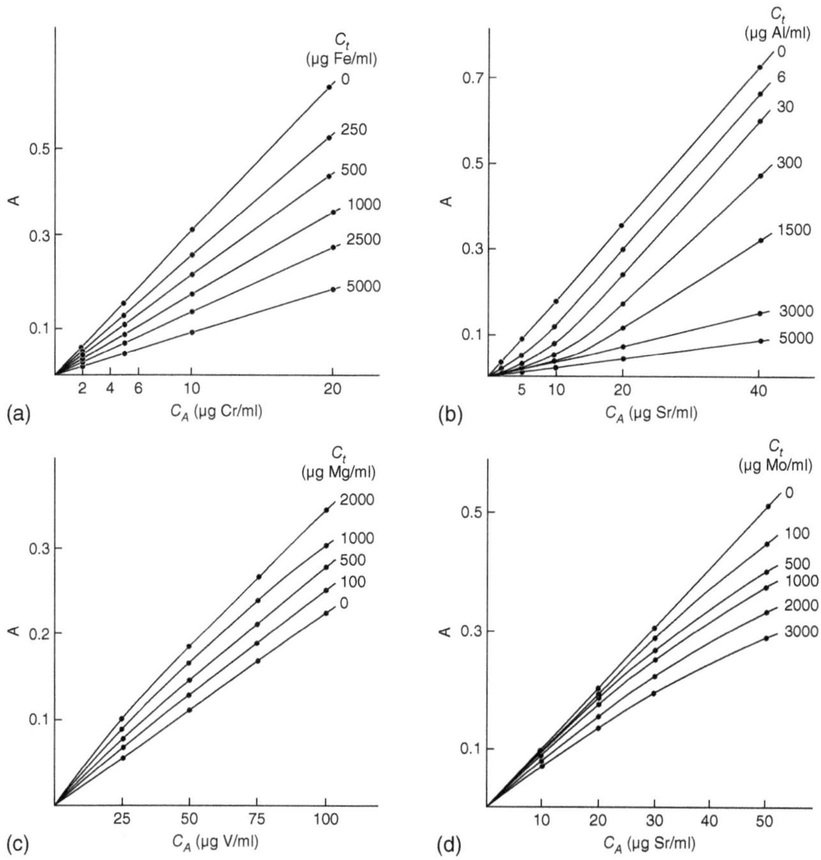

Figure 4.7 Simulated real functions obtained by flame atomic absorption spectrometry (FAAS) for chromium (a), strontium (b, d), and vanadium (c) influenced in terms of shape and position by interferents: iron (a), aluminum (b), magnesium (c), and molybdenum (d). Source: Adapted from Pszonicki and Skwara [9].

The universal way to eliminate interferences is to remove the interferents from the sample (**matrix isolation**) or to isolate the analyte from the sample matrix (**analyte isolation**). Both objectives are achieved by appropriately selected laboratory techniques, e.g. extraction, precipitation, gaseous diffusion, etc. However, any such procedure is fraught with the risk of losing some analyte from the sample, i.e. viewed from the calibration side – the formation of a negative additive preparative effect – in relation to the true function.

Often interference effects are eliminated by chemical means by adding appropriately selected special reagents to the sample. By their nature, these are used primarily when chemical interference is present. They usually act via **matrix modification** usually in such a way that they release the interferent from the compound with the analyte (releasing reagents), or they mask the analyte or interferent preventing the compound formation reaction between them. This action is selective, i.e. it depends on the type of analyte and interferent as well as on the measurement method used.

It is sometimes necessary to add them to the sample in large excess relative to the analyte concentration, because of the danger of introducing into the sample some of the analyte present in the reagent. For this reason, it is usually recommended to add the special substance to the sample and to the standard solutions in equal concentration.

Special substances can also be used to compensate for interference effects provided that they have the ability to create equal measuring conditions for the analyte in the sample and in the standards. This is called **matrix buffering**. A typical example of such a buffering substance is an easily ionized element (e.g. cesium) that, when added in excess to sample and standards introduced into the flame of an emission spectrometer, equalizes the ionization conditions of other elements of the type (e.g. potassium and sodium) whose reaction without the presence of a buffer is the source of their mutual influence.

Another way to compensate for interferences is to arrange for the same chemical composition in the standards as in the sample. This process, called **matrix matching**, usually involves adding interferents of the type and concentration found in the sample to a series of standard solutions. Proper selection of the added interferents theoretically provides effective compensation for any type of interference effects. In practice, however, this is usually a very difficult and laborious task because it requires complete information about the interferents present in the sample. Most often, therefore, only those components that are present in the sample in the highest concentrations and can therefore act as interferents are placed in standard solutions.

The easiest matrix matching procedure to use is when the interference is caused by one known component present in the sample in such excess that the effect caused by it is by far the dominant one (this is the case, for example, when determining trace impurities in a reagent). In such a situation, not only it is known what the interferent is, but also what the concentration (approximately) of this component in the sample is.

The matrix matching procedure is also particularly effective in such exceptional situations where the analyst has a natural sample containing no analyte available (e.g. blood free of psychotropic substances determined in the blood of patients addicted to these substances). Such a matrix with added analyte of known concentration provides a standard that almost perfectly matches the sample under analysis in terms of chemical composition.

It should be noted that a standard prepared in this way, when subjected in the course of analysis to the same preparative procedures as the sample, provides an opportunity to compensate not only for interference effects but also for other potential uncontrolled effects occurring in a given analytical method. It is thus a kind of "ideal" calibration procedure, leading to an accurate analytical result regardless, at least in theory, of the sources, type, and magnitude of these effects. Not surprisingly, then, the search for and creation of opportunities to prepare such complex-matrix standards (i.e. sources of strong and complex interferences) suited to various types of quantitative analyses is one of the popular trends in modern analytical chemistry.

A special role in the matrix-matching procedure play the certified reference materials (CRMs). If CRM is suited to the sample analyzed with regard to its composition

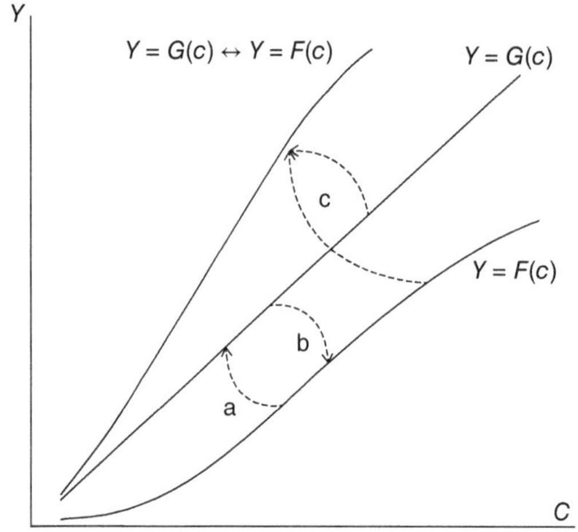

Figure 4.8 The paths to matching the real function, $Y = F(c)$, by the model function, $Y = G(c)$, as a result of application of different ways of elimination and compensation of interference effects: matrix isolation and matrix modification (a), matrix matching (b), and matrix buffering (c).

it can serve itself as the standard. If, in addition, CRM is free of analyte, it can be added to the set of standards considered as the matrix of the sample analyzed. As the most CRMs are available in the solid state, they are especially useful for matching the matrix of solid samples (in particular, using laser-ablation ICP-MS).

The choice of the appropriate way to eliminate and/or compensate for interference effects depends on many factors that characterize a particular analytical situation, and the decision to make this choice is entirely up to the analyst. From a calibration point of view, it is important that any correctly chosen and properly implemented method of eliminating or compensating for interference effects leads to an accurate representation of the real function by the model function (provided that there are no other uncontrolled effects). Depending on the procedure used, this can occur by different paths, as shown in Figure 4.8: the model function (calibration plot) can be adjusted in shape and position to the real function and vice versa, or both functions can be moved to the same position and assume the same shape. In either of these situations, there is a chance to avoid interference effects and obtain an accurate analytical result.

The final analytical process that can contribute to compensating for interference effects is an appropriate calibration procedure. The capabilities of different calibration approaches in this regard will be specifically highlighted in Chapter 6 of this book.

4.4 Mathematical Modeling of Real Function

When formulating a model function that approximates the form of the real function in an empirical calibration, the analyst does not inquire into exactly what uncontrolled effects caused the form of the real function to be as it is and what actually justifies the use of a model function of a particular form. For example, when formulating a nonlinear model function, it does not really matter to him whether the nonlinearity of the real function that he expects is due to a detection effect or an interference effect. And indeed, practice shows that such an approach is completely justified and sufficient in empirical calibration.

Sometimes, however, if the analyst wants to be able to observe how the elimination or compensation of uncontrolled effects changes the form and position of the model function toward a better fit with the real function, then he should have a function of more complex form. Such control, however, is possible only if the mathematical description of the model function takes into account the existence of uncontrolled effects of various kinds or at least those whose occurrence is most likely and most common. It should be emphasized, however, that it is not a question of creating a model on the basis of physicochemical processes and phenomena simulating the preparation of the sample for measurement and the conditions of its measurement, because such modeling is the domain of theoretical calibration, not empirical.

When attempting to formulate such an empirical model, however, some compromise should be found between overcomplicating its form by attempting to include as many effects as possible, regardless of their practical significance, and simplifying it too far. Therefore, it is here proposed to assume that in the absence of any uncontrolled effects the real function, $Y = F(c)$, is considered in the linear range of dependence of the analytical signal, Y, on the analyte concentration, c, and is described by the formula:

$$F(c) = B_1 \cdot c \tag{4.9}$$

It is therefore assumed that the coefficient B_0 (see Eq. (4.2)) takes a zero value for zero analyte concentration, and all deviations from the zero value are random and can be neglected.

It is further assumed that as a result of the presence of these effects, the linear real function changes its form to one consistent with the following model:

$$F(c) = A(c_p, c_n) + B_1 \cdot c + B_2 \cdot c \cdot \frac{c_m}{1 + a \cdot c + b \cdot c_m} \tag{4.10}$$

or else:

$$F(c) = A(c_p, c_t) + B_1 \cdot c + B_2 \cdot c \cdot H(c, c_m) \tag{4.11}$$

where coefficients A, B_1, B_2, a, and b define various forms, linear or nonlinear, that the real function may take after taking into account effects of different nature. The hyperbolic form of the function $H(c,c_m)$ results from the fact that, as it has been repeatedly stated and proved, the calibration graph of this form is best fitted to the measurement points in the case of interference effects [9, 10]. This gives reason to assume that the real function also most often takes this or similar mathematical form.

Model (4.10) has the following properties:

- it accounts for an additive interference effect, $A(c_p, c_t)$, caused by the presence of reagents added to the sample during sample processing with a total concentration of c_p (blank effect) and native interferents with a total concentration of c_t,

- takes into account the multiplicative interference effect ($B_2 \cdot H$) with linear ($a = 0$) or hyperbolic ($a \neq 0$) character, caused by native interferents with total concentration c_m (that is, it gives the real function a nonlinear character caused by the interference effect),
- gives the possibility to take into account a speciation effect of linearly multiplicative character by changing the measurement sensitivity of the analyte expressed by the coefficient B_1 ($B_1' \neq B_1$),
- takes into account the linear ($b = 0$) and nonlinear ($b \neq 0$) dependence of the analytical signal on the concentration of native interferents.

At the same time, model (4.10) does not take into account the presence of other effects that are much less likely to occur, e.g.:

- an additive effect originating from the residual amount of analyte added to the sample (e.g. as a result of a chemical reaction or sample processing by a specific extraction technique), assuming that this amount is negligibly small,
- an additive effect of a different origin (e.g. resulting from baseline shifts), assuming that such an effect is relatively rare and, if present can be eliminated or compensated for beforehand,
- occurrence of interference effect of complex mechanism (e.g. causing change of signal proportional to the ratio of analyte and interferents concentration).

Model (4.10) will be used in Chapters 5 and 6 primarily to assess the robustness of the discussed empirical calibration methods in quantitative analysis to uncontrolled effects (in particular, to interference effects). It will also help in demonstrating the specificity of each method. However, it should be emphasized that this model, like any other empirical model, will not be able to capture, as will be shown, all the natural characteristics of the real function and predict all the real changes that may occur in the sample during its processing and measurement.

References

1 Thompson, S.L.R., Ellison, S.L.R., and Wood, R. (2002). Harmonized guidelines for single-laboratory validation of methods of analysis. *Pure and Applied Chemistry* 74 (5): 835–855.

2 Bates, D.M. and Watts, D.G. (1988). *Nonlinear Regression Analysis and Its Applications*. New York: Wiley.

3 Miller, J.N. (1991). Basic statistical methods for analytical chemistry. Part 2. Calibration and regression methods. A review. *Analyst* 116 (1): 3–14.

4 Theil, H. (1950). A rank invariant method of linear and polynomial regression analysis, I, II, III. *Proceedings of the Koninklijke Nederlandse Akademie Wetenschappen, Series A Mathematical Sciences* 53: 386–392, 521–525, 1397–1412.

5 Kościelniak, P. (1993). Non-linear robust regression procedure for calibration in flame atomic absorption spectrometry. *Analytica Chimica Acta* 278 (1): 177–187.

6 Knecht, J. and Stork, G. (1974). Percentage and logarithmic procedures for calculation of calibration curves (in German). *Fresenius' Zeitschrift für Analytische Chemie* 270 (2): 97–99.

7 Massart, D.L., Vandeginste, B.G.M., Buydens, L.M.C. et al. (1997). *Handbook of Chemometrics and Qualimetrics. Part A*. Amsterdam: Elsevier.

8 Van Loco, J., Elskens, M., Croux, C. et al. (2002). Linearity of calibration curves: use and misuse of the correlation coefficient. *Accreditation and Quality Assurance* 7 (7): 281–285.

9 Pszonicki, L. and Skwara, W. (1989). The standard addition and successive dilution method for evaluation and verification of results in atomic-absorption analysis. *Talanta* 36 (12): 1265–1276.

10 Kościelniak, P. and Parczewski, A. (1985). Efficient application of non-linear transformation of factors in empirical modelling based on experimental designs. *Analytica Chimica Acta* 177: 197–202.

5

Comparative Calibration Methods

As in qualitative analysis, the group of comparative methods includes those methods in which the sample and standards are prepared for measurement separately from each other. These methods can be divided into:

- **external calibration methods**, in which no other substances are used as additional standards to aid the calibration process, and
- **internal calibration methods**, in which additional substances are added in equal concentrations to the sample and standards.

5.1 External Calibration Methods

The group of external calibration methods includes the **external standard method**, which is the most basic and widely used method in quantitative analysis. It is conceptually and experimentally the simplest method because it does not require any additional calibration procedures or substances to support the calibration process.

External calibration methods also include the method where the sample prepared for measurement is only gradually diluted in parallel with the standard. This calibration approach is called the **dilution method**.

5.1.1 External Standard Method

In its basic form, the external standard method (ESM) on the preparative step involves preparing a sample and two standard solutions containing the analyte itself at concentrations c_1 and c_2, with the concentration of the analyte in the sample, c_0, being greater than c_1 and less than c_2. This procedure is shown in Figure 5.1.

In the measurement step, the signal intensities are measured for the sample, Y_0, and for the standard solutions, Y_1, Y_2. Assuming that the analyte in the sample is not subject to uncontrolled effects, i.e. that the real function is described by a linear function of Eq. (4.9), on the basis of the signals obtained for the standard solutions a linear model function of equation:

$$Y = B_1 \cdot c \tag{5.1}$$

Calibration in Analytical Science: Methods and Procedures, First Edition. Paweł Kościelniak.

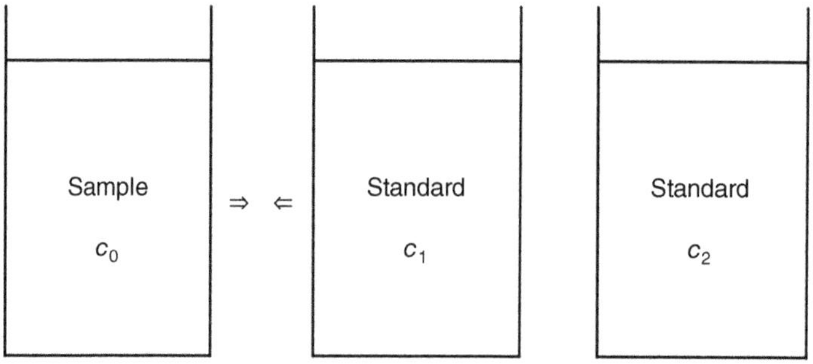

Figure 5.1 Scheme of the basic form of the external standard method at the preparative stage of the quantitative analysis.

where B_1 is a factor that determines the slope of the calibration graph, respectively. Assuming that the intercept B_0 is random, the concentration of analyte in the sample, c_x, is determined by the formula:

$$c_x = \frac{Y_0}{B_1} \tag{5.2}$$

or

$$c_x = \frac{Y_0 \cdot (c_2 - c_1) - Y_1 \cdot c_2 + Y_2 \cdot c_1}{Y_2 - Y_1} \tag{5.3}$$

The calibration procedure of the method in the measurement and calculation steps is shown in Figure 5.2a. As seen, in the external standard method the concentration of the analyte in the sample is determined in an interpolative way.

In some cases, the external standard method is performed using two standards, one of which contains no analyte ($c_1 = 0$). It is then assumed (after experimental

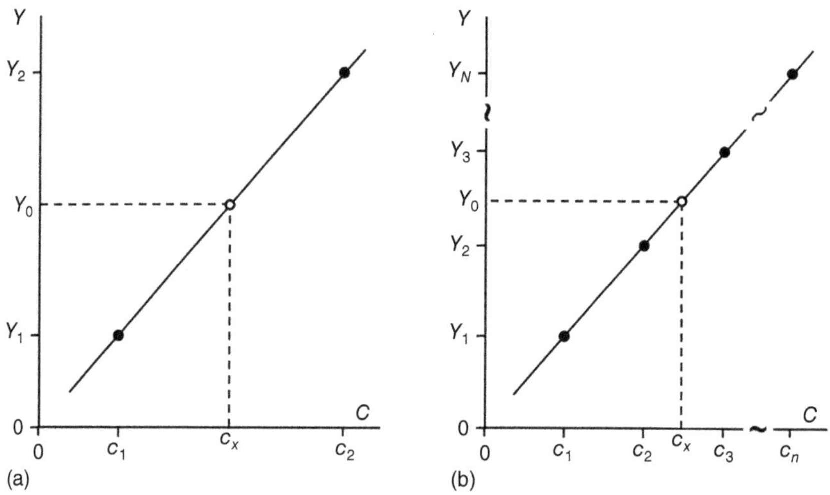

Figure 5.2 Calibration in accordance with two-standard (a) and multipoint (b) forms of the ESM method.

verification) that the linear range of the model function can reach zero analyte concentration and the signal Y_1 corresponding to this standard is equal to zero (within the random error limit). The analytical result is then determined from the formula:

$$c_x = \frac{Y_0}{Y_2} \cdot c_2 \tag{5.4}$$

Another, even more procedurally simplified approach is the so-called single standard external method. In this case, a single standard of nonzero analyte concentration is used and the zero signal is determined (usually instrumentally) without introducing any standard solution into the measurement system. The concentration of the analyte in the sample is then also determined from formula (5.4). However, this procedure is criticized mainly from the point of view of potentially increased errors in analytical results [1]. It is assumed that the method in this version can be applied only when the analyzed sample contains the analyte in the concentration similar to the analyte concentration in the standard solution. This version can also be applied as a fast and convenient recalibration method for cyclically checking the magnitude of changes in analytical signal intensity under changes in experimental conditions. It must be said, however, that because of its extreme simplicity it is quite often used in laboratories for serial routine analyses (e.g. clinical analyses).

Usually, in the external standard method, the calibration graph is constructed in a multipoint version (Figure 5.2b), i.e. not by means of two, but of several prepared standard solutions containing the analyte at different, known, and well-defined concentrations. The method in this version is shown in Figure 5.3b. In this case, a linear function is fitted to the experimental points and the analytical result, c_x, is determined from the formula:

$$c_x = \frac{\hat{Y}_0}{B_1} \tag{5.5}$$

where $\hat{Y}_0$ is the value of the linear function corresponding to the intensity of Y_0, and the coefficient B_1 is calculated according to the chosen method of fitting the linear function.

The random error of the analytical result in the external standard method depends on several factors. If it is assumed that the random errors of the coefficient A and the preparation of the standard solutions are negligible and that the random error of the intensities of the measured signals for all standard solutions is the same, independent of the values of these intensities, then the value of the standard deviation of the analytical result, can be approximately calculated from the following formula valid for a linear multipoint calibration graph [2]:

$$s_{c_x} = \frac{s_{Y_x}}{B} \cdot \sqrt{\frac{1}{m} + \frac{1}{n} + \frac{(Y_0 - \overline{Y})^2}{B_1^2 \sum_i (c_i - \overline{c})^2}} \tag{5.6}$$

where m is the number of repetitions of measurements for each sample and standard solution, n – the number of standard solutions, and s_{Y_x} – the random error of the

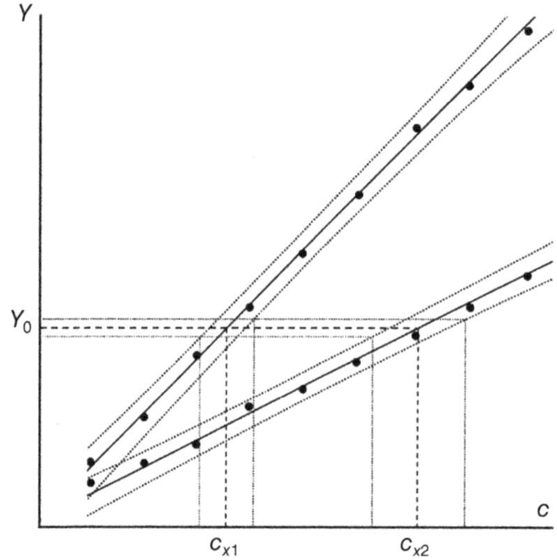

Figure 5.3 Random errors of analytical results, c_{x1} and c_{x2}, obtained by the ESM method on the basis of the signal intensity, Y_0, and of two calibration graphs of the same confidence limits prepared under different conditions in terms of sensitivity.

signal measured for the analyte in the sample, expressed with the formula:

$$s_{Y_x} = \sqrt{\frac{\sum_i (Y_i - \hat{Y}_i)^2}{n-2}} \tag{5.7}$$

From Eq. (5.6), it is possible to determine the confidence limit (shown in Figure 5.3) within which the calibration graph falls under certain conditions. Equation (5.6) shows that the precision of the analyte determination is greater than the number of times the signal intensity of each standard and sample solution is measured, and the number of measurement points, n, equals the number of standard solutions. The random error of the analytical result also depends strongly on slope B_1, which is a measure of measurement sensitivity. Figure 5.3 clearly shows that higher the measurement sensitivity, the more precise the determination of an analyte based on signal intensity with specific random fluctuations is likely to be (this is why care is taken to perform analyses under conditions of the highest possible sensitivity). Figure 5.3 also shows that the analytical result should be determined on the basis of a signal as close as possible to the central part of the calibration graph, although of course this condition cannot be treated as definitive for both theoretical and practical reasons.

The systematic error in the analytical result results from the presence of uncontrolled effects that change the position and shape of the real function. As a result, the linear calibration graph becomes an inaccurate representation of the real function. The errors that can then be made when uncontrolled effects of different nature are present are shown in Figure 5.4.

What then are the conditions for obtaining an accurate analytical result in the external standard method? This is best demonstrated mathematically.

If we assume that the calibration graph is described by a linear function, and the real function "deformed" by uncontrolled effects is described by the hyperbolic model adopted earlier (4.10), the fit of the two functions at the point corresponding

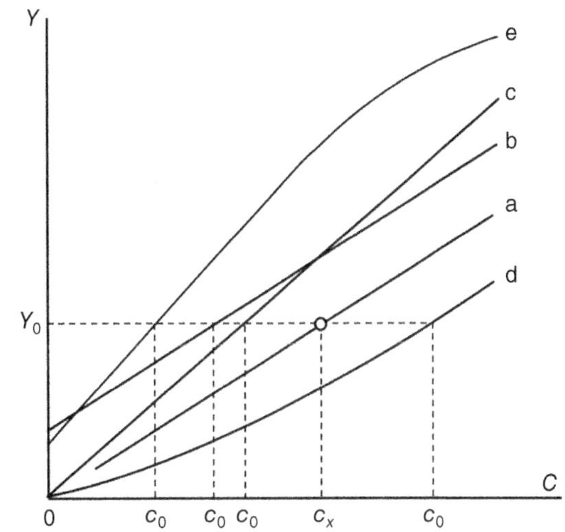

Figure 5.4 Systematic errors of the analytical result, $|c_0 - c_x|$, committed in the ESM method when the linear model function (a) maps the real function in the forms corresponding to additive (b), linearly multiplicative (proportional) (c), nonlinearly multiplicative (d) or complex (e) effect.

to the signal Y_0 is expressed by the equation:

$$B_1 \cdot c_x = A(c_p, c_t) + B_1' \cdot c_0 + B_2H(c_0, c_m) \cdot c_0 \tag{5.8}$$

where, as a reminder, the coefficient $A(c_p, c_t)$ expresses the magnitude of the additive effect caused by the components c_p added to the sample during processing and the native components, c_t, present in the sample, and $B_2H(c_0, c_m)$ is the value of the multiplicative effect of linear or hyperbolic nature determined by the function $H(c, c_m)$ (Eq. (4.10)). The coefficients $B_1 \neq B_1'$ express the possibility of the analyte being present in the sample (B_1') and in the standards (B_1) in different chemical forms, i.e. the presence of a speciation effect.

It follows from Eq. (5.9) that the ESM method provides an exact fit of the model function to the real function, i.e. reaching the state $c_x = c_0$, only if:

- additive and multiplicative interference effects are absent or eliminated (i.e. $A = B_2 = 0$), and the speciation effect is compensated ($B_1' = B_1$), or
- all these effects are compensated, which corresponds to taking the model function in exactly the same form as the real function:

$$A(c_p, c_n) + B_1 \cdot c_x + B_2H(c_x, c_m) \cdot c_x = A(c_p, c_n) + B_1 \cdot c_0 + B_2H(c_0, c_m) \cdot c_0 \tag{5.9}$$

It follows that the **ESM method is generally not resistant to uncontrolled effects** and, if these effects do occur, requires the use of combined processes to eliminate and compensate for them.

Certainly, the blank effect (BE) is the easiest to deal with. If the concentration of the native interferents causing the additive effect is equal to zero ($c_t = 0$), the effect can be eliminated by taking a measurement for the reagents used to prepare the sample or standards alone and subtracting this signal from the signals obtained for the sample or standards, respectively. This treatment is shown in Figure 5.5a in the case where the effect is only for the analyte in the sample. If the sample and standard

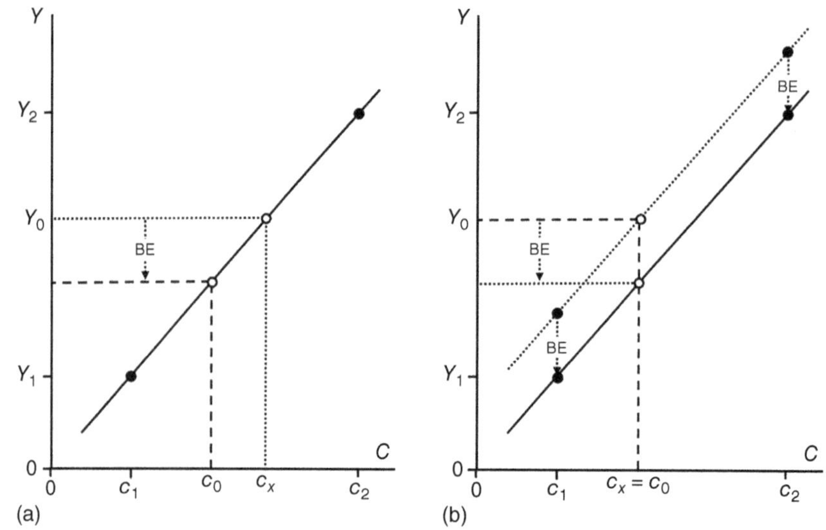

Figure 5.5 Blank effect (BE) eliminated (a) and compensated (b) when the sample alone (a) or together with standards (b) is affected in the ESM method; c_x, c_0 – the analyte concentrations obtained before and after correction, respectively.

were prepared in the same way and with the same reagents (and this is very often the case) and these reagents are the only source of the additive effect, then the calibration will spontaneously compensate for it, which is consistent with Eq. (5.9). The process of compensating for the blank effect is shown in Figure 5.5b.

Some practical considerations arise in this connection. To avoid the necessity of eliminating the blank effect, it is well to add the same reagents to the sample and to the standards regardless of whether the presence of some of them is necessary in one or the other solution. Care should also be taken to ensure that the concentrations of these reagents in the sample and standards are equal. If a series of standard solutions is made up by diluting a stock solution, then a solution containing these reagents in the concentrations in which they are present in the sample shall be used for this purpose.

A much more serious problem arises when the additive effect is caused by components naturally present in the sample ($c_t \neq 0$). To eliminate it in a way such as a blank effect, the analyst must know the type and amount of components responsible for the effect, which is rarely possible. Additive effects from unknown interferents, or known but present in the sample in unknown amounts, can only be eliminated by separate procedures requiring special conditions and handling. Chapter 9 will be devoted to this issue.

The role of speciation effect in external standard method calibration is relatively rarely discussed. As is well known, it is revealed when the analyte present in the sample has a different chemical form than the analyte in the standards and the two forms produce an analytical signal with different measurement sensitivity. If the analyte in the sample has a higher sensitivity than the analyte in the standard solutions, this

Figure 5.6 Systematic positive and negative errors, $(c_0 - c_x)$, caused by the speciation effect expressed by different measurement sensitivity of the analyte in the sample. B_1, than in the standards, (B_1') (compare Eq. (5.9).

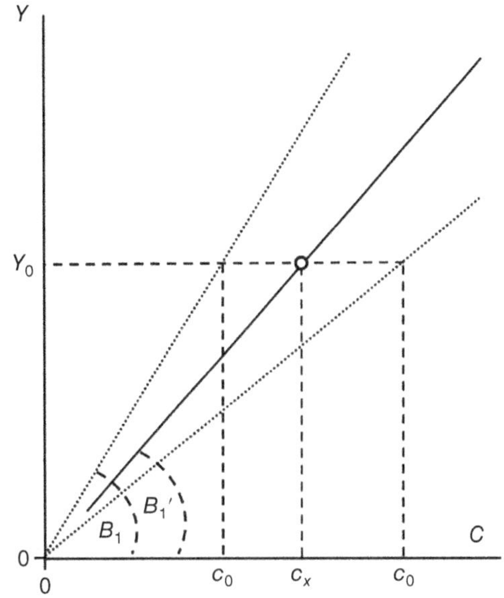

effect causes the analyte to be determined with a positive bias and vice versa. This is illustrated in Figure 5.6.

Table 5.1 shows the significance of the speciation effect on the example of the results of selenium determination by hydride generation – atomic fluorescence spectrometry (HG-AFS) [3]. Selenium was present in the calibration solutions in organic (selenomethionine) or inorganic (Se) form. As can be seen, the relative errors of the analytical results reached 50%, and increased when the speciation effect was superimposed on the interference effect. These results clearly demonstrate the potentially very large negative role of the speciation effect in the external standard method. In this case, the effect was effectively eliminated by bringing the organic form of selenium in the sample to the inorganic form by sample mineralization (using UV lamp).

As already stated, the analyst tries to perform determinations in the linear range of the calibration function. Sometimes, however, the problem arises that the experimental point system is nonlinear over the entire range of analyte concentrations considered and, therefore, the calibration graph is also nonlinear over this range. In this situation, the so-called **bracketing technique** can be used. This is a version of the two-point calibration using the external standard method illustrated in Figure 5.7. It involves setting the analyte concentration of the two standards at levels c_1 and c_2 that are as close as possible to each other, while including the analyte concentration, c_x, in the sample. The concentration of analyte in the sample is calculated according to Eq. (5.3). The better the above condition is satisfied, the more accurately the linear calibration graph reflects the nonlinear nature of the real function.

From the formal point of view the ESM method in the bracketing technique version is not correct, because if the real function is nonlinear in a wide range of analyte concentrations, it remains nonlinear in any narrower concentration range. By approximating it with a linear function one must therefore expect to make

Table 5.1 Systematic errors (expressed by relative error $[RE] = [(c_0 - c_x)/c_0] \cdot 100\%$), caused by the speciation and interference effects in the ESM method when selenium determined by HG-AFS was present in the sample and standards in the same concentration ($20\,\mu g\,l^{-1}$) but in different chemical forms; Se-Met – organic form (selenomethionine), Se – inorganic form, Cu – interferent, UV lamp – the digestion tool.

Sample	Analyte in sample	Analyte in standards	Effect	UV lamp	RE (%)
1	Se-Met	Se	Speciation	+	−1.7
2	Se-Met	Se	Speciation	−	−48.7
3	Se-Met + Cu	Se	Speciation + interference	+	−45.0
4	Se-Met + Cu	Se	Speciation + interference	−	−84.3
5	Se-Met + Se	Se-Met	Speciation	+	0.8
6	Se-Met + Se	Se-Met	Speciation	−	−25.8

Source: Adapted from Stafiński et al. [3].

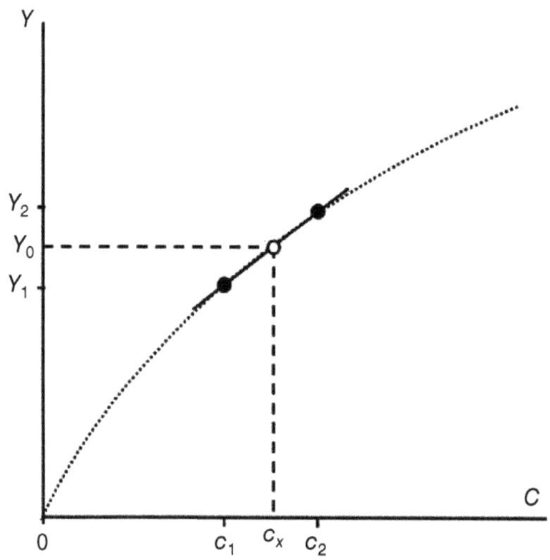

Figure 5.7 Bracketing technique: analyte concentration in a sample, c_x, is found on the basis of linear model function (solid line) approximating nonlinear real function (dotted line) in a narrow concentration range (between c_1 and c_2).

systematic errors. These errors, however, may be very small and, on the other hand, the method may improve the precision of analytical results.

The bracketing technique can be used for batch analyses only if the series of samples contains similar analyte concentrations, falling within the range determined by c_1 and c_2 concentrations. But even if this is not the case, because of the simple and fast procedure, it can be used, for example, to recalibrate determinations to compensate for temporal variation in signal intensity (time drift).

5.1.1.1 Modified Procedures

A number of modifications to the basic procedure of the external standard method can also be found in the literature that adapt the method to calibration under a

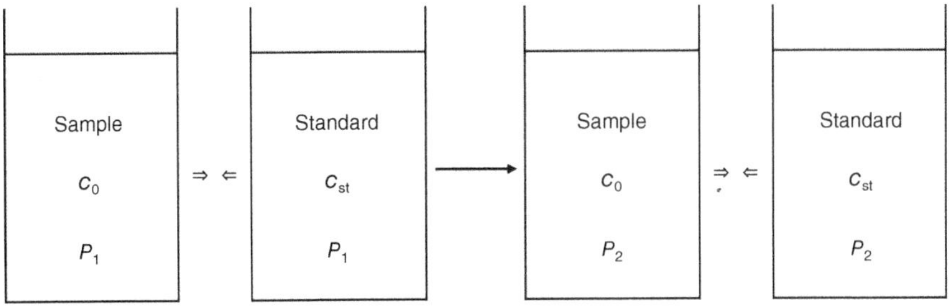

Figure 5.8 Scheme of the parametric ESM method at the preparative stage.

variety of specific conditions. Of these, two approaches, although occasionally used in practice, are nevertheless universal in the sense that they can be used successfully in conjunction with a variety of analytical methods and under a variety of analytical conditions. In both cases, it is assumed (which must be verified experimentally) that the intensity of the signal measured for a standard solution containing no analyte is equal to zero. The calibration procedure then requires a single standard with analyte concentrations of c_{st} (greater than c_0).

One of these methods, which can be called the **parametric ESM method**, consists in its basic version of subjecting these solutions to measurements under conditions defined by two different values P_1 and P_2 of a selected experimental parameter P (see Figure 5.8). This parameter is usually of an instrumental nature (e.g. radiation wavelength) and should be chosen such that a change in its value results in a clear change in signal intensity.

The measurement results obtained can be represented as shown in Figure 5.9: two two-point calibration graphs are constructed, based on the origin of the coordinate system and on the measurement points obtained for both values of parameter P. Then, the signal intensities measured for the sample under the two different conditions are related to the respective calibration graphs, determining two estimates (apparent), c_{x1} and c_{x2}, of the analytical result c_x' from the formula:

$$c'_{xi} = \frac{Y_{0i}}{Y_i} \cdot c_{st} \tag{5.10}$$

where Y_{0i} and Y_i ($i = 1, 2$) are the intensities of the signals measured for the sample and the standard solution, respectively, under conditions of the i-th value of parameter P. If the values of the apparent concentrations obtained differ only by chance, the final analytical result shall be calculated as the arithmetic mean of these values.

The calibration procedure described applies to the situation where the sample and standard do not contain any blanks. If the sample and standard contain such components of equal type and concentration, the blanking effect will be "automatically" compensated for under both conditions specified by the P parameter values. Otherwise, it requires elimination by correcting the signals measured for the sample and standard by the signals measured for the blank solutions. Note that this correction must be made separately under both measurement conditions.

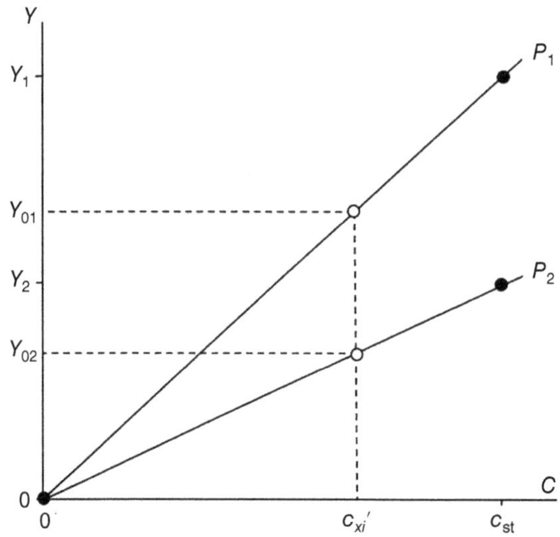

Figure 5.9 Parametric ESM method: two calibration graphs obtained for a single standard solution (c_{st}) at different conditions defined by values P_1 and P_2 of parameter P allow for estimation of the analytical result, c_x, by different values of apparent concentrations $(c_x')_1$ ($i = 1, 2$).

It can be expected that the precision of determining the concentration of an analyte in a sample according to formula (5.4) is not high due to the small number of measurement data. If the change of instrumental conditions can occur rapidly and with good reproducibility, then using the same individual sample and standard solutions, measurements can be made for them under conditions defined by a larger number of values of the parameter P, which will obviously increase the precision of the analytical result. It should be noted, however, that changing the conditions under which measurements are made is coupled with a gradual decrease in the optimum measurement sensitivity, which is associated with an increase in the random error of the analytical result. Another possibility is to increase the number of standard solutions subjected to measurements under different conditions, although then the method partially loses its characteristic features.

A valuable advantage of the parametric ESM method is the possibility to handle a single standard solution. This facilitates compliance with the condition that sample and standard are processed in the same way and consequently compensates for instrumental and preparative effects. The fast procedure also allows the calibration to be performed on a sample-by-sample basis. This individual treatment of the samples during the calibration process is always beneficial to minimize accidental and systematic errors.

An obvious limitation of the method is the assumption of the linear character of the real function under different conditions defined by the values of the parameter P. If the values of the apparent concentrations vary only by chance, this proves the validity of this assumption. It may also testify to the absence of other uncontrolled effects causing a systematic error in the analytical result, but is no conclusive evidence of the absence of such effects. Indeed, there is a very high probability that a change in the value of an instrumental parameter will induce a proportional change in any effect, resulting in no significant change in the value of the apparent concentration. This creates the danger of the psychological trap of overconfidence and misplaced trust in the accuracy of the method in a situation where nearly identical analyte determination results are obtained.

However, if the obtained values of apparent concentrations differ from each other statistically significantly, it can be said with high confidence that uncontrolled effects are the cause of this situation. In the first place, one should expect the presence of a nonlinear interference effect or a speciation effect. This information, although essentially negative from the point of view of the aim of the analysis, is very important for the further course of the analysis, because it indicates the necessity to make the appropriate correction of the analytical procedure used so far. This is a great advantage of the described calibration approach over the traditional approach, which does not offer any possibility of diagnosing interference and speciation effects.

A special laboratory process that can be used in the external standard method is the controlled dilution of solutions. Calibration solutions are prepared by diluting the pretreated sample and standard solutions with an appropriately selected diluent. If a blank is present, the diluent should be a blank solution as the calibration procedure used will then compensate for the blank. After dilution, the initial analyte concentrations, c_0 and c_{st}, in the sample and standard solution decrease proportionally to the values of $k_v \cdot c_0$ and $k_v \cdot c_{st}$. The parameter k_v is the so-called dilution degree, given by the formula:

$$k_v = \frac{V_0}{V_1 + V_0} \tag{5.11}$$

where V_0 is the *initial* volume of the solution diluted (in this case the sample and standard) and V_1 is the volume of the diluent. Note that the parameter k_v takes values from 0 to 1 for the infinitely diluted and undiluted solutions, respectively. The preparative procedure of the method is schematically shown in Figure 5.10.

Assuming that the degree of dilution is an experimental parameter, the **dilution ESM method** is basically similar to the parametric ESM method discussed previously. However, the difference is that the concentration of the analyte in the standard solution is known before and after changing its dilution, i.e. after changing the value of the parameter k_v. After measuring the analyte signal intensities for all four calibration solutions (Figure 5.10), the measurement results can therefore be presented as a "traditional" single calibration graph, as shown in Figure 5.11.

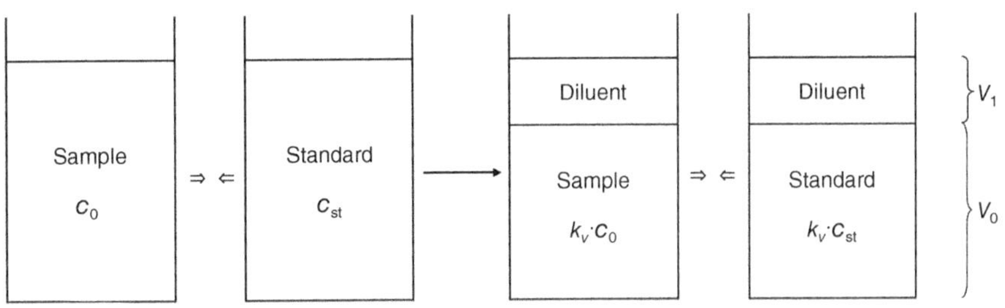

Figure 5.10 Scheme of the dilution ESM method at the preparative stage.

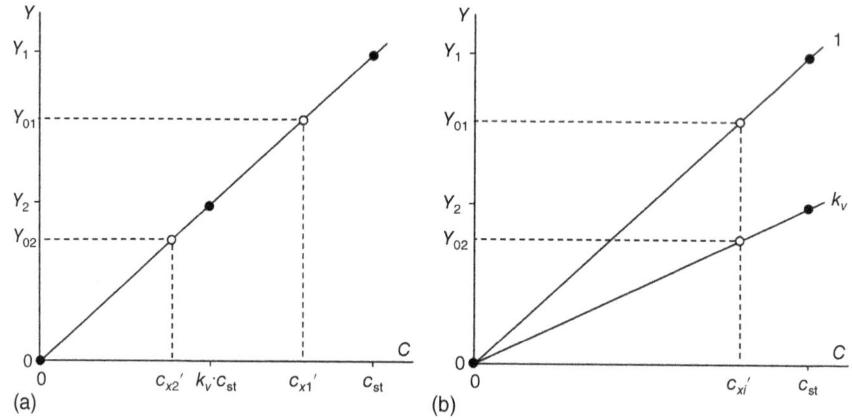

Figure 5.11 Dilution external standard method: the same measurement results obtained for a single sample and standard are presented in the forms of two-point calibration graph (a) and two-line calibration graphs (b), both allowing for determination of an analyte in a sample on the basis of two apparent analyte concentrations c_{xi}' ($i = 1, 2$) obtained for undiluted sample and standard ($k_v = 1$) and both diluted with factor k_v.

Based on the graph, two values of apparent concentrations are determined from the formula:

$$c_{xi}' = \frac{Y_{xi}}{Y_i \cdot k_{vi}} \cdot c_{st} \tag{5.12}$$

where Y_{xi} and Y_1 ($i = 1, 2$) are the signal intensities measured for the sample and for the standard solution diluted with degree equal to 1 and k_v, respectively. If the values of the apparent concentrations differ only randomly, the final analytical result is calculated, as before, as the arithmetic mean of these values.

Assuming an interpretation of the measurement results as shown in Figure 5.11a, the precision of the analytical result is increased – compared to the traditional approach in its basic version (see Figure 5.2a) – by measuring the signal intensity for the sample twice before and after dilution. As before, this precision can be increased by increasing the number of measurement points by taking measurements for the sample and standard at several dilution steps of both solutions.

The method has similar advantages to the previous method: calibration information is provided by a single standard, sample and standard can be treated in the same way, and the calibration process can be applied to each of a number of samples individually without significant time and effort. However, calibration by this method requires great care, since any systematic error made in the preparation of the sample and standard solutions as well as in their progressive dilution will be transferred to subsequent diluted solutions.

If one assumes that the degree of dilution is an experimental parameter, then the measurement results shown in Figure 5.11a can also be interpreted similarly to the previously described parametric calibration approach and presented in the analogous manner shown in Figure 5.11b. The question that arises, however, is whether this manipulation of the data makes sense and provides any benefit. Well, it is worth noting that if the measurement results obtained for the calibration solutions form a single calibration graph, it gives the possibility to identify the effect of nonlinearity of the real function. It will be manifested not only by nonlinear arrangement of measurement points but also by variation of values of apparent concentrations. Larger variability of those values after presentation of measurement results in the form of several two-point calibration graphs will prove that apart from the nonlinearity effect there is another uncontrolled effect, most probably interference or speciation.

It should be noted that the manifestation of an interference effect by changing the degree of sample dilution is much more likely than by changing any instrumental parameter. This is because diluting a sample is equivalent to making an equal change in the concentration of all its components, including of course the analyte and the interferents. Thus, this disturbs some states of equilibrium between them, which generally induces a change in the degree of interaction between the interferents and the signal produced by the analyte. The situation with the speciation effect may be similar, although it must be said that it is very poorly studied in this respect.

The ESM method in the form presented above is even more diagnostic of potential uncontrolled effects than in its previous, parametric form. However, it is very rarely used in practice. This approach, like the previous one, is mainly seen as a fast and efficient calibration method, which is expected to determine an analyte in the form of two or more randomly different estimates of its concentrations unambiguously indicating the final value of the analytical result.

However, the conventional ESM method is mainly used in its unmodified version, although paradoxically this version – due to potential uncontrolled effects impossible to diagnose – carries the highest risk of systematic errors and analytes determination with high inaccuracy. The method in this version is also commonly recommended in the vast majority of international standards and various other analytical regulations. In scientific publications, it appears as a standard, somehow universally applicable way of performing analytical calibration.

Situating this external standard method among other possible calibration approaches, one doubts whether it should really be so widely used and recommended in quantitative analysis. Its undoubted advantage is its conceptual clarity and procedural simplicity, which is of great importance from the point of view of uncertainty of analytical results obtained with this method. A major practical advantage is also the possibility to determine the analyte concentration by this method in multiple samples using a single calibration graph. However, these advantages should only be considered when potential uncontrolled effects as sources of analytical error are very well recognized and eliminated or compensated for.

5.1.2 Dilution Method

The possibility of using the sample dilution process in the calibration process has been recognized for quite a long time. The first calibration method based on this process was developed in 1959 by P.T. Gilbert Jr. [4] and applied "to circumvent interferences" of aluminum, iron, copper, and calcium with cadmium by flame photometry. Originally called by him the method of successive dilution", already a year later it appeared under the name of "dilution method" given by J.A. Dean [5], and under this name it appears most frequently in the literature. Initially arousing great interest, among others, thanks to the works of A. Shatkay [6, 7], however, in later years it did not gain much popularity in analytical practice.

Underlying the concept of the **dilution method** is the assumption that **progressive dilution of the sample results in sufficient modification of the sample matrix so that uncontrolled effects, in particular the interference effects, diminish with the dilution process and are then fully compensated for in a hypothetical sample of infinite dilution**. This approach thus differs from the previously discussed modification of the external standard method in that it is expected not only to determine the concentration of the analyte in the sample, but also, as an added value, to compensate for uncontrolled effects and consequently improve the accuracy of the analytical result. In this sense, the dilution method becomes a separate calibration method.

At the preparative stage, the procedure of the dilution method is similar to that in Figure 5.11. The difference is that the individual solutions of sample and standard, according to the theoretical concept of the dilution method, are not diluted once or even several times, but the dilution process is carried out until a dilution degree value k_{vi} close to zero is reached, i.e. as much dilution of sample and standard as practical considerations allow. This is schematically shown in Figure 5.12.

At each dilution step (i.e. for successive values of dilution steps k_{vi}, where $i = 1, 2, \ldots, n$) the signal intensities for the standard, Y_i, and for the sample, Y_{xi}, are measured, each time taking into account possible signals from the blanks and establishing a zero signal value for zero analyte concentration. The measurement results can be interpreted as shown in Figure 5.13. At each dilution step of the sample and standard, a series of apparent concentration values c_{xi}' of the analyte in the sample is determined from the formula:

$$c_{xi}' = \frac{Y_{xi}}{Y_i} \cdot c_{\text{st}} \tag{5.13}$$

The apparent concentrations can then be represented as a function of the degree of dilution of the sample, as shown in Figure 5.13. If there are no uncontrolled effects during dilution, then all values of the apparent analyte concentrations should be equal (within random error). If, on the other hand, effects are present, then the apparent concentration obtained for the undiluted sample ($k_v = 1$) is assumed to be

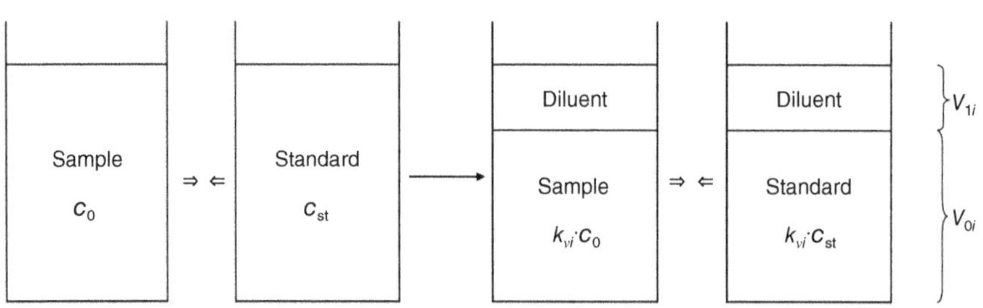

Figure 5.12 Scheme of the dilution method at the preparative stage.

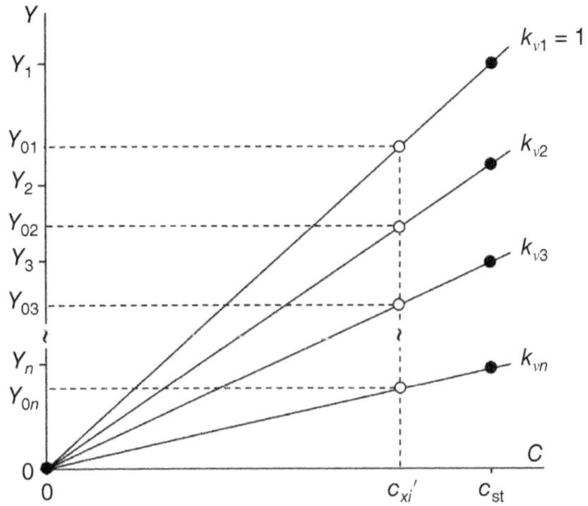

Figure 5.13 Dilution method: a set of two-point calibration graphs allowing for determination of an analyte in a sample on the basis of a series of apparent analyte concentrations c_{xi}' ($i = 1, 2, \ldots, n$) obtained in the interpolative way at different relative dilution values, k_{vi}.

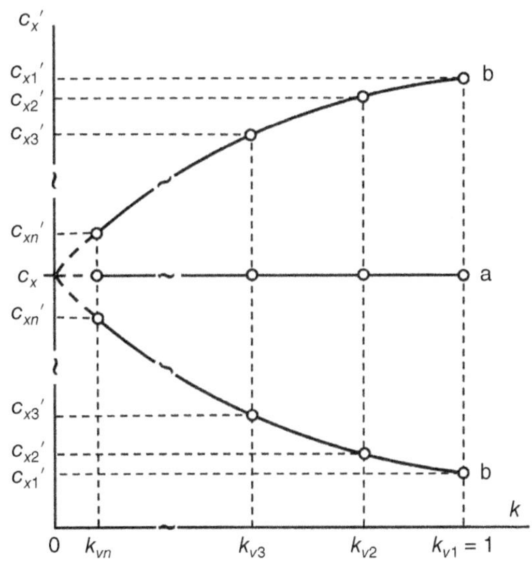

Figure 5.14 Dilution method: final analytical result, c_x, is calculated after extrapolation of a function fitted to the values of apparent analyte concentrations, c_{xi}', obtained at different relative dilution values k_{vi}, in cases of the absence (a) and presence (b) of uncontrolled effects.

different from the true concentration, but as dilution proceeds, subsequent values of the apparent concentrations will approach the true value, i.e.:

$$c_0 = \lim_{k_{vi} \to 0} c'_{xi} \tag{5.14}$$

To determine the analytical result, c_x, which is a measure of the true concentration, c_0, a function (linear or nonlinear) must therefore be fitted to the experimental points $[(c'_x)_i, k_i]$ (c'_{xi}, k_{vi}) and extrapolated to the value $k_{vi} = 0$. This is illustrated in Figure 5.14.

Note that in the dilution method, changes in the analytical signal in the sample are compared to changes occurring in the standard solution. Thus, regardless of whether or not the analyte in the standard undergoes certain effects, the effects acting on the analyte in the sample are potentially compensated for rather than eliminated.

To make the compensation process more efficient, the standard solution can be made somewhat similar to the sample (e.g. contain interferents), but from a theoretical and practical point of view, this is not necessary.

The question arises: however, how valid and plausible is the assumption of gradual compensation of effects as the sample is diluted and their complete compensation in an infinitely dilute sample? Approaching this question from the theoretical side, one can once again rely on the general mathematical form of the real function (4.10).

When the measurement results are interpreted as in Figure 5.14, it can be assumed that the initial slope B_1 of the calibration graph (for $k_v = 1$) decreases according to the value of the sample dilution step. Then the linear model function at each dilution step specified by the dilution ratio k_{vi} takes the form:

$$Y_{0i} = k_{vi} \cdot B_1 \cdot c'_{xi} \tag{5.15}$$

If a real function is nonlinear, its representation by a linear function is expressed by the equation:

$$k_{vi} \cdot B_1 \cdot c'_{xi} = A(k_{vi}c_p, k_{vi}c_n) + B'_1 \cdot k_{vi}c_0 + B_2H(k_{vi}c_0, k_{vi}c_m) \cdot k_{vi}c_0 \tag{5.16}$$

Hence:

$$c_x = \lim_{k_{vi}\to 0} c'_{xi} = \lim_{k_{vi}\to 0} \left(\frac{A(k_{vi}c_p, k_{vi}c_n)}{k_{vi} \cdot B_1} + \frac{B'_1 \cdot c_0 + B_2H(k_{vi}c_0, k_{vi}c_m) \cdot c_0}{B_1} \right) \tag{5.17}$$

where coefficients B_1 and B_2 express possible occurrence of speciation effect. It can be assumed that as the concentrations of interferents causing the additive interference effect decrease to zero, the coefficient A (the signal caused by the presence of these interferents) will also tend to zero, i.e.:

$$\lim_{k_{vi}\to 0} \frac{A(k_{vi}c_p, k_{vi}c_n)}{k_{vi} \cdot B_1} = \lim_{k_{vi}\to 0} \frac{k_{vi} \cdot A(c_p, c_n)}{k_{vi} \cdot B_1} = \frac{A(c_p, c_n)}{B_1} \tag{5.18}$$

It is also seen that as the sample is diluted, the function describing the multiplicative interference effect goes to zero regardless of whether it is linear ($a = 0$) or nonlinear ($a \neq 0$) with respect to the analyte concentration:

$$\lim_{k_{vi}\to 0} H(k_{vi}c_0, k_{vi}c_m) = \lim_{k_{vi}\to 0} \frac{k_{vi}c_m}{1 + a \cdot k_{vi}c_0 + b \cdot k_{vi}c_m} = 0 \tag{5.19}$$

Thus:

$$c_x = \lim_{k_{vi}\to 0} \frac{A(c_p, c_n) + B'_1 \cdot c_0}{B_1} \tag{5.20}$$

Thus, from the theoretical point of view, the dilution method allows to obtain the real result ($c_x = c_0$) when the additive and speciation effects do not occur or the first one has been eliminated ($A = 0$) and the second one compensated ($B_1 = B_1'$). Under these conditions, the **dilution method gives a chance to compensate for linear and nonlinear multiplicative effects**. The method also allows for **compensation of the complex effect**, occurring in the form of linear and nonlinear multiplicative effects in different ranges of analyte concentration.

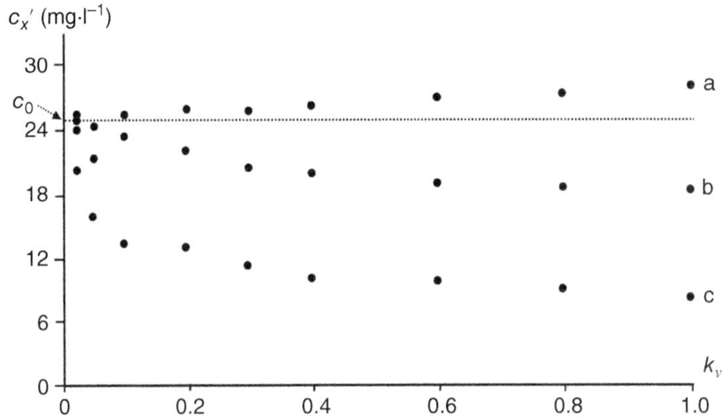

Figure 5.15 Calibration by dilution method: results obtained for FAAS determination of calcium in the synthetic samples containing 25 mg l^{-1} Ca with 0 (a), 25 (b), and 100 (c) mg l^{-1} Si; concentration of Ca in the standard solution: 4 mg l^{-1}; maximum dilution: $k_v = 0.025$ (from own collection).

How are the theoretical assumptions of the dilution method fulfilled in practice? Figure 5.15 shows as an example the results of the dilution method for the determination of calcium obtained with flame atomic absorption spectrometry (FAAS).

The results obtained for the analyte alone (line *a*) indicate that there is a small positive effect occurring in the undiluted sample ($k_v = 1$), which decreases with dilution and basically compensates for the infinitely diluted sample. The effect is due to the high concentration of analyte in the sample (25 mg l^{-1}), beyond the linear range of the real function. The presence of silicon in the undiluted sample additionally causes a large interference effect, the larger the interferent concentration (which is obvious). Regardless of the interferent concentration, the effects (nonlinearity and interference) are gradually compensated as the sample dilution process proceeds.

Thus, as can be seen, calibration by the dilution method provides the possibility to compensate for interference effects. Moreover, it also allows compensation for the effect of nonlinearity of the real function whether it is due to interference effects or detection effects. This is of great importance because the analyst when undertaking a calibration with this method, does not have to pay particular attention to whether he is working in the linear or in the nonlinear range of the calibration graph.

The experiments also revealed important limitations of this calibration method. First of all, the experimental points obtained for a sample of high dilution ($k_v < 0.2$) are characterized, as expected, by a relatively large random scatter. This is due to the large fluctuations in the intensity values of the signals measured for small concentrations of analyte in the sample. It is therefore very important to determine under each new analytical condition the optimum value of the dilution ratio of sample and standard, such that on the one hand, it corresponds to small measurement fluctuations, but on the other hand, ensures that a small area is left for the extrapolation process.

Another problem is the mutual position of the experimental points that determines the form of the function fitted to those points (c_{xi}' vs. k_v). While the nonlinear

Table 5.2 Results obtained by the dilution method for FAAS determination of calcium in synthetic samples after fitting nonlinear functions: parabolic (a), exponential (b), hyperbolic (c), and logarithmic (d) to the same set of experimental points (compare Figure 5.11) [8].

	Sample composition ($mg\,l^{-1}$)		Calculated analyte concentration, c_x ($mg\,l^{-1}$)			
c_{st} ($mg\,l^{-1}$)	$c_{Ca} = c_0$	c_{Si}	a	b	c	d
4	25	25	25.9	26.6	27.4	50.6
4	25	50	24.2	25.3	27.0	58.5
4	25	100	18.0	20.9	25.2	56.4
25	25	25	24.8	25.0	25.1	35.2
25	25	50	26.0	27.2	28.3	32.4
25	25	100	18.4	21.7	24.5	58.6

detection effect does not fundamentally change the linear arrangement of the points, the interference effect makes the points arrange themselves strongly nonlinear, contrary to expectations, the more the concentration of the interferent in the sample is higher. This forces one to fit a nonlinear function to these points and to extrapolate this function, which is riskier from the point of view of the accuracy of the analytical result than if these operations involved a linear function.

If the nonlinearity is evident, the accuracy of the final result may depend on the type of nonlinear function fitted. Table 5.2 presents results that demonstrate this [8]. These results also confirm that a simple hyperbolic function (with the general formula given earlier to the function $H(c_0, c_m)$) by far best represents the point system especially when it is strongly nonlinear. It is also worth noting that the final analytical results do not really depend on whether there is a nonlinearity effect or whether it is compensated by equal analyte concentrations in the sample and in the standard solution.

So, can it be said that the dilution method meets the expectations placed in it? Based on the experimental example presented, we can say rather yes. Although the data in Table 5.2 (column *c*) show that the analytical results, c_x, obtained with this method are subject to errors sometimes exceeding 10% of the expected value, c_0, a large contribution is made by random errors, also reaching several percent of relative standard deviation (RSD). Furthermore, as shown in Figure 5.15, the results obtained are much more accurate than they would be if no sample dilution process was used during calibration, but the analyte was determined in the undiluted sample, i.e. by the external standard method. Note that this applies to results obtained under strong interference effects in the sample.

Experiments performed by the author's team with various analytical methods have shown that the method is similarly effective in many other analytical situations differing in the type of analyte and interferent, their mutual concentration in the sample, and the type of measurement method used. In all such cases, it allows the more accurate analytical results to be obtained the smaller the effects that occur.

Of course, it always leads to correct results when there are no uncontrolled effects, but in this case, it makes more sense to use the external standard method in the multipoint version for calibration.

Unfortunately, however, in practice, one can find situations where the dilution method practically fails in the sense that the dilution of the sample does not lead to compensation of uncontrolled effects. In Figure 5.16 it is apparent, for example, that the interference effect caused by the presence of aluminum in the sample during the determination of calcium by flame atomic absorption spectrometry does not decrease as the sample is diluted, but increases unexpectedly (line *a*). As a result, the analytical result obtained is much more inaccurate than it would be if it were determined in an undiluted sample.

This seemingly surprising result can be easily understood by realizing that in the FAAS technique, interferents generally do not interact with the analyte in the sample solution, but after the sample enters the flame. Both interferences, i.e. caused by silicon and aluminum, are related to the size of the dried aerosol particles in the flame, which is related to the concentrations of the of analyte and interferent in solution. During the dilution process, the particles gradually decrease in size, and the chance of the interferent binding an adequate amount of analyte to the refractory compound becomes smaller and smaller, especially if only one type of compound (e.g. $CaO–SiO_2$) is formed.

In the case of aluminum various refractory compounds of different ratios of Al to Ca atoms, (e.g. $3CaO–Al_2O_3$, $5CaO–3A_{12}O_3$, $3CaO–5Al_2O_3$) are exceptionally formed during volatilization process. Therefore, even where the sample is very diluted and the aerosol particles are very small, aluminum atoms are apparently able to find an adequate amount of calcium atoms to bind them in the form of a selected refractory compound. As a result, the effect of aluminum on calcium may become greater as the sample is diluted.

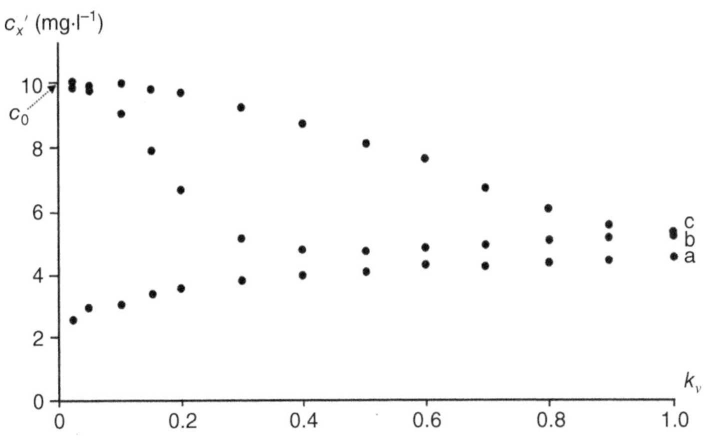

Figure 5.16 Calibration by dilution method: results obtained for FAAS determination of calcium in the synthetic sample contained 10 mg l^{-1} with 80 mg l^{-1} Al and diluted with lanthanum solution of concentrations 0 (a), 20 (b), and 100 (c) mg l^{-1} La: concentration of Ca in the standard solution: 10 mg l^{-1}; maximum dilution: $k_v = 0.025$ (from own collection).

It seems unlikely, however, that any interference effect (other than that of an additive nature) would remain significant in an "infinitely" dilute sample. It can be assumed that the type of interference that occurs between calcium and aluminum in the flame is nevertheless gradually eliminated but under conditions of such high sample dilution that it is not possible to verify this phenomenon experimentally (with the current degree of sophistication of the measuring apparatus).

From a practical point of view, however, the problem is that if even one experimental example contradicts the basic assumptions of the dilution method, the method in its basic version cannot be used without knowledge or prior experimental verification of these assumptions. This greatly reduces its practical usefulness. Therefore, can any specific procedure be proposed for the dilution method that could increase its calibration versatility?

One such approach is to use a selected special reagent to dilute the sample [9]. The action of the reagent is then different from that of the external standard method in that as the sample is diluted, i.e. the concentration of the analyte and interferent decreases, the concentration of the reagent increases simultaneously. To compensate for the possible effects of this reagent on the analyte the reagent should be used concurrently with the dilution of the standard solution.

An example of this procedure is shown in Figure 5.16 (lines b and c). As can be seen, the interference effect is eliminated when the ratio of the concentration of the special reagent to the concentration of the analyte in the sample is sufficiently large, and this occurs in that area of the dependence of the apparent analyte concentration on the degree of dilution of the sample that is crucial to the accuracy of the analytical result. It is also worth noting that this can be achieved with a reagent of relatively low concentration. In the example presented here, the complex effect of aluminum on calcium was effectively eliminated with lanthanum at an initial concentration of $20\,\mathrm{mg\,l^{-1}}$ (line b), which is several tens of times lower than the concentration usually established in the external standard method. Obviously, the effectiveness of lanthanum in a higher, but still relatively small concentration ($100\,\mathrm{mg\,l^{-1}}$), was even greater (line c) and was already revealed in the initial dilution phase, which influenced such an arrangement of experimental points that was more favorable for the approximation and extrapolation processes. As a result, an accurate analytical result could be achieved in both cases after fitting the nonlinear function to only a few final (for $k_v < 0.15$) experimental points.

The underlying process of diluting the sample and standard can also be effectively used to effectively compensate for interference effects using interferents (!) [9]. To do this, sample and standard solutions should be diluted in parallel with a solution containing the same, or even a different interferent than the one that dominates the sample. As the dilution proceeds, the concentration of the interferent naturally present in the sample decreases, and the concentration of the interferent added with the dilution solution increases equally in the sample and standard solutions. Consequently, the overall concentration of the added interferent gradually equalizes in the two solutions and there is a buffering effect of gradually equalizing effects of the interferent on the analyte in these solutions. Even if the natural interferent effect

Table 5.3 Results obtained for FAAS determination of calcium in concentration of $c_0 = 10\,\mathrm{mg\,l^{-1}}$ in the presence of various interferents by ESM method and by dilution method using (or not) different interferents in the role of buffers; concentration of Ca in the standard solution: $10\,\mathrm{mg\,l^{-1}}$.

		Calculated analyte concentration, c_x (mg l^{-1})		
Interferent (mg l^{-1})	**Buffer (mg l^{-1})**	**ESM method**	**Dilution method without buffer**	**Dilution method with buffer**
80 Al	40 Al	5.98 ± 0.13	2.38 ± 0.15	9.65 ± 0.32
80 Al	10 Al	5.78 ± 0.11	2.38 ± 0.15	9.87 ± 0.28
20 Ti	10 Ti	5.45 ± 0.08	6.78 ± 0.43	8.42 ± 0.32
10 Ti	40 Al	7.01 ± 0.15	8.52 ± 0.33	10.15 ± 0.55
10 V	40 Al	3.21 ± 0.04	10.19 ± 0.34	9.92 ± 0.43
10 V	5 V	4.09 ± 0.17	10.19 ± 0.34	9.82 ± 0.24
20 Ti + 10 V	10 Al	0.76 ± 0.03	2.43 ± 0.36	10.51 ± 0.79
20 Ti + 80 Al	40 Al	4.43 ± 0.09	2.89 ± 0.44	9.72 ± 0.67

Source: Kościelniak [9], table 1 (p. 108)/Elsevier.

does not tend to diminish as the sample is diluted, the buffering effect becomes increasingly dominant and determines the accuracy of the analytical result.

The results of this procedure, obtained for the determination of calcium in the presence of various interferents present in the sample and added to it, are shown in Table 5.3 (the samples were diluted using the flow injection technique, which will be approximated in a Chapter 7). As can be seen, the dilution method with buffer leads in most cases to more accurate results than the method without buffer and too far better results in this respect than the ESM method. What is also important, the concentration of the buffer interferent can be small, even smaller than the concentration of the interferent present in the sample (for the same reasons that the concentration of the special reagent can also be small).

However, the precision of the results obtained in this way is lower than without the use of a buffer, especially when the buffering interferent is selected in such a way that it gradually reduces the intensity of the signal measured for the analyte in the sample and in the standard (as in the case presented in Table 5.3). It is therefore much better to select for buffering an interferent that, under the given analytical conditions, has the ability to increase the intensity (although this is certainly not easy since there are far fewer such interferents than those that act inversely).

The ability to use a buffer interferent instead of a special reagent is convenient in that the analyst does not need to know the qualitative composition of the sample to select this reagent. The buffer interferent is selected relative to the analyte regardless of which sample component is the interferent or whether there is one or more interferents in the sample. The high efficiency of the dilution method using a buffer interferent to compensate for interferences in natural samples is shown in Table 5.4.

Table 5.4 Results obtained for FAAS determination of calcium in the iron ore samples (certificated reference materials) using ESM method and dilution method with the use of buffer (10 mg l^{-1} Al).

Sample code	Main sample components (%)					Analyte amount found (%)	
	Ca	Al	Fe	Si	Mg	ESM	Dilution method
2.53	0.286	0.37	66.1	1.90	0.17	0.163 ± 0.009	0.272 ± 0.022
2.54	2.423	2.97	41.5	6.07	2.21	1.715 ± 0.045	2.531 ± 0.098
2.55	0.529	0.51	64.8	2.45	0.89	0.382 ± 0.007	0.552 ± 0.018
2.57	0.051	0.71	83.9	1.15	0.02	0.013 ± 0.002	0.055 ± 0.005

Source: Kościelniak [9], table 2 (p. 109)/Elsevier.

Certainly, the precision of the analytical results remains a weakness of the dilution method. As Tables 5.3 and 5.4 show, one has to reckon with the fact that the random scatter of the results can be from two to several times larger than that obtained under similar analytical conditions by the external standard method. This obviously has a considerable influence on the uncertainty of the value of the determined analyte concentration in the sample. The dilution method may therefore be recommended (preferably in versions using a special reagent or buffer interferent) in cases where the uncontrolled effects in the sample are of unknown nature and are so large that the systematic errors in the analytical result due to them may be much larger than the random errors due to the nature of the method itself.

Compared to the external standard method, additional random and systematic errors in the analytical result may also result from the dilution process itself. If a sample and standard with different dilution rates are prepared by stepwise dilution of two initial solutions, errors made at each step of this process may accumulate and the actual dilution rate values may differ significantly from those predicted. There may also be situations where the sample and standard are not diluted to the same extent. The process must therefore be carried out with great care and attention.

However, it seems that the role of the dilution process in calibration is underestimated and therefore too rarely used. This process is the simplest additional procedure that can be used to support the ESM method to increase its utility. It does not require the use of additional chemical reagents, and if desired, the concentration of these reagents, as shown, can be relatively small, which is very important from a green chemistry policy perspective. Although multiple dilution is a rather laborious and time-consuming process, it can be performed fully automatically (e.g. by liquid injection technique), which reduces these practical limitations of the dilution method while increasing the precision and accuracy of sample and standard preparation for measurement. This issue will be discussed in more detail in Chapter 7.

5.2 Internal Calibration Methods

In internal calibration methods, as in external calibration methods, the sample and analyte standards are prepared for measurement separately from each other. Therefore, they can be classified as comparative calibration methods. What characterizes them is, as in qualitative analysis, the addition of other substances in known, strictly fixed amounts to solutions of the sample and standard containing a known amount of analyte. The role of these "internal standards" (IS) is so important that their action adds new values to the analytical process. Regardless of its specificity, each internal calibration method obviously fulfills the basic task of determining the analyte with satisfactory accuracy and retains the characteristics that determine its practical utility.

Depending on the role of the substance to be added, calibration is performed using either the **internal standard method** or the **indirect method**.

5.2.1 Internal Standard Method

In the preparative step of the **internal standard method** a substance selected to be an internal standard (IS) is added to the sample with analyte concentration c_0 and to the standard solutions with analyte concentrations c_1 and c_2 ($c_2 > c_1$) in such a quantity that its concentration (c_w) in these solutions is equal (Figure 5.17). This substance shall not be a constituent of the sample to be analyzed, shall not react with constituents of the sample, and shall not give rise to a signal under measurement conditions characteristic of the analyte.

In such prepared solutions, both the analyte and the internal standard are measured under two measurement conditions that are optimal for both substances. The response of a measuring instrument to analyte concentration is expressed in terms of the ratios of the intensities of the signals obtained for the analyte in the sample, Y_0, and in the standards, Y_1, Y_2, to the intensities of the signals obtained for the internal standard in these solutions, Y_{w0}, Y_{w1}, and Y_{w2}. Assuming that the dependence of the signal Y measured for on the concentration c of the analyte and the dependence of the signal Y_w measured for the internal standard on the concentration c_w of

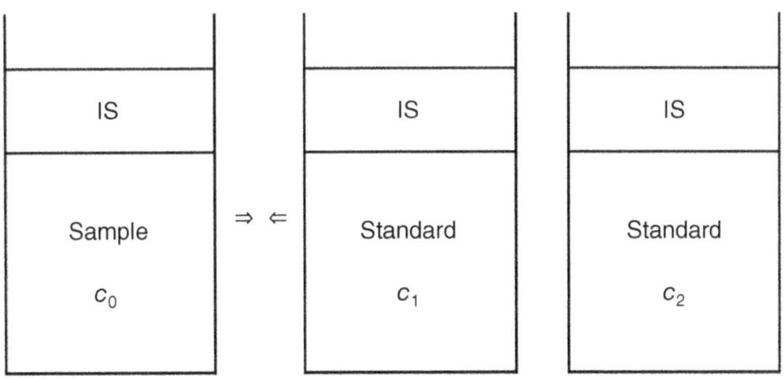

Figure 5.17 Scheme of the basic form of the internal standard method at the preparative stage of the calibration process.

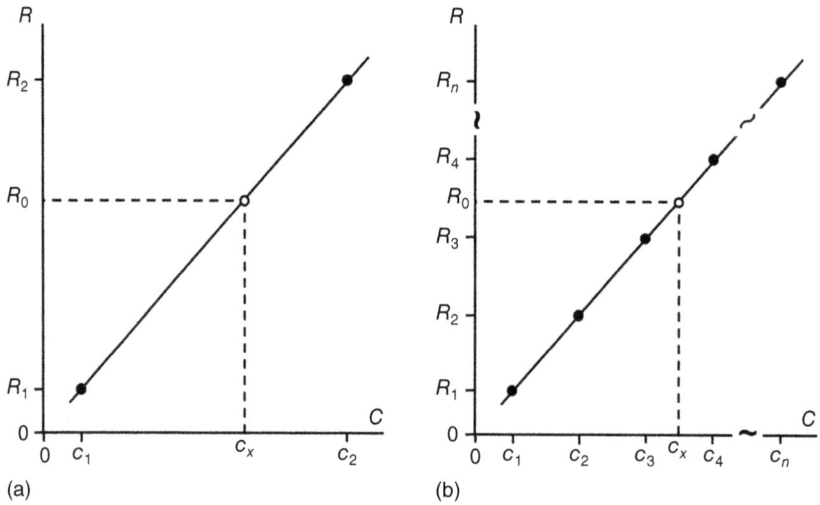

Figure 5.18 Calibration in accordance with internal standard method: interpolative transformation of the ratio of analyte to internal standard signals obtained for the sample, R_x, and for the analyte standards, $R_1, \ldots, R_n$, to the analytical result, c_x, using two-point (a) and multipoint (b) calibration graphs.

that standard are linear, and assuming that this model function is described by the formula:

$$R = \frac{B_1 \cdot c}{B_{w1} \cdot c_w} \tag{5.21}$$

where $R = Y/Y_w$. Assuming that value $B_{w1} \cdot c_w$ is constant (within the limits of random measurement error), then formula (5.21) can be presented in the form:

$$R = B_{r1} \cdot c \tag{5.22}$$

The image of the function (5.22) is the linearly increasing calibration graph shown in Figure 5.18a.

The intensity ratio value, $R_0 = Y_0/Y_{w0}$, measured for the sample is then related to the model function and the concentration c_x of the analyte in the sample is determined by interpolation from the formula:

$$c_x = \frac{R_x \cdot (c_2 - c_1) - R_1 \cdot c_2 + R_2 \cdot c_1}{R_2 - R_1} \tag{5.23}$$

where $R_1 = Y_1/Y_{w1}$ and $R_2 = Y_2/Y_{w2}$.

If the procedure of the method is implemented in a multipoint version (see Figure 5.18b), the ratios of the measured signals for analyte and internal standard are determined for all standard solutions and, after fitting a linear function to the measurement points, the analytical result is calculated from the formula:

$$c_x = \frac{\hat{R}_0}{B_{r1}} \tag{5.24}$$

where $\hat{R}_0$ is the value of the function (5.22) corresponding to the intensity ratio R_0, and the value of the coefficient B_{r1}' is determined by the chosen method of fitting the linear function.

As in quantitative analysis, the main objective of the internal standard method is to compensate for measurement errors made during analysis. The basic assumption made in this method is that a change $Y \pm E$ in the intensity of the signal Y measured for the analyte can be compensated for by a similar change in magnitude and direction in the signal $Y_w \pm E_w$ measured for the internal standard when the signal Y is measured relative to the signal Y_w, i.e.:

$$\frac{Y \pm E}{Y_w \pm E_w} \approx \frac{Y}{Y_w} \quad \Leftrightarrow +E \approx +E_w \vee -E \approx -E_w \tag{5.25}$$

It is obvious that the condition (5.25) can be satisfied the more easily the errors E and E_w are smaller. Thus, **the internal standard method serves primarily to compensate for random measurement errors** and thus contributes to improving the precision of analytical results obtained without the use of an internal standard (i.e. those obtained by the external standard method, for example).

This property of the internal standard method has been noticed for a long time. The first reports on it date back to the nineteenth century, but its systematic application for analytical purposes was initiated in the 1920s by the work of W. Gerlach [10], which concerned the reduction of errors caused by excitation source instability and variable photographic in arc and spark emission spectrometry. The popularity of the method in the past is evidenced by the fact that shortly after the introduction of commercial recording flame photometers, a photometer was constructed that allowed simultaneous measurement of signals from two elements to perform calibrations using the internal standard method [11]. Currently, the method is quite often used, especially in analyses by chromatographic techniques and inductively coupled plasma-mass spectrometry (ICP-MS) method.

The key elements for successful calibration with the internal standard method are the maintenance of appropriate measuring conditions and the proper selection of the substance serving as internal standard. Measurements for the analyte and the internal standard should be performed with the same measurement method and at the same time or shortly after each other. Preference is therefore given to methods that allow simultaneous measurements for at least two substances (and these are the separation methods and ICP-MS). This is of course a certain limitation of the wide applicability of the method.

The choice of internal standard should follow the generally recognized general principle that an **internal standard should resemble as closely as possible the analyte in its chemical and physical properties**. The idea is that due to their mutual similarity, both substances are similarly susceptible to any changes occurring during the analytical process. Observance of the above conditions usually (but not always) allows for a reduction of random errors in the analytical result.

However, the question arises whether and to what extent the internal standard method is robust to the occurrence of statistically significant systematic errors. In other words, the question is whether and to what extent the internal standard can contribute to strengthening the robustness of the method to the occurrence of uncontrolled effects compared to when the standard is not included in the calibration process, i.e. compared to the external standard method. Mathematical approaches may provide some answers to this question.

The relative measurement error, D, caused by uncontrolled effects committed in the external standard method can be expressed by the formula:

$$D = \frac{Y_q - Y_p}{Y_p} \cdot 100\% \tag{5.26}$$

where Y_p *i* Y_q are the intensities measured for the analyte before and after the effects occur, respectively. If, due to the occurrence of the same effects, the intensity of the internal standard signal changes from the value Y_{wp} to Y_{wq}, then the relative error made in the internal standard method, D_w, is given by the formula:

$$D_w = \frac{\frac{Y_q}{Y_{wq}} - \frac{Y_p}{Y_{wp}}}{\frac{Y_p}{Y_{wp}}} \cdot 100\% \tag{5.27}$$

when Eqs. (5.26) and (5.27) are taken into account, the relationship between the quantities D and D_w takes a form that depends only on the ratio Y_{wp}/Y_{wq}:

$$D_w = \left[\left(\frac{D}{100} + 1\right) \cdot \frac{Y_{wp}}{Y_{wq}} - 1\right] \cdot 100\% \tag{5.28}$$

The relationship (5.23) is illustrated in Figure 5.19 for several values of D errors made by the external standard method. From Eq. (5.28) and Figure 5.19, it follows that:

- if uncontrolled effects do not change the signals of the internal standard ($Y_{wp} = Y_{wq}$), then systematic errors made in the calibration with the participation of this standard are the same as those without its participation (i.e. external standard method, $D = D_w$),
- the contribution of the internal standard may increase systematic errors made without its contribution if $Y_{wp}/Y_{wq} > 1$ or may partially compensate for them if $Y_{wp}/Y_{wq} < 1$,
- the contribution of the internal standard can fully compensate for systematic errors made in the external standard method if its response to uncontrolled effects is the same as that of the analyte, i.e. $Y_{wp}/Y_{wq} = Y_p/Y_q$.

Thus, it can be generally said that:

- the internal standard method can contribute to compensating for uncontrolled effects acting on the analyte only if the internal standard is also subject to these effects, and
- the robustness of the internal standard method to uncontrolled effects is determined by the sensitivity of the internal standard compared to the sensitivity of the analyte to the presence of these effects.

To investigate in more detail the robustness of the internal standard method to uncontrolled effects, it is again necessary to use the general empirical formula (4.11) simulating the form of the real function taking into account effects of different types and nature. According to the above conclusions, it should be assumed a priori that

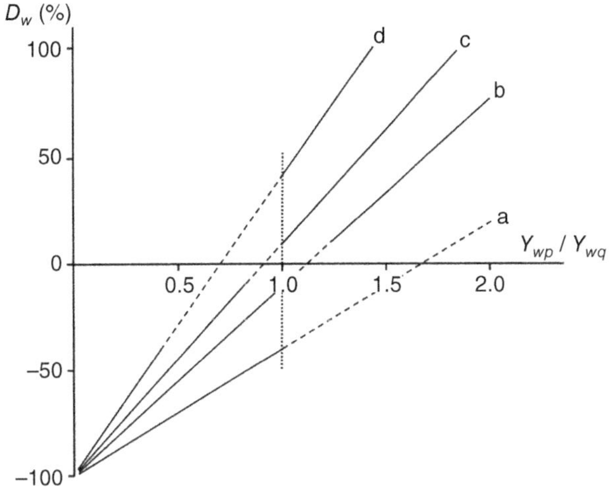

Figure 5.19 Relationship between the errors D_w, and D committed in the internal and ESM methods, respectively, and the ratio Y_{wp}/Y_{wq} of the signal intensities measured for the internal standard when D is equal to −40% (a), −10% (b), 10% (c), and 40% (d); dependence ranges corresponding to $D_w < D$ are denoted by dashed lines.

the occurrence of uncontrolled effects concerns not only the analyte but also the internal standard present in the sample and in the standards, i.e. it manifests itself by a change in the intensity of its signals. Assuming that the same sample components, c_p, c_t, and c_m, cause the additive and multiplicative interference effects with respect to the internal standard and to the analyte, the representation of the real one by means of a model function is expressed by the equation:

$$\frac{B_1' \cdot c_x}{B_{1w} \cdot c_w} = \frac{A(c_p, c_t) + B_1 \cdot c_0 + B_2 \cdot c_0 \cdot H(c_0, c_m)}{A_w(c_p, c_t) + B_{1w} \cdot c_w + B_{2w} \cdot c_w \cdot H_w(c_w, c_m)} \tag{5.29}$$

where coefficients B_1 and B_2 again express the possible presence of a speciation effect. Equation (5.29) illustrates additive effects acting on the analyte, $A(c_p, c_t)$, and on the internal standard. $A_w(c_p, c_t)$, cannot be compensated even when they are caused by the same interferent and when they are not accompanied by multiplicative effects ($B_2 = B_{2w} = 0$). The compensation of blank effects in the sample, $A(c_p)$ can only occur if the components causing these effects are also added in equal concentrations to the sample and to the standard solutions, and at the same time the internal standard is free of multiplicative effects, for then Eq. (5.29) takes the form

$$\frac{A(c_p) + B_1 \cdot c_x}{B_{1w}} = \frac{A(c_p) + B_1' \cdot c_0 + B_2 H(c_0, c_m) \cdot c_0}{B_{1w}} \tag{5.30}$$

It may be noted that Eq. (5.30) is analogous to Eq. (5.8) for calibration by the external standard method. Thus, this confirms the observation that **when the internal standard is not subject to any uncontrolled effects, the internal standard method is just as irresistible to uncontrolled effects as the external standard method.**

If additive effects are not present or have been eliminated, and the speciation effect does not exist or has been compensated for, $B_1 = B_1'$, then Eq. (5.29) takes the form:

$$\frac{B_1 \cdot c_x}{B_{1w} \cdot c_w} = \frac{B_1 \cdot c_0 + B_2 H(c_0, c_m) \cdot c_0}{B_{1w} \cdot c_w + B_{2w} H_w(c_w, c_m) \cdot c_w} \tag{5.31}$$

It follows from this equation that $c_x = c_0$ only if the ratio is valid:

$$\frac{B_1}{B_{1w}} = \frac{B_2 H(c_0, c_m)}{B_{2w} H_w(c_0, c_m)} \tag{5.32}$$

It follows from Eq. (5.31) that, when uncontrolled effects are present, the internal standard method can lead to an accurate analytical result ($c_x = c_0$) only when:

- the analyte and the internal standard are subject to multiplicative effects of the same linear or nonlinear form from the same interferents, and
- the ratio of the effects acting on the analyte and the internal standard is equal to the ratio of the measurement sensitivities of the analyte and the internal standard.

It is worth noting that the accuracy of the analytical result does not depend on the concentration of the internal standard, although, of course, the concentration should be high enough that the intensities of the measured signals are sufficiently greater than the measurement noise.

It should be emphasized that the above conditions are much more restrictive than the one that is quite common and can be found in publications, namely, that the internal standard method allows compensation for interference effects as long as they are linear. It should also be realized that these restrictive conditions apply to uncontrolled effects regardless of their magnitude, so they also apply to near-random effects. The reason for this is that, as can be seen from numerous literature studies, the elimination of errors, even of a random nature, by means of an internal standard occurs in quantitative analysis with varying degrees of success and not always as expected.

One of the reasons for this problem may also be, paradoxically, the increasingly complex design of modern measurement instruments, which makes the type and number of various sources of random and systematic variations in signal intensity larger and more difficult to control. Even if the internal standard is able to compensate for errors coming from some sources, other, uncompensated errors may dominate and decisively influence the resultant value of precision and accuracy of analytical results. The possibility cannot be ruled out that the internal standard reacts to some sources by changing the signal in the direction opposite to that of the analytical signal, which may adversely affect the precision and accuracy of the results even if errors from other sources are compensated for.

The key element determining the efficiency of calibration by the internal standard method remains the choice of the internal standard so that it is as similar as possible to the analyte in terms of chemical and physical properties. Having no strict rules for this choice, the analyst is guided mainly by literature reports and his own statement. This procedure, however, often proves unreliable because, despite the high degree of similarity, the two substances may not be susceptible to accidental or systematic effects uncontrolled in the manner specified above.

The selection of the internal standard in *ICP-MS* analysis by inductively coupled plasma-mass spectrometry (ICP-MS) is particularly difficult and unreliable. Although there is a generally accepted recommendation that the internal standard should be selected as close in mass number as possible to that of the analyte, numerous examples show that this criterion is not always confirmed in practice. The situation is best captured by the following conclusion: "... there is no single physical or chemical parameter that reliably allows a priori selection of a 'good' internal standard for any given analyte" [12].

A good confirmation of this thesis are the results presented in Table 5.5 for the determination of selenium in different isotopic forms by ICP-MS [13]. Despite the use of elements such as germanium (with mass numbers similar to those of the analyte) and rhodium recommended in this study, no statistically significant improvement in the precision of the results was observed compared to the results obtained without the use of internal standards. The method also did not prove to be fully effective in eliminating the inteferents introduced into the sample, and only in the case of the ^{82}Se isotope determination did it contribute to a clear improvement in the precision of the determination.

In other individual cases, however, experimental studies fully support the above criterion for internal standard selection. As an example, the very good, thoroughly verified effectiveness shown by bismuth as an internal standard in improving the precision and accuracy of lead determination in a variety of materials with complex matrix by various ICP techniques [14].

The efficiency of the internal standard method looks best in analysis by chromatographic and electrophoretic techniques. For the determination of organic compounds, the common practice is to use the structural analogs, with consideration of the structural similarities between the internal standard and the analyte. A properly chosen internal standard allows for effective elimination of errors from sources such as viscosity, pressure and temperature variation, spontaneous injection sample losses between injection and detection, and detector response variation. In the

Table 5.5 Results of the ICP-MS determination of selenium (20 mg l^{-1}) in synthetic samples without and with sodium and potassium (320 and 280 mg l^{-1}) as interferents with use of the ESM method (with no internal standard) and the internal standard method with ^{72}Ge and ^{103}Rh as internal standards.

		No		^{72}Ge		^{103}Rh	
Analyte	Interferents	c_x (mg l^{-1})	RSD	c_x (mg l^{-1})	RSD	c_x (mg l^{-1})	RSD
^{77}Se	No	19.68	2.40	19.61	2.15	19.47	3.05
^{78}Se	No	19.86	2.75	19.74	1.70	19.21	5.40
^{82}Se	No	19.62	2.25	19.65	2.65	19.45	3.35
^{77}Se	Na + K	19.31	3.05	24.68	3.80	25.05	1.50
^{78}Se	Na + K	16.25	7.25	25.89	4.45	26.68	4.95
^{82}Se	Na + K	16.04	6.25	20.90	6.70	20.72	4.00

case of separation methods combined with mass spectrometry, stable isotope labeled standards are often used as internal standards, i.e. compounds with the same chemical structure as the analytes in which one or more atom is replaced with an isotopic analog (commonly ^{2}H, ^{13}C, ^{15}N, ^{17}O). An alternative (e.g. in the case of quantitative metabolite profiling) is the strategy named isotope-coded derivatization where the sample is derivatized with naturally labeled reagent, while a standard solution is separately derivatized with isotopically labeled reagent and spiked into the sample solution as the internal standard. However, the practical limitation is that isotopically labeled compounds are not always available and are usually very expensive.

Theoretical and experimental studies have shown that random and systematic changes in the migration velocity of the sample and standard solutions occurring in the analysis by capillary electrophoresis technique affect the position of the peaks to a quite similar extent as their intensity (area) and in both cases, these effects can be compensated equally effectively by the internal standard method [15]. The effectiveness of this compensation in both cases is greater the closer the peak coming from the internal standard is to the peak coming from the analyte, as shown by the data in Table 5.6 (mode a). This is understandable since the mutual position of the peaks is generally the closer the corresponding substances are to each other. Further improvement of the results can be obtained by performing time correction of peak area ratios, or, alternatively, transformation of electropherograms from the time-related scale into the electrophoretic mobility-related scale (modes b–d).

There is no doubt that the internal standard method is a very valuable calibration method. It is a kind of supplement to other calibration methods in quantitative analysis because, unlike them, it serves to a large extent to compensate accidental measurement errors and to improve precision of analytical results. Its limitations result from the lack of unambiguous criteria for the selection of an internal standard suitable for a particular analyte determined by a given measurement method. Attempts to solve this problem are being made in numerous scientific centers, e.g.

Table 5.6 The relative standard deviation values (%, $n = 30$) obtained for the peak areas of three different internal standards (ISs) differently situated in relation to the analyte peak, calculated using various modes.

	Mode	No IS	IS		
			IS1	IS2	IS3
a	Peak areas measured in the migration time scale	53.7	6.2	12.2	23.8
b	Peak areas measured as divided by migration time in the migration time scale	10.4	3.0	5.9	9.9
c	Peak areas measured in the electrophoretic mobility scale neglecting the voltage ramping effect	40.2	5.9	6.4	15.5
d	Peak areas measured in the electrophoretic mobility scale with the voltage ramping effect correction	41.3	9.9	6.8	16.7

Source: Adapted from Nowak et al. [15].

by searching for "universal" internal standards. However, as it seems, this issue is so difficult that it also requires extended, systematic theoretical research to better define the connection between physicochemical properties of elements and groups of organic compounds with their potential action as internal standards.

5.2.2 Indirect Method

The **indirect method** in quantitative analysis, as in qualitative analysis, is characterized by the fact that the calibration process is aided by an additional substance reacting with the analyte. The intensity of the analytical signal is measured for a substance other than the analyte involved in this reaction. The value of the measured signal gives an indirect indication of the concentration of the analyte in the sample and hence the indirect nature of the method.

The name "indirect method" can be misleading, as empirical calibration methods are sometimes considered "indirect", as opposed to "direct" non-calibration methods. This type of approach can be found in some papers [16]. In general, however, the indirect method is considered a calibration method according to the convention adopted in this paper.

The question naturally arises: what is the purpose of calibration by the indirect method in quantitative analysis? Well, the examples of using a chemical reaction in this way are very numerous, which makes the indirect method in quantitative analysis much more frequently used than in qualitative analysis. The "exchange" of an analyte for another chemical compound gives, among other things, the following advantages:

- determination of an analyte in a form that does not produce a measurement signal with an available measurement method (e.g. determination of organic compounds after conversion to organometallic compounds),
- increasing accuracy of analyte determination by transferring it into a form more stable under given experimental conditions (e.g. into a stable complex),
- increasing the measurement sensitivity of an analyte by transferring it to another form triggering a signal with higher sensitivity,
- compensation of the speciation effect by transferring two different chemical forms of an analyte to a single form of another substance,
- performing two-component analysis by transferring two analytes to another single substance.

There are many examples in the literature of this and other analyte handling to achieve specific positive effects.

At this point, however, a certain doubt may arise: after all, chemical reactions involving an analyte are a natural part of sample preparation for analysis regardless of the calibration method used, so why to give these processes special importance in determining the type of calibration method. Therefore, it should be clearly emphasized: only those chemical reactions are of such importance in which a well-defined substance is formed that is capable of releasing a measurement signal of sufficient intensity that clearly corresponds to the amount of analyte that has been reacted.

The conditions for a chemical reagent in the indirect method are actually the same as in qualitative analysis, namely:

- it must not be a natural component of the sample to be analyzed,
- it should react with the analyte selectively, sufficiently fast and with high efficiency,
- it should lead to a product that does not react with other components of the sample and whose signal is clearly separated from the signals of these components.

It is also worth noting that precise knowledge of the stoichiometry of the reaction underlying the indirect method is not necessary (although obviously desirable). From a formal point of view, the reaction does not contribute theoretical elements to the indirect method that "violate" its empirical calibration character.

The indirect method in quantitative analysis is realized in two different variants depending on the type of chemical reaction product. If the selected substance, R, reacting with the analyte, An, remains in excess to the amount of analyte that results from the stoichiometry of the reaction, the reaction can be represented in the following general way:

$$\mathrm{An} + R = \mathrm{AnR} + R_{\mathrm{r}} \tag{5.33}$$

where R_{r} is the reactant remaining after the reaction has taken place. It follows that the measurement of the signal representing the analyte can be made both for the reaction product, AnR, and for the reactant in amount R_{r}. In the former case, the signal intensity increases with increasing analyte concentration as the amount of AnR product gradually increases, while in the latter case, the signal intensity decreases because increasing analyte concentration leaves a smaller and smaller concentration of unreacted reagent after the reaction. While in qualitative analysis these procedural differences are of little calibration significance, in quantitative analysis the direction of signal change determines the shape and type of the actual and model functions, and to some extent the analytical properties of the indirect method.

In a situation where the signal intensity is measured for the reaction product, AnR, the calibration procedure of the indirect method in its basic version at the preparative stage involves, as shown in Figure 5.20, the preparation of a sample solution with analyte concentration c_0 and two standard solutions with analyte concentrations c_1 and c_2 covering their c_0 concentrations ($c_1 < c_0 < c_2$). To all these solutions, the chosen reagent R is added in equal volume and in such quantity that it is capable of reacting with the total amount of analyte contained in the sample and standard solutions.

After the chemical reaction is complete, the intensities of the Y_{r1} and Y_{r2} signals for the AnR reaction product in the model solutions are measured and presented as a function of analyte concentration, as shown in Figure 5.21a. The model function has the form:

$$Y_r = B_1 \cdot c \tag{5.34}$$

where B_1 is a factor that determines the measurement sensitivity of the analyte. After measuring the signal intensity Y_{r0} for a sample and relating it to a calibration

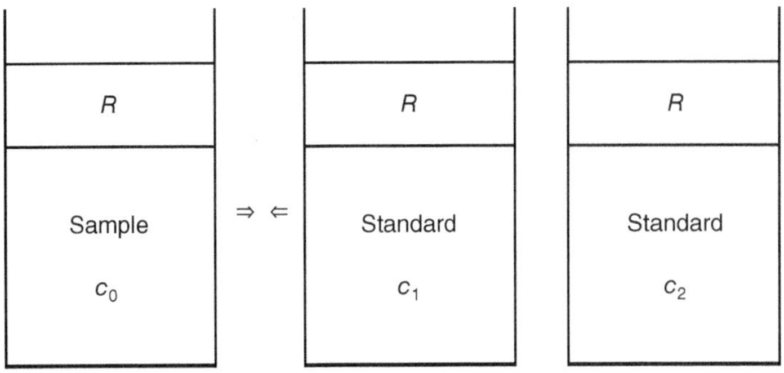

Figure 5.20 Scheme of the preparative scheme of the indirect method when the signal has to be measured for the reaction product AnR.

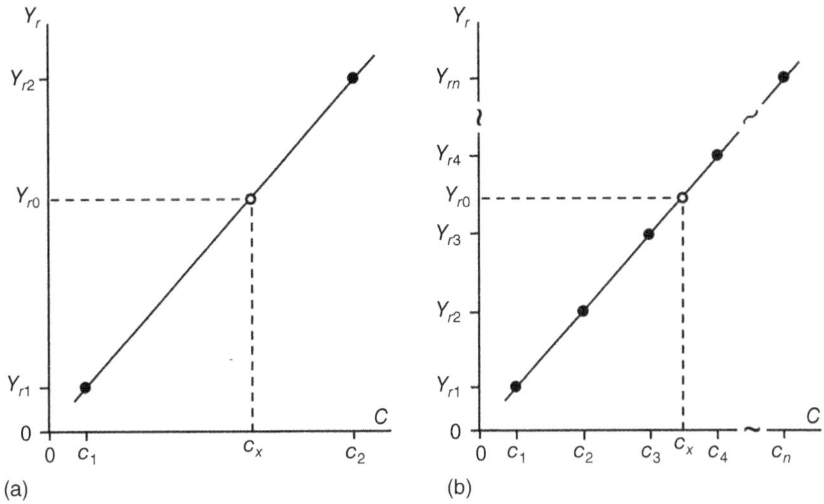

Figure 5.21 Calibration in accordance with two-point (a) and multipoint (b) versions of the indirect method when the signal Y_r is measured for the reaction product AnR (see reaction (5.33).

graph, the concentration of the analyte in the sample, c_x, can be determined using the formula:

$$c_x = \frac{Y_{r0} \cdot (c_2 - c_1) - Y_{r1} \cdot c_2 + Y_{r2} \cdot c_1}{Y_{r2} - Y_{r1}} \tag{5.35}$$

The indirect method in this form can, of course, also be implemented in a multitemplate version, using a series of several standard solutions containing analyte at concentrations lower and higher than the putative analyte concentration in the sample (see Figure 5.21b). After taking measurements and fitting a linear function to the experimental points using the same calculation methods as for the extended version of the external standard method, the analytical result is determined from the formula:

$$c_x = \frac{\hat{Y}_{r0}}{B_1} \tag{5.36}$$

where $\hat{Y}_{r0}$ is the value of the linear function corresponding to the intensity of Y_{r0}, and the coefficient B_1 is calculated according to the chosen method of fitting the linear function to the experimental points.

As can be seen, the indirect method in this form is very similar to the external standard method (ESM) from the calibration side. This similarity is particularly evident in the last steps of the calibration procedure: the construction of a linear, increasing model function and the path to determine the value of the analyte concentration in the sample. Thus, it is not surprising that in many cases, without noticing the chemical reaction transforming the analyte into another chemical form, the indirect method thus implemented is not considered a separate calibration method, but is treated as an external standard method.

The specificity of the indirect method manifests itself to a much greater extent when signal intensity measurements are made for the reagent R_r remaining after reaction with the analyte (see reaction (5.33)). The preparative step then involves the preparation of a sample (with analyte concentration c_0) and two standard solutions, the first of which contains no analyte ($c_1 = 0$) and the second of which contains analyte at a concentration c_2 greater than the analyte concentration in the sample ($c_2 > c_0$). To all these solutions, an equal, known, and well-defined amount of the selected analyte-reactive substance is added in excess. This procedure is shown in Figure 5.22.

The characteristic of the indirect method thus considered is the decreasing form of the real function, as shown in Figure 5.23a and the formula of the calibration line graph:

$$Y_r = Y_{r1} - B_1 \cdot c \tag{5.37}$$

The concentration of the analyte in the sample, c_x, can be determined using the formula:

$$c_x = \frac{Y_{r0} - Y_{r1}}{Y_{r2} - Y_{r1}} \cdot c_2 \tag{5.38}$$

where Y_{r0} is the signal intensity measured for the analyte in the sample.

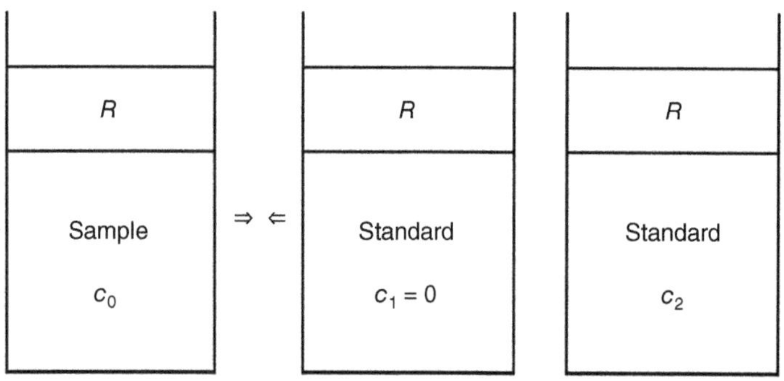

Figure 5.22 Scheme of the preparative scheme of the indirect method when the signal is caused by an unreacted amount R_r of reagent (see reaction (5.33)).

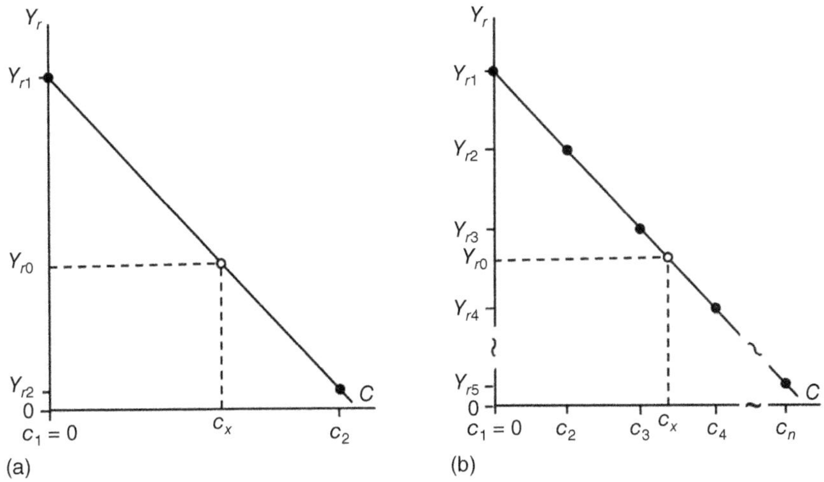

Figure 5.23 Calibration in accordance with two-point (a) and multipoint (b) versions of the indirect method when the signal Y_r is measured for an unreacted amount R_r of reagent R (see reaction (5.33)).

If the indirect method in this form is implemented in the multipoint version (see Figure 5.23b), the analytical result is determined by the formula:

$$c_x = \frac{Y_{r1} - \hat{Y}_{r0}}{B_1} \tag{5.39}$$

where values Y_{r1} *i* B_1 are calculated according to the chosen method of fitting a linear function to the experimental points.

An interesting aspect of the indirect method is that it can be used for the determination of substances that are interferents in relation to the analyte in the external standard method. The roles are then exchanged: such an interferent becomes the analyte in the indirect method, while the analyte - the substance that causes the measurement signal. This approach opens the way for the determination of substances that are not determined by a given measurement method because, for example, their measurement sensitivity is too low. Standards containing varying concentrations of interferents (e.g. aluminum in FAAS) are then prepared and the reagent (e.g. calcium) added to the sample and standard solutions in an equal, constant concentration is measured after reacting with the interferent (see Figure 5.24). However, one should then expect a nonlinear form of the real function and the need to approximate it with a nonlinear calibration graph.

The precision of the analytical result obtained by the indirect method generally depends on the same factors that play a role in the external standard method: on the measurement sensitivity (absolute value of the coefficient B_1), the number of standard solutions, and the number of repetitions of signal intensity measurements measured for these solutions, and on the random error of the signal intensity measured for the sample and the position of this intensity value relative to the calibration graph.

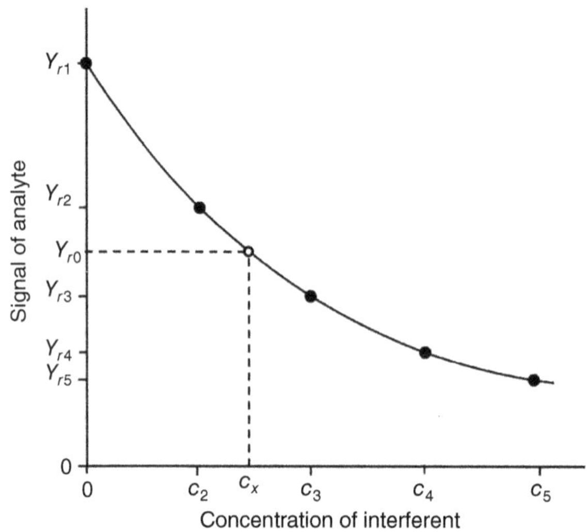

Figure 5.24 Typical calibration graph obtained in the direct method for the FAAS determination of an interferent (e.g. aluminum) using an analyte (e.g. calcium).

It should be pointed out, however, that there is a special property of the indirect method which distinguishes it from the external standard method, but which is not always recognized or exploited. This is due to the fact that, in the case of a decreasing model function, only the reactant is present in the first standard solution, since it produces a signal of maximum intensity with a relatively small random error. The concentration of the analyte in the next calibration solution (c_2) may be large enough so that the reduced signal also has a small random error. Thus, the portion of the calibration graph bounded by the concentrations $c_1 = 0$ and c_2 can be used to determine the analyte present in the sample at any small concentration with very small measurement uncertainty (which the external standard method does not offer). The random error of the analytical result then depends more on the error of preparation of the sample containing the analyte in low concentration than on the measurement error of the corresponding signal.

The indirect method in both its forms is also similar to the external standard method in its lack of robustness to random and systematic errors due to uncontrolled effects. It can be easily proven mathematically in the same way as before that the linear model function allows in the indirect method to determine the true analyte concentration ($c_x = c_0$) only if multiplicative linear and nonlinear effects have been eliminated and additive and speciation effects have been compensated.

What is the significance of the chemical reaction in the context of analytical errors made in indirect method calibration? Apart from the other mentioned conditions it should fulfill, it is important that the degree of analyte reaction is constant over time, independent of the analyte concentration and independent of the chemical environment of the analyte (i.e. the same in the sample as in the standards). Any random or systematic deviation from these conditions during the course of the analysis alters the concentration of the signal-causing substances and, therefore, the intensity of the signal. In the analytical and calibration process, the chemical reaction should therefore be considered as a potential source of preparative effects.

The chemical reaction, by leading to the appearance of an additional substance involved in the calibration process, also potentially reduces the resistance of the indirect method to interference effects. This is because sample components can induce changes in the analytical signal not only by directly interacting with the form of the reactant for which this signal is measured but also by interacting with other components of the chemical reaction of which they are in proper equilibrium with each other. This is shown schematically in Figure 5.25.

In particular, by acting on the analyte, An, the interferent may reduce the amount of analyte available to the reactant, such as by forming a more stable compound with the analyte, resulting in not all of the analyte reacting with the reactant. Simultaneously with the formation of a stable compound of the interferent with the analyte, the equilibrium of the reaction may shift back, causing additional amounts of the reactant to be released into solution, resulting in an increase in the intensity of the measurement signal. Since a fixed amount of interferent in solution will always react with the same amount of analyte, it is reasonable to assume that the increase in the concentration of the reactant in the system after the reaction (and thus the increase in the analytical signal) will be constant regardless of the analyte concentration, i.e. the interference effect will be additive in nature.

Also, the possibility cannot be excluded that the interferent will enter into a competitive reaction with the added reagent, R, in parallel with the analyte and change the amount of reagent, causing the signal sent to not tell the true amount of analyte in the sample. In particular, in the case of a reaction leading to the formation of a sparingly soluble compound between the analyte and the reagent, the interferent may form other stable compounds with the reagent, which will be removed from solution along with the actual product. An even more complex situation can arise when the reactant itself does not trigger the analytical signal, but must be carried out into another compound for this purpose (e.g. with the correct color in spectrophotometric analysis). The interferent can then interact not only directly with the reactant, but also with the other components of this additional reaction.

The effect of the interferent on the reaction product, AnR, can also be to shift the equilibrium of the reaction, which contributes to a change in the amount of unbound reactant. When the analyte-reagent reaction occurs in solution, this type of effect can occur by changing the pH or ionic strength of the solution, which is often caused by the presence of a sample component. Such interactions will tend to produce additive effects since changes in the reaction environment induced by a fixed amount of interferent will have a fixed.

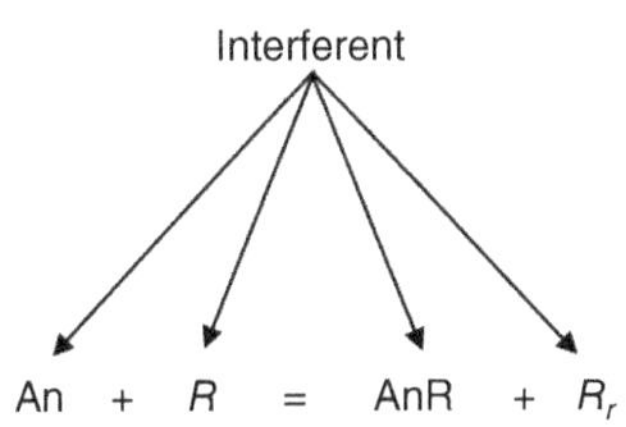

Figure 5.25 Scheme of possible pathways of the interferent influence on individual reaction components in the indirect method.

In many cases, such as in analysis performed using flame techniques, the interferent has the ability to affect the reactant R introduced into the solution prior to separation of the reaction product, but also to affect in the flame the reactant remaining after the reaction, R_r, for which the signal is measured. If, in addition, the interferent affects other components of the chemical reaction in the analytical system, the end result is difficult to predict. This is exemplified by the results shown in Figure 5.26 [17]. As can be seen, in a particularly unfavorable case, the presence of the interferent can even completely neutralize the reaction of the analyte with the reagent and make the determination of the analyte practically impossible.

It should be noted that such an effect of the interferent on the analyte, which manifests itself as a multiplicative effect in the external standard method, changes its character when, in the indirect method, the analyte acts as a reactant and the signal is measured for that residue of that reactant, R_r, after the reaction. Since the amount of reactant R_r decreases as the concentration of the analyte in the calibration solutions increases, the interference effect, although still proportional to the concentration of the analyte, becomes smaller and smaller. The position of the calibration graphs a and b in Figure 5.26b is probably an example of such a phenomenon.

Another problem is the additive blank effect, which in the indirect method can have two different sources: it can come from such components present in the sample or added to the sample and standards that react with the added reagent together with the analyte, or from other components that trigger the signal under conditions characteristic of the AnR and R_r reaction products. If the signal is measured for the AnR compound, then in both cases there is an additive increase in the signal measured for the analyte in the sample and standards. If, on the other hand, the signal is measured for the reagent R_r remaining after the reaction, then the co-reagents "consuming" the reagent, in addition, will cause a negative blank effect, and the

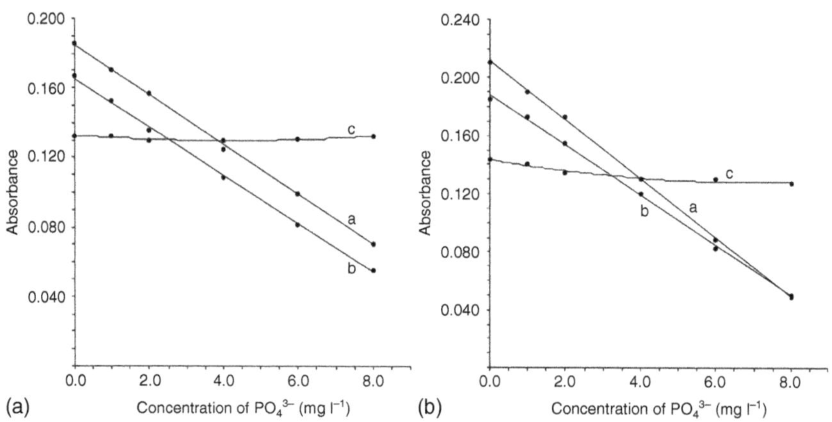

Figure 5.26 Calibration graphs constructed in the direct method for phosphate determined by the FAAS method when strontium (a) and calcium (b) were used as reagents producing signal and no additional substances (a), silicon (b), and vanadium (c) were additionally present in the standards. Source: Kościelniak and Wieczorek [17], fig 5 (p. 431)/with permission of Taylor & Francis.

substances releasing their own signal will cause a positive effect. Under particularly favorable circumstances, these effects can therefore completely or partially compensate for each other without any intervention by the analyst.

In any case, it is possible to eliminate the blank effect in the way described for the external standard method, i.e. by taking separate measurements and making appropriate corrections. In any case, it is possible to eliminate the blank effect in the way described for the external standard method, i.e. by making separate measurements and making appropriate corrections. However, because of the possibility of the origin of this effect from different sources, its complete elimination in the indirect method is certainly more difficult. If the effect is caused by equal amounts of the same blank-inducing substances in the sample and in the standards, it will be compensated for regardless of the form in which the indirect method is implemented.

Thus, a chemical reaction in an indirect method is a potential source of uncontrolled effects, especially interference effects, of various, often difficult to predict, nature. On the other hand, a well-chosen chemical reaction can be a factor in reducing some effects. For example, with appropriate reaction selectivity, a reagent reacting only with the analyte provides a good chance of freeing the analyte from the possible influence of interferents present in the sample. Furthermore, when the analyte is present in the sample and in the standards in a different chemical form, the reagent can be selected so that a chemical reaction with both forms of an equal amount of analyte leads to an equal loss of that reagent after the reaction. In this respect, the indirect method is thus superior to the external standard method.

In summary, the indirect method is an extremely useful and necessary calibration method, largely complementary to the external standard method. However, it should be used with great attention paid to possible uncontrolled effects, especially interference effects. A helpful tool that can help to control these effects is the induced chemical reaction. Thus, it can be said that the choice of a suitable reagent reacting with the analyte is a key element of the indirect method, largely determining the precision of the accuracy of the analytical result obtained by this method.

References

1 ISO Guide 35:2006 (2006). *Reference materials. General and statistical principles for certification*, the International Organization for Standardization, Geneva.

2 Miller, J.N. and Miller, J.C. (2005). *Statistics and Chemometrics for Analytical Chemistry*, Ve. Essex: Pearson Education Limited.

3 Stafiński, M., Wieczorek, M., and Kościelniak, P. (2013. Influence of the species effect on trueness of analytical results estimated by the recovery test when determining selenium by HG-AFS. *Talanta* 117: 64–69.

4 Gilbert, P.T. Jr. (1959). Determination of cadmium by flame photometry. *Analytical Chemistry* 31 (1): 110–114.

5 Dean, J.A. (1960). *Flame Photometry*. New York: McGraw-Hill.

6 Shatkay, A. (1968). Photometric determination of substances in presence of strongly interfering unknown media. *Analytical Chemistry* 40 (14): 2097–2106.

7 Shatkay, A. (1970). Dilution as the changing parameter in photometry. *Applied Spectroscopy* 24 (1): 121–128.
8 Kościelniak, P. (1993). Calibration by the dilution method in flame atomic spectrometry. *Universitatis Iagiellonicae Acta Chimica* 1092 (36): 27–38.
9 Kościelniak, P. (1998). Calibration procedure for flow injection flame atomic absorption spectrometry with interferents as spectrochemical buffers. *Analytica Chimica Acta* 367 (1–3): 101–110.
10 Gerlach, W. (1925). Zur Frage der richtigen Ausführung und Deutung der "quantitativen Spektralanalyse". *Zeitschrift für anorganische und allgemeine Chemie* 142: 383–398.
11 Heidel, R.H. and Fassel V.A. (1950). An Instrument for Internal Standard Flamephotometry and Its Application to the Determination of Calcium in the Rare Earths. *Iowa State College, Ames Laboratory ISC Technical Reports*, ISC-107.
12 Finley-Jones, H.J., Molloy, J.L., and Holcombe, J.A. (2008). Choosing internal standards based on a multivariate analysis approach with ICP(TOF)MS. *Journal of Analytical Atomic Spectrometry* 23 (9): 1214–1122.
13 Wieczorek, M., Tobiasz, A., Dudek-Adamska, D. et al. (2017). Analytical strategy for the determination of selenium in biological materials by inductively coupled plasma – mass spectrometry with a dynamic reaction cell. *Analytical Letters* 50 (14): 2279–2291.
14 Bechlin, M.A., Ferreira, E.C., Gomes Neto, J.A. et al. (2015). Contributions on the use of bismuth as internal standard for lead determinations using ICP-based techniques. *Journal of the Brazilian Chemical Society* 26 (9): 1879–1886.
15 Nowak, P., Woźniakiewicz, M., and Kościelniak, P. (2018). Flow variation as a factor determining repeatability of the internal standard-based qualitative and quantitative analyses by capillary electrophoresis. *Journal of Chromatography A* 1548: 92–99.
16 El-Azazy, M.S. (2018). Analytical calibrations: schemes, manuals, and metrological deliberations, Chapter 2. In: *Calibration and Validation of Analytical Methods: A Sampling of Current Approaches* (ed. M. Stauffer), 17–34. London: IntechOpen.
17 Kościelniak, P. and Wieczorek, M. (2010). Extrapolative version of the indirect calibration method. *Analytical Letters* 43 (3): 424–435.

6
Additive Calibration Methods

6.1 Basic Aspects

Almost all the procedures of calibration methods presented and discussed so far, both theoretical and empirical, have been based on the basic principle of analytical calibration. It says, as is well known, that the sample and the standard should be drawn up separately and made as similar to each other as is possible and reasonable. This ensures an accurate representation of the real function by the model function, and consequently the determination of an analytical result close to the real result. This is, however, as already mentioned, very difficult and often impossible in practice.

The simplest way to make the sample and the standard similar is to add the standard to the sample. From the calibration side, this is a very logical and rational action. This procedure makes the difficulty of accurately reproducing the composition of the sample in the standard largely disappear. The standard need not, and should not, contain the components of the sample matrix, because when combined with the sample, all of its components also become the matrix of the standard.

An example of such a calibration procedure is the standard addition method in qualitative analysis, although it is not very important there and is rarely used as a separate calibration method for analyte identification. In quantitative analysis, it is quite different. This is because the change in signal intensity caused by adding a standard to a sample can provide more potentially useful information about the amount of analyte in the sample than about the type of analyte.

Over the years, various calibration approaches based on the process of adding analyte to the sample, or additive calibration methods, have emerged in quantitative analysis. In general, they involve combining a sample with standards, observing the effects of this procedure with a measuring instrument, interpreting the measurement data accordingly, and converting this data into the concentration of the analyte in the sample. They can be divided into three categories (using generally accepted terminology):

Calibration in Analytical Science: Methods and Procedures, First Edition. Paweł Kościelniak.

- **standard addition method** – when the standard containing an analyte of the same isotopic form as the analyte in the sample is added to the sample,
- **titration method** – when the standard other than the analyte, which reacts with the analyte in the sample, is added to the sample, and
- **isotopic dilution method** – when the standard to be added contains an analyte of a different isotopic form than that of the analyte in the sample.

From a calibration point of view, each of these methods is not so much about fitting the model function to the real function, but more about reproducing the real function with the model function. Thus, one might think that doing so should allow one to compensate for at least some uncontrolled effects and increase the accuracy of the analytical result. The following sections will show whether and to what extent analytical practice confirms the theoretical supposition.

6.2 Standard Addition Method

The **standard addition method** involves combining a sample with a standard of that analyte whose amount in the sample is being determined. This procedure can be performed in a variety of ways. The standard may be added to the sample so as to cause or not cause dilution, or the sample may be diluted only after the standard has been added to it. As will become apparent, these different techniques for preparing calibration solutions used in the preparative stage of the calibration procedure further determine, in the measurement stage, the form of the calibration graph, and in the transformation stage, the mathematical way in which the analytical result is determined. In other words, the preparative stage largely determines the specific variant of the SAM and its calibration characteristics.

The question arises – how to distinguish nomenclaturally between these variants of the method without going into the details of their preparation? It seems that the most substantive criterion for the division is the mathematical way of **transformation** of the analytical signal into analytical result and such a criterion has been adopted in the present study. Thus, further on we will talk about the **SAM** in its variants: **extrapolative standard addition method (E-SAM)**, **interpolative standard addition method (I-SAM),** and **indicative standard addition method (In-SAM)**.

6.2.1 Extrapolative Variant

The standard addition method came into the "analytical arena" rather late. This method was used in 1935 by Foster et al. [1] for the spectrographic determination of lead in cerebrospinal fluid. The authors described the procedure as follows: "The spectrum of the fluid to be analyzed is photographed before and after the addition of a known lead, and the intensity of a lead line with reference to an 'internal standard' is measured in each exposure." Interestingly, the first approach was thus somehow combined with the internal standard method. Shortly thereafter, the SAM method was introduced to polarographic analysis independently by H. Hohn [2] and E.N. Varasova [3], and in combination with this analytical technique and emission spectrometry techniques, it was mostly used in the following years. The term "standard addition" was first introduced by J.J. Lingane and H. Kerlinger in 1941 [4], although still several years later some authors gave the SAM method the name "internal standard method" [5]. The first use of the multipoint method was described by Polish scientist W. Kemula in 1966 [6] using a "hanging drop mercury electrode" he had invented.

The pioneering variant of the SAM method was the extrapolation method (E-SAM) and in this form, the method is still most commonly used today. The character of the E-SAM variant is a consequence of the way the calibration solutions are prepared, as shown in Figure 6.1. The sample is divided into two equal portions of volume V_0 containing analyte at an equal, unknown concentration c_0. One portion is then diluted with a diluent (containing no analyte) of volume V_1, and the other portion is diluted with a standard solution of known analyte concentration c_1 and the same volume V_1. In the calibration solutions prepared in this way, the concentration of analyte derived from the sample is equal to $k_v{\cdot}c_0$, and the concentration of analyte derived from the addition of the standard in the second solution is equal to $(1-k_v){\cdot}\Delta c_1$, where $k_v = V_0/(V_0+V_1)$ is the degree of dilution of the sample.

For both solutions, the intensities of the Y_0 and Y_1 signals are then measured and assigned in the coordinate system shown in Figure 6.2a to zero concentration and Δc_1 concentration of the analyte in the added standard, respectively. If the Y_0 and Y_1 signals are not subject to systematic uncontrolled effects and the real function over these signals is linear, then the model function expresses the dependence of these signals on the concentration of analyte in the sample, c_x, and in the standard solution, Δc_1, by the formula:

$$Y_0 = B_1 \cdot k_v c_x \tag{6.1}$$

$$Y_1 = B_1 \cdot [k_v c_x + (1-k_v)\Delta c_1] \tag{6.2}$$

where B_1 is a factor that determines the measurement sensitivity of the analyte. Equations (6.1) and (6.2) give the formula for the analytical result, c_x, which is a

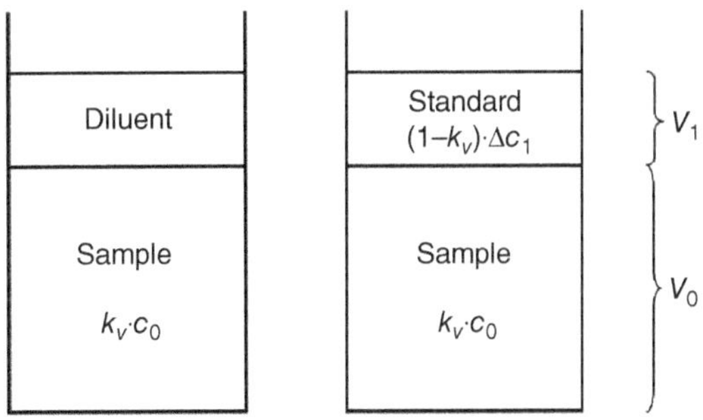

Figure 6.1 Scheme of the basic form of the E-SAM method at the preparative stage of the calibration process.

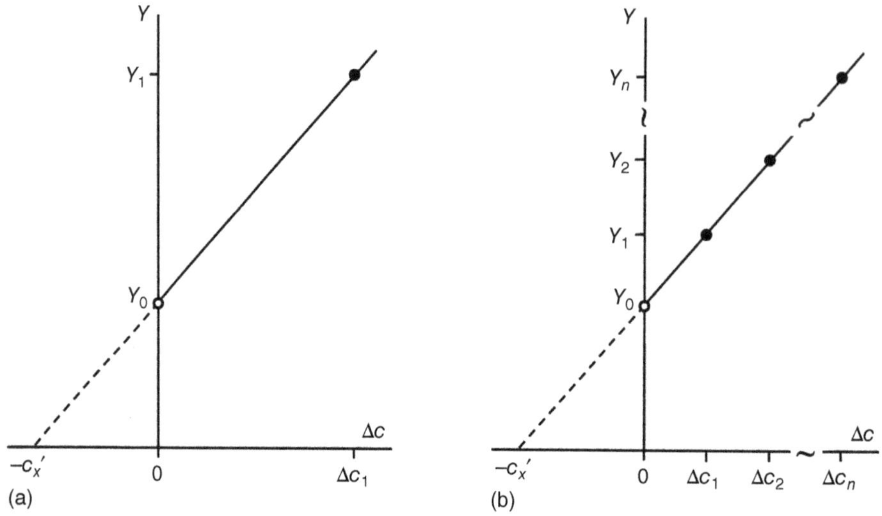

Figure 6.2 Extrapolative standard addition method (E-SAM) : extrapolative transformation of signal Y_x obtained for analyte in the sample and signals Y_i ($i = 1, \ldots, n$) obtained for the sample with standard additions, Δc_i, to the analytical result, c_x', using two-point (a) and multipoint (b) calibration graphs.

measure of the concentration c_0 of the analyte in the sample:

$$c_x = \frac{Y_0}{Y_1 - Y_0} \cdot \frac{1 - k_v}{k_v} \cdot \Delta c_1 \tag{6.3}$$

For the multipoint version of the extrapolation method, standards with increasing, well-defined analyte concentrations are added to successive portions of the sample. If the sample solution with the standard with the highest analyte concentration, Δc_n, is prepared as in Figure 6.1 (without diluent), the constructed calibration graph (Figure 6.23b) allows the analytical result to be determined from the formula:

$$c_x = \frac{\widehat{Y}_0}{\widehat{Y}_n - \widehat{Y}_0} \cdot \frac{1 - k_v}{k_v} \cdot \Delta c_n \tag{6.4}$$

where $\widehat{Y}_0$ and $\widehat{Y}_n$ are the signal intensities determined for the sample and for the sample with the standard with the highest analyte concentration, Δc_n, respectively, resulting from the model function fitted (e.g. by the least squares method) to the measurement points.

The characteristics of the E-SAM method are as follows:

- calibration solutions are prepared so that **the concentrations of the native analyte and all other sample components are equal in all solutions,**
- **the calibration graph covers a limited range of analyte concentration**: from the concentration of the analyte in the sample to the concentration of the analyte in the sample plus the highest concentration of the analyte in the standard,
- **the concentration of an analyte in a sample is determined by extrapolating the calibration graph to a zero signal.**

It is important to pay closer attention to the graphical illustration of the E-SAM method. The assignment of negative values to the analytical results, c_x, in Figure 6.2 is a purely formal procedure, resulting from the convention adopted for scaling the abscissa axis: not based on the total analyte concentrations in the sample (because these are unknown), but on the concentrations of analyte added to the sample (because these are known). The calculated analytical results should of course be given positive values.

From the graph shown in Figure 6.2a, the following formula is used to determine the analytical result:

$$c_x' = \frac{Y_0}{Y_1 - Y_0} \cdot \Delta c_1 \tag{6.5}$$

The difference between Eqs. (6.3) and (6.5) is due to the fact that the concentration of the analyte in the sample taken for calibration decreases as a result of diluting the sample with the standard added to it (see Figure 6.1) and the measure of its value is then the apparent concentration calculated according to Eq. (6.5). To obtain a c_x result that is a valid measure of c_0 concentration, formula (6.3) must be used that takes into account the current value of the dilution factor k_v. This remark obviously also applies to the multipoint standard version of E-SAM.

Thus, it can be said that the dilution of the sample changes the position of the calibration graph with respect to the position of the true function, but this change can be easily accommodated by taking into account the corresponding dilution of the sample. This is easily seen in Figure 6.3, where the position of the true function (line a) is shown relative to the position of the calibration lines formulated after diluting the sample with the standard to varying degrees (lines b–d). The determined concentration of an analyte in a diluted sample is a direct measure of the true concentration only when the dilution factor, k_v, is equal to 0.5, or otherwise when the volumes V_0 and V_1 are equal.

Another natural consequence of adding a standard to the sample is that the sensitivity of the analyte is reduced compared to the sensitivity that can be achieved when the calibration is performed by another method that does not require dilution of the sample (e.g. external standard method). This obviously has a negative effect on the precision of the analytical results obtained.

As can be seen in Figure 6.4, the precision of the results obtained by the E-SAM method depends, as in the external standard method, primarily on the measurement sensitivity of the analyte and on the number of replicate measurements performed for the sample and the sample with the standard added. However, if both methods are performed under the same measurement conditions, the random scatter of analytical results obtained each time is theoretically larger for the E-SAM method. In Figure 6.4, it is evident that this is due to the need to extrapolate the calibration graph: the random error in the position of the calibration graph in the measurement area increases quite significantly in the extrapolation area, which directly affects the error in the analytical result.

The precision of the results obtained by the E-SAM method also depends on the increment size parameter, P, which determines the ratio of the highest concentration of analyte in the standard solution, c_n, to the concentration of native analyte

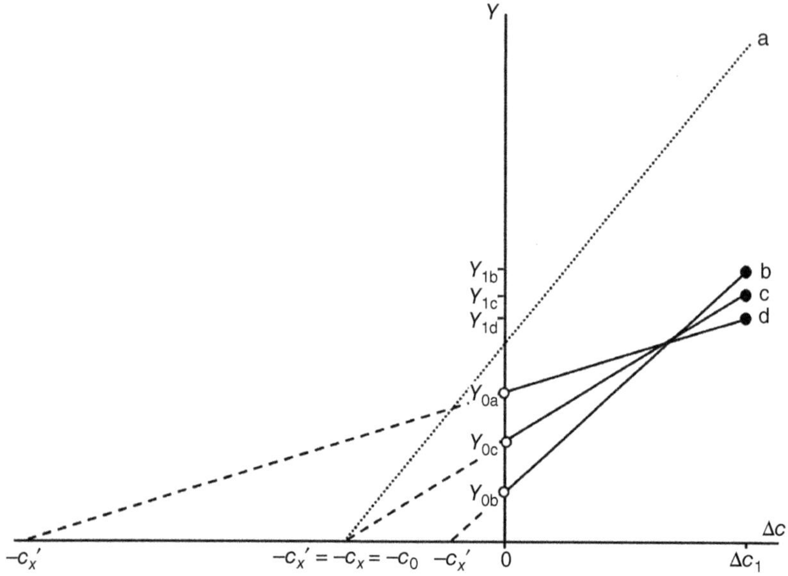

Figure 6.3 Real function (a) and calibration graphs obtained for $k_v = 0.25$ (b), 0.50 (c), and 0.75 (c) leading in SAM to apparent analyte concentrations, c_x'.

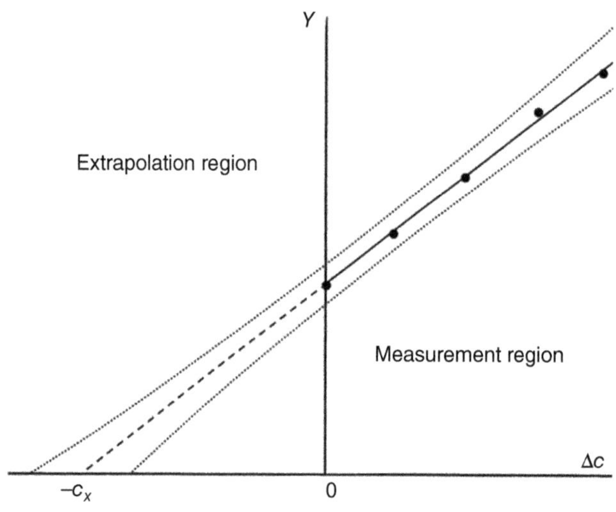

Figure 6.4 Confidence limit of the calibration graph affecting the precision of the analytical, c_x, obtained by the SAM method in the extrapolative way.

in the sample, c_0, after addition of the standard to the sample ($P = \Delta c_n/c_0$) and on the number of measurement points, N (i.e. the number of standards added minus 1). This relationship is shown in Figure 6.5. As can be seen, it is very important that **the concentration of analyte in the last addition is greater than the concentration of native analyte in the sample** ($P > 1$). Interestingly, the random error only

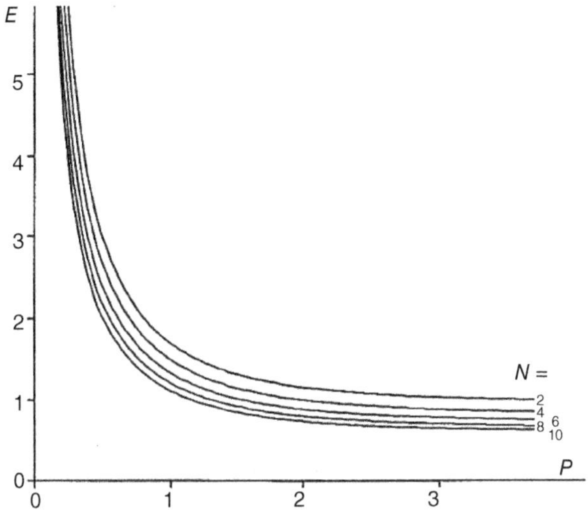

Figure 6.5 Dependence of the random error, E (in terms of variance), of the analytical result on the increment size, P, and the number of measurement points, N. Source: Kościelniak [7], fig 3 (p. 279)/with permission of Elsevier.

depends very little on the N number. Therefore, one should agree with the opinion expressed by many authors that the use of several additions (i.e. the application of the E-SAM method in the multipoint version) is more necessary to verify the linearity of the model function than to improve the precision of the determination.

What is the robustness of the E-SAM method to uncontrolled effects of a systematic nature? The introduction of the standards into the natural environment of the sample, i.e. the application of the "ideal" matrix matching procedure, should, as it seems, favor the compensation of effects caused by the components present in the sample, i.e. mainly interference effects. Let us try to prove it mathematically.

By adding a standard to the sample and thus constructing a model function based on the real function, in the basic version of the E-SAM method the two functions are matched over the range of concentrations 0 and Δc_1 of the added analyte and over the range of signals Y_0 and Y_1 corresponding to these concentrations (see Figure 6.2). If the real function is subject to uncontrolled effects as expressed by Eq. (4.10), it takes on the following values at these points:

$$Y_0 = A + B_1 \cdot k_v c_0 + B_2 H_1 \cdot k_v c_0 \tag{6.6}$$

$$Y_1 = A + B_1 \cdot k_v c_0 + B_1{}' \cdot (1 - k_v)\Delta c_1 + B_2 H_1 \cdot k_v c_0 + B_2{}' H_2 \cdot (1 - k_v)\Delta c_1 \tag{6.7}$$

where $A = A(k_v c_p, k_v c_t)$ is the magnitude of the additive effect from components of the diluted sample that are naturally present in the sample, c_t, or were added to it during sample processing, c_p, coefficients B_1 and $B_1{}'$ determine the measurement sensitivity of the analyte in the sample and in the standard, respectively, and $B_2 H_1$ and $B_2{}' H_2$ are measures of the magnitude of the multiplicative effect taking into account the different linear ($B_2 \neq B_2{}'$) and nonlinear ($H_1 \neq H_2$) nature of this effect in the sample and in the standard, respectively. The speciation effect is assumed to be responsible for the differences $B_1 \neq B_1'$ and $B_2 \neq B_2'$. The values of H_1 and H_2 are given by the formulas:

$$H_1 = \frac{k_v c_m}{1 + a_1 \cdot k_v c_0 + b_1 \cdot k_v c_m} \tag{6.8}$$

$$H_2 = \frac{k_v c_m}{1 + a_2 \cdot (1 - k_v)\Delta c_1 + b_2 \cdot k_v c_m} \tag{6.9}$$

It can be seen that when the interference effect is linear ($a_1 = a_2 = 0$) and acts in the same way on the analyte in the sample and in the standard ($b_1 = b_2$), the values of H_1 and H_2 are equal to each other.

By substituting Eqs. (6.6) and (6.7) into Eq. (6.3), one obtains a relation specifying the conditions that should be satisfied for the analytical result obtained to be a valid measure of the analyte concentration in the sample ($c_x = c_0$) when uncontrolled effects are present:

$$c_x = \frac{A + (B_1 + B_2 H_1)}{(B_1' + B_2' H_2)} \cdot c_0 \tag{6.10}$$

It follows from Eq. (6.10) that $c_x = c_0$ when:

- additive effect does not occur or eliminated ($A = 0$),
- the speciation effect does not occur or compensated ($B_1 = B_1'$, $B_2 \neq B_2'$),
- multiplicative interference effect does not occur ($B_2 = B_2' = 0$) or when it occurs but is linear and does not change this functional form before and after the addition of the standard to the sample ($H_1 = H_2$).

Thus, under certain conditions, **the extrapolative standard addition method, E-SAM, is capable of compensating for proportional (linearly multiplicative) interference effects.**

The condition for compensation of interference effect is absence of additive and speciation effects. The additive effect caused by components added to the sample during sample processing (blank effect) can be eliminated as in the external standard method, namely by correcting measured signals for the sample and the sample with standard additions by the value of signal intensity caused by these components, as shown in Figure 6.6a. Blank effect (BE) compensation is practically not possible. Theoretically, it can only occur if, as a result, the slope of the calibration graph changes in proportion to the total analyte concentration in the sample (see Figure 6.6b), but this requires that the concentration of analyte in the standard solution is matched precisely to the concentration of analyte in the sample, which is obviously impossible in practice.

Additive effects from unknown interferents, or those known but present in the sample in unknown amounts, can only be eliminated as in the external standard method – by separate procedures described in Chapter 4.

The speciation effect is also an important source of systematic error in the E-SAM method. As Figure 6.7 shows, depending on whether the analyte present in the sample in the triggering form has a greater signal than the analyte of the same concentration in the form added to the sample, or vice versa, the analytical results

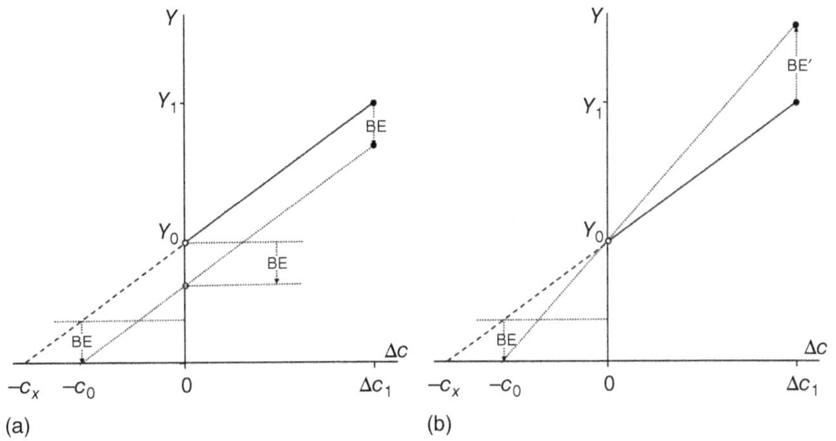

Figure 6.6 Blank effect (BE) eliminated (a) and compensated (b) in the E-SAM method; c_x and c_0 – the analyte concentrations obtained before and after correction, respectively.

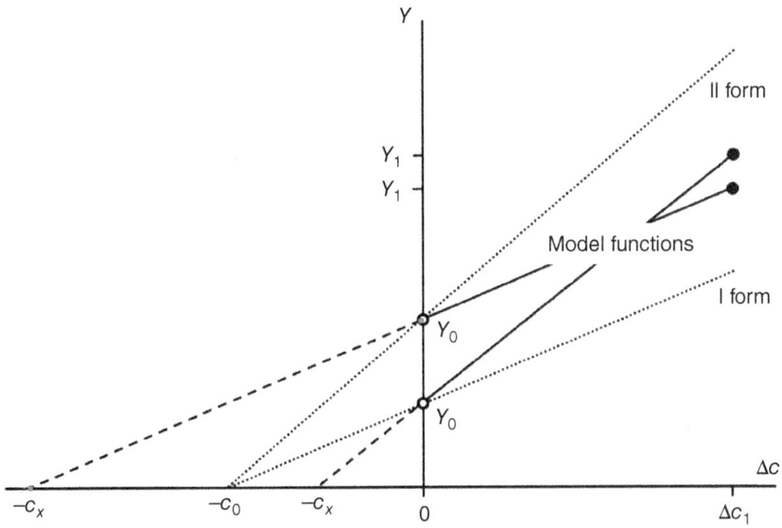

Figure 6.7 Systematic errors caused in the E-SAM method by speciation effect when the analytes of less (I form) and greater (II form) sensitivity are in the sample or in the standards.

obtained can be very different from each other and greater or less than the value that would be obtained if the two forms were equal. As can be seen, unlike the multiplicative nature of the speciation effect in the external standard method, in the E-SAM method it manifests itself as a composite of additive and proportional effects. The effects of its occurrence are therefore also more difficult to predict.

The great importance of the speciation effect in the E-SAM method is shown by the results in Table 6.1 obtained under the same conditions under which the external standard method was tested in this respect (see Table 5.1). It can be seen that when the analyte in the sample was present in organic form and in the standards in inorganic form, the analyte determination errors of both methods were similar. It

Table 6.1 Systematic errors (RE = $[(c_0 - c_x)/c_0]\cdot 100\%$), caused by the speciation and interference effects in the E-SAM method when selenium determined by hydride generation atomic fluorescence spectrometry (HG-AFS) was present in the sample and standards in the same concentration (20 µg l^{-1}) but in different chemical forms; Se-Met – organic form (selenomethionine), Se – inorganic form, Cu – interferent, UV lamp – the digestion tool.

Sample	Analyte in sample	Analyte in standards	Effect	UV lamp	RE (%)
1	Se-Met	Se	Speciation	+	−1.9
2	Se-Met	Se	Speciation	−	−47.7
3	Se + Cu	Se	Interference	−	3.7
4	Se-Met + Cu	Se	Speciation + interference	+	−6.1
5	Se-Met + Cu	Se	Speciation + interference	−	−70.8

Source: Adapted from Stafiński et al. [8].

was also shown that, as before, the effect could be effectively eliminated by digestion of the samples. It is also worth noting that regardless of the test conditions, the interference effect was effectively eliminated by using the E-SAM.

Systematic errors in the analytical result can also occur in the E-SAM method when the interference effect is not uniform over the full range of analyte concentrations. For example, the real function may be linear over the concentration range of the analyte being measured and nonlinear in the extrapolation region. When constructing a calibration graph, i.e. fitting a linear function with a specified slope to the measurement points, the analyst extrapolates according to that slope because he is unaware of the change in the nature of the real function over that range. The error he may make in this situation is shown in Figure 6.8A. It can sometimes be reduced by adding a special reagent to each portion of the sample, but this method is not always effective.

Another problem is the choice of the model function to approximate the real function when the experimental point system is not explicitly linear and can be treated as nonlinear or vice versa, as shown in Figure 6.8B. While in the external standard method this dilemma in the situation of such a distribution of measurement points is not of great importance, in the E-SAM method the analytical results obtained after linear and nonlinear approximation can be, as can be seen, significantly different due to the extrapolation process.

This problem is even more clearly highlighted by the results in Table 6.2 relating to an ambiguous distribution of measurement points such as that shown in Figure 6.8. It can be seen that in such cases accurate results can be obtained either by linear or nonlinear approximation, depending on the nature, linear or nonlinear, of the real function. Since the appropriate type of approximation is difficult to predict or evaluate visually, the statistical tests mentioned earlier must be used. The data in Table 6.2 further show that, from the point of view of the accuracy of analytical results, the decision between a linear and a nonlinear function is much more important than which type of nonlinear function should be used.

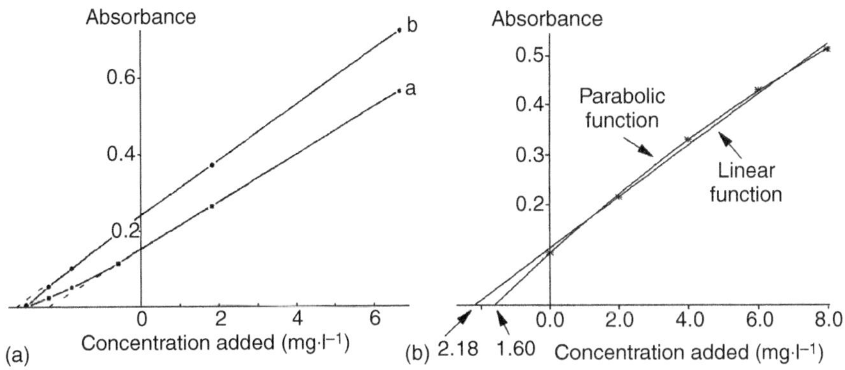

Figure 6.8 Typical systematic errors committed in the E-SAM method by (A) – linear extrapolation of the linear calibration graph instead of nonlinear extrapolation (FAAS determination of Ca in the presence of Al and Si with (a) no releasing reagent, (b) 0.2% lanthanum releasing agent. Source: Hosking et al. [9], fig 7 (p. 308)/American Chemical Society), and (B) – fitting linear function instead of nonlinear one to the same measurement points (FAAS determination of Cd. Source: Kościelniak [7], fig 7 (p. 282)/with permission of Elsevier.)

Table 6.2 Results of the FAAS determination of Zn, Fe, Mn, and Ni in a sample of aluminum alloy (reference material) using the SAM method with various functions approximating the real function (in all cases the quality coefficient QC was less than 3%).

Analyte	Expected content, c_0 (%)	Obtained content, c_x (%), with approximation				
		Linear	Parabolic	Exponential	Hyperbolic	Logarithmic
Zn	0.26	0.257	0.204	0.200	0.198	0.195
Fe	0.47	0.408	0.478	0.472	0.469	0.467
Mn	0.30	0.289	0.261	0.259	0.258	0.257
Ni	0.37	0.305	0.406	0.392	0.387	0.383

Source: Kościelniak [7], table 3 (p. 283)/with permission of Elsevier.

In view of the above problems, it is commonly assumed that calibration by the E-SAM method can be performed only under conditions that give full confidence in the linear nature of the real function. It is difficult to disagree with this assumption, taking into account additionally that the method leads to compensation of interference effects only if these effects are linear. While this does not mean that the nonlinear real function cannot be accurately fitted after compensation for linear interference effects, this is certainly much less likely than for a linear function.

Nevertheless, it is worth looking at the interesting results that were obtained in the work [7] with respect to the problem of nonlinear approximation of the real function in the E-SAM method. Namely, it was shown that the applicability of this type of modeling depends on the type and degree of curvature of the model function fitted to the measurement points. This degree was determined by Q parameter of the formula:

$$Q = \frac{\hat{Y}_{n/2} - \hat{Y}_n - \hat{Y}_0}{\hat{Y}_n - \hat{Y}_0} \tag{6.11}$$

where $\hat{Y}_{n/2}$ is the signal determined from the pattern of the nonlinear model function at the midpoint of the measurement range. The value of the Q parameter is positive or negative depending on whether the function is convex or concave, respectively, with respect to the linear function connecting the points $\hat{Y}_0$ and $\hat{Y}_n$. It has been proved that the possibility appears of reducing the random error with increasing convex curvature of the calibration line, i.e. when positive value of parameter Q is increased, as shown in Figure 6.9. Comparing this figure with Figure 6.5, one can see that the nonlinear (e.g. parabolic) regression can lead to even better precision than the linear regression (!). provided that the analysis is performed under conditions allowing the calibration line to be explicit of convex shape. On the other hand, from the point of view of precision of analytical results, one should definitely not undertake SAM analysis when the model function takes even the slightest concave shape ($Q < 0$).

When the system of measurement points is distinctly nonlinear ($0.2 < Q < 0.5$), the problem becomes not whether a linear or nonlinear function should be fitted to these points, but what kind of nonlinear function should be used. Table 6.3 presents the results of this type of study. They clearly show that as the degree of curvature of the real function increases, different nonlinear functions lead to increasingly divergent results. In the flame photometry method, the hyperbolic function proved to be

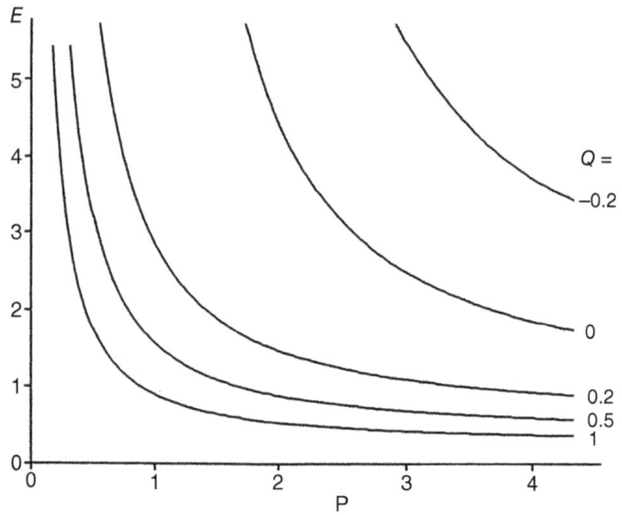

Figure 6.9 Random error, E, committed in the E-SAM method when parabolic function of various curvature degrees Q is used for approximation of nonlinear real function; P increment size, number of standard additions $N = 4$. Source: Kościelniak [7], fig 4 (p. 279)/with permission of Elsevier.

Table 6.3 Results of the determination of sodium (10 mg l^{-1}) in the presence of calcium (200 mg l^{-1}) by flame photometry using linear and various nonlinear model functions for E-SAM calibration.

	Analyte concentration, c_x (mg l^{-1}), found with approximation				
Parameter Q	**Linear**	**Parabolic**	**Exponential**	**Hyperbolic**	**Logarithmic**
0.437	45.17	18.53	13.62	10.07	5.38
0.350	31.39	14.64	11.90	9.89	7.10
0.272	24.72	13.57	12.05	10.97	9.52
0.173	18.30	11.77	10.80	10.14	9.26
0.094	12.95	9.97	9.79	9.69	9.57
0.051	11.43	9.77	9.69	9.64	9.58

Source: Adapted from Kościelniak [7].

the most appropriate once again, although it seems that recommending its use in various other analytical methods is too risky and should be reviewed.

Thus, it can be said that **nonlinear modeling should not be avoided at all costs in the E-SAM method**, especially when the system of measurement points is definitely nonlinear and gives confidence in the existence of a real function of this nature.

It should be emphasized, however, that the greatest benefit an analyte can gain by using E-SAM calibration is the ability to obtain an accurate analytical result when a linear interference effect is present. This is an even greater benefit since this type of effect is most common across analytical methods. The analyst has a chance to compensate for this effect without knowing the type and concentration of the interferents present in the sample. In addition, the method can be used preventively, that is, when he or she is unsure whether or not the interference effect threatens the accuracy of the analytical result. It should also be noted that the method does not require the use of separate chemical reagents and follows a relatively simple procedure.

In the light of these observations, the question arises as to why the E-SAM method is not used far more frequently in analytical laboratories, replacing the external standard method more commonly in such situations where, for example, samples with a complex matrix representing a high potential hazard from interference effects are analyzed. This is undoubtedly influenced by the limitations of the method outlined above, which are mainly related to the necessity of determining the analytical result by extrapolation. Practical considerations also play an important role, as it should be noted that the E-SAM method – unlike the external standard method – provides a single calibration graph for each sample analyzed. All these limitations have led to many attempts to modify the E-SAM method to improve its performance and analytical capability.

6.2.1.1 Modified Procedures

Since the beginning of the introduction of the standard addition method into analytical practice, attempts have been made to modify it in various ways. These concerned all stages of the calibration procedure of this method – preparation of calibration solutions, interpretation of measurement results, and the method of transformation of signals to the concentration of analyte in the sample. For example, very early on, the first attempts were also made to combine the process of adding a standard to a sample and diluting this solution in the so-called additive and partitioning method [10]. Calibration by this method on the basis of the results obtained by the internal standard method was also proposed [11], and strong interference effects were proposed to be minimized mathematically by the so-called double coefficient method [12]. These and other proposals, however, passing the test only in isolated cases, did not find wider practical application in later years.

As mentioned, a certain drawback of the E-SAM method is its reduced precision, which is due to the extrapolation process of the calibration graph. To improve this limitation of the method, several attempts have been made to transform its character from extrapolation to interpolation. The most interesting proposal of this kind concern changes in the calibration procedure at the stage of interpretation of measurement results.

The concept is very simple: it is proposed to transform the measurement data obtained by the E-SAM method in such a way that, pictorially, the coordinate system is shifted along the signal intensity axis (Y) by the value of the signal intensity obtained for the sample, Y_0 [13]. This procedure, shown in Figure 6.10, makes it possible to relate the signal Y_0 to the calibration graph in its measurement part and to determine the analyte concentration in the sample, c_x, by interpolation.

Experimental studies have shown that this approach does not significantly improve the precision of the E-SAM method, as shown by the results presented in Table 6.4. However, it should be emphasized that the described modification is

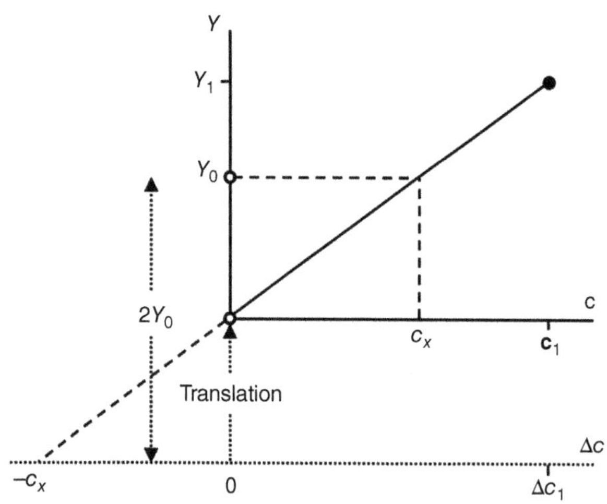

Figure 6.10 Transformation of measurement data obtained in the E-SAM method, allowing the analytical result, c_x, to be obtained in the interpolative way.

Table 6.4 Random errors in the ET-AAS determination of Ni in fuel oil, obtained in the SAM method in the extrapolative and interpolative ways; increment size ≈1.5.

Signal change (%)	Confidence interval of the analytical result (ppb)	
	Extrapolative way	**Interpolative way**
0	10.26	8.48
+10	11.68	9.85
−10	16.51	13.32
+20	16.44	14.09
−20	28.77	22.53

Source: Adapted from Andrade et al. [13].

worth further experimental verification because it does not require any procedural or measurement changes in the calibration process, and thus retains the basic advantages of the method.

Another approach is to determine the sum of the analyte concentrations in the sample under analysis and in the standard solutions by the external standard method (i.e. by the interpolation method) and then to determine the analyte in that sample by the E-SAM method using the same series of standard solutions added to the sample [14]. The results obtained in the interpolative way (i.e. sums of the analyte concentrations in the sample and in the standard additions) are then presented as correlated with the analyte concentrations added to the sample (as shown in Figure 6.11). The analytical result is determined by extrapolating the obtained calibration graph to zero analyte concentration.

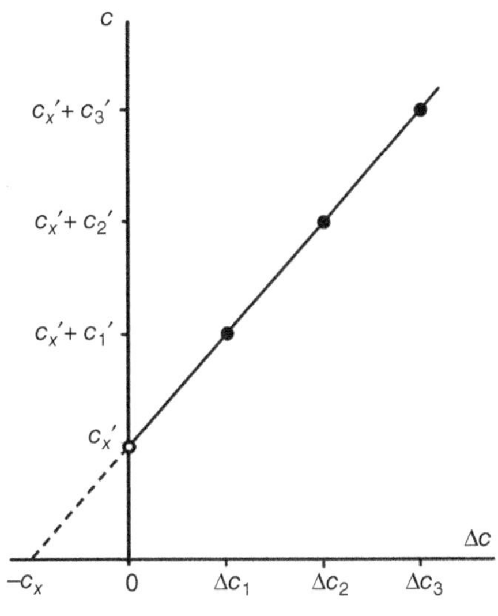

Figure 6.11 Correlated procedure: apparent analyte concentrations obtained for a sample and standards in the interpolative way ($c_x' + c_i'$) are correlated with the analyte concentration in the standard added to the sample (Δc_i) leading to the analytical result (c_x) in the extrapolative way.

Thus, the presented approach avoids the process characteristic of the E-SAM method, i.e. extrapolation of the calibration graph to the zero value of the analytical signal, and thus it can be helpful when this process is difficult (e.g. due to large measurement fluctuations) or even impossible (e.g. in potentiometric measurements) to perform. Furthermore, in some cases, this way of interpreting the results allows the transformation of the nonlinear system of measurement points revealed by the E-SAM method to a linear system. It should also be noted that this method is also useful for detecting the linearly multiplicative interference effect in the analyzed sample since the measure of this effect is the angle of slope of the graph shown in Figure 6.11 (an angle equal to 45° indicates the absence of this effect, an angle statistically different from 45° indicates its presence).

In contrast to the above approach, one that can be termed **the consecutive one-point calibration** is focused exclusively on the interpretation and transformation stages of the E-SAM method [15]. It consists of extrapolative calculation of the apparent concentrations of the analyte in the sample based on measurements performed for the sample and for each consecutive solution of the sample with standard addition (see Figure 6.12a). The apparent concentrations are then expressed as a function of the added analyte concentrations, and the final result is calculated by extrapolation of this function to zero added concentration (Figure 6.12b). Such an approach gives a chance, in particular, for determination of the analyte with improved accuracy when the model function obtained in the E-SAM method is nonlinear (even when it is of concave shape).

The simplest and most obvious modification of the E-SAM method at the preparative step is to add standards to the sample in a sequential manner (one after the other) without making up successive solutions to an equal volume. The E-SAM based on such a procedure has been described in the literature several times, and the most complete characterization of it is contained in a series of papers by R.J.C. Brown et al. initiated by the article [16].

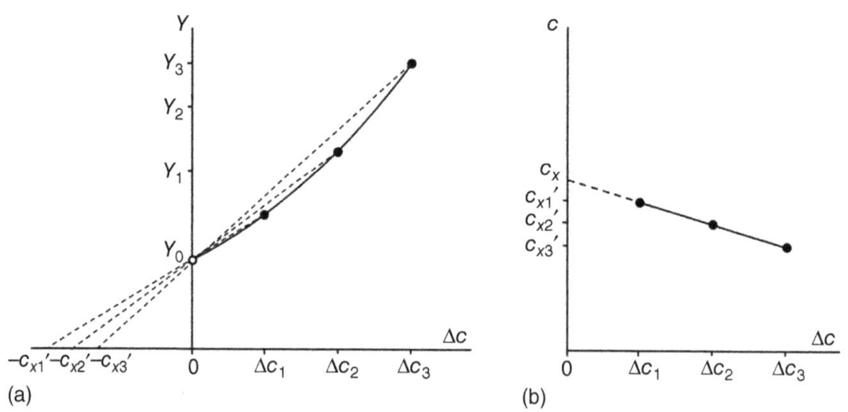

Figure 6.12 Principle of the consecutive one-point calibration procedure in the first (a) and second (b) step.

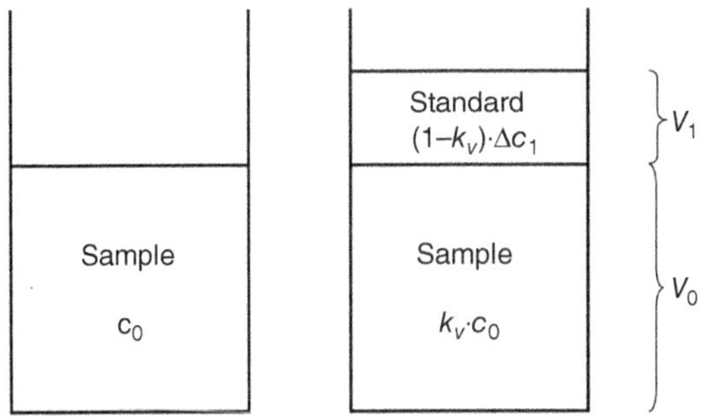

Figure 6.13 Scheme of the basic form of the sequential E-SAM method at the preparative stage of the calibration process.

In the basic version of **sequential SAM method,** an initially prepared, undiluted sample of volume V_0 and concentration c_0 is then diluted with the addition of a standard of volume V_1 and analyte concentration Δc_1. When the standard is added to the sample, the analyte concentration derived from the sample is, therefore, $k_v \cdot c_0$, and the analyte concentration derived from the standard addition is equal to $(1 - k_v)\cdot\Delta c_1$. This procedure is shown in Figure 6.13.

The intensity values Y_0 and Y_1 measured for both solutions are related to the analyte concentrations in the prepared calibration solutions as follows:

$$Y_0 = B_1 \cdot c_x \tag{6.12}$$

$$Y_1 = B_1 \cdot [k_v c_x + (1 - k_v)\Delta c_1] \tag{6.13}$$

where B_1 is a factor that determines the measurement sensitivity of the analyte. From these equations, a formula is derived to estimate the actual concentration of an analyte in a sample taking into account the current dilution of the sample k_v:

$$c_x = \frac{(1 - k_v) \cdot Y_0}{Y_1 - k_v \cdot Y_0} \cdot \Delta c_1 \tag{6.14}$$

The measurement results can be represented on the same coordinate system as in the conventional E-SAM method (Figure 6.2a) and the value of the apparent analyte concentration, c_x', can be determined by extrapolation. The concentration c_x is then calculated from the formula:

$$c_x = \frac{(1 - k_v) \cdot c_x'}{\Delta c_1 + (1 - k) \cdot c_x'} \cdot \Delta c_1 \tag{6.15}$$

Thus, correction of the c_x' value, in this case, requires consideration of the current maximum concentration of the analyte added to the sample with the last standard solution.

The multipoint version of the sequential method involves continuing to add a standard solution to a progressively more dilute sample. If the same standard is added in equal volumes, the response of the measuring instrument as a function of analyte concentration in successive additions is nonlinear, as shown in Figure 6.14a.

To avoid approximation of the real function by a nonlinear model function, the concentration of the analyte in the sample can be calculated as shown in Figure 6.14a, i.e. by extrapolation of linear functions considering the signal intensity obtained for the analyte in the sample, Y_0, and the intensity Y_n measured

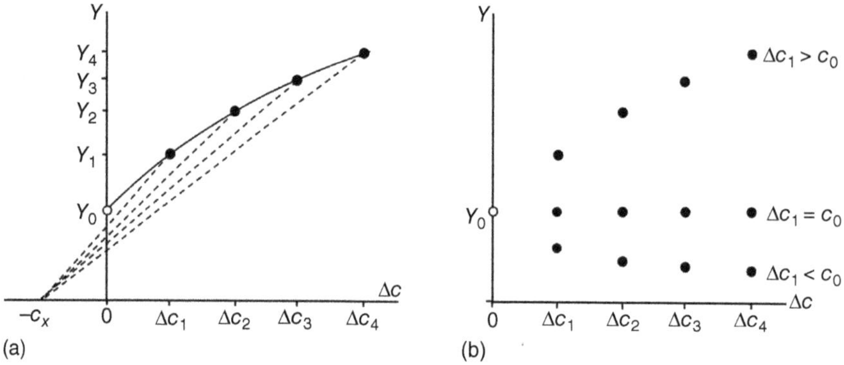

Figure 6.14 Multipoint sequential E-SAM method: possible way for calculation of analytical result, c_x (a) and different distribution of the measurement points depending on the concentration ratio of the analyte added, c_1, to the analyte present in the sample, c_0 (b).

after the nth addition of a standard solution of concentration Δc_1 and volume V_1 to a sample of initial volume V_0 (i.e. after the nth dilution of the sample with the standard solution). These values are determined using the formula:

$$c_x = \frac{k_{vn} \cdot Y_0}{Y_n - k_{vn} \cdot Y_0} \cdot n \cdot \Delta c_1 \tag{6.16}$$

where:

$$k_{vn} = \frac{V_0}{V_0 + n \cdot V_1} \tag{6.17}$$

If there are no uncontrolled effects in the sample, then the c_x values calculated in successive sample dilution sequences can be assumed to vary randomly and the measure of the final analytical result is then the arithmetic mean of these values.

Note that the calibration graph in the sequential SAM method is increasing when the analyte concentration in the added standard is greater than the analyte concentration in the sample ($\Delta c_1 > c_x$), as seen in Figure 6.14b. Only in this situation does the addition of the standard solution overcompensate for the loss in total analyte concentration caused by diluting the solution with this addition. Otherwise ($\Delta c_1 < c_x$) the graph is nonlinear and decreasing, and when $\Delta c_1 = c_x$ the graph is constant. In both cases the determination of the analyte concentration in the sample by the sequential method becomes impossible.

As can be seen in Figure 6.14a, the concentration of analyte as it is diluted is determined with decreasing precision as the "local" measurement sensitivity of the analyte decreases. The situation can generally be improved by gradually increasing the analyte concentration in successive portions of the added standard, but this is an impractical procedure and requires very tight control.

The initial measurement sensitivity can be increased by setting a higher concentration of analyte in the standard relative to the concentration of analyte in the sample. This is best done by fixing the volume of standard solution with a given mass of analyte at a very small level because then the degree of dilution of the sample is also small (according to Eq. (6.17)) and the tendency of the sensitivity to gradually decrease is weaker (the calibration graph becomes more and more linear). In other words, **the smaller the volume of solution added at a sufficiently high concentration, the sequential method becomes more similar to the E-SAM method.**

However, there are two dangers of handling a standard solution of high concentration and small volume. A large concentration of standard may cause the linear range of the true function to be rapidly exceeded after successive sample dilution steps and superimpose this nonlinearity on the naturally nonlinear nature of the calibration graph caused by the dilution process. Errors in the apparent analyte concentrations obtained in this way shown in Figure 6.11 will then become systematic. On the other hand, the determination of a very small volume of standard may have a relatively large error, which will accumulate and increasingly affect the error of the analytical result as standard is added to the sample. Thus, from the point of view of the accuracy of analytical results, the concentration of analyte in the added standard and the volume of this standard are key parameters of the sequential method, requiring particularly careful and accurate optimization.

In view of these limitations of the method, a fundamental problem becomes its resistance to uncontrolled effects. In particular, the question arises as to whether the sequential method of adding standard to the sample causing progressive dilution maintains the ability of the method to compensate for linear interference effects.

Approaching the issue in the same mathematical way as used for the E-SAM method, the conditions for obtaining an accurate estimate of the analyte concentration in a sample when uncontrolled effects are present are given by the equation:

$$c_x = \frac{(1-k_v)\cdot[A+(B_1+B_2H_0)\cdot c_0]}{(A'-k_v\cdot A)+B_2\cdot(H_1-H_0)\cdot k_v c_0+\left(B_1'+B_2'H_2\right)\cdot(1-k_v)\Delta c_1}\cdot\Delta c_1 \tag{6.18}$$

where $A = A(c_p, c_t)$, $A' = A(k_v c_p, k_v c_t)$. Value of function H_0 is given by:

$$H_0 = \frac{c_m}{1+a_0c_0+b_0c_m} \tag{6.19}$$

and the values of the functions H_1 and H_2 are calculated from Eqs. (6.8) and (6.9). If the additive effect is absent or eliminated ($A = A' = 0$), and the speciation effect is absent or compensated ($B_1 = B_1'$, $B_2 = B_2'$), then Eq. (6.18) is simplified to the form

$$c_x = \frac{(B_1+B_2H_0)\cdot(1-k_v)c_0}{B_2\cdot(H_1-H_0)\cdot k_v c_0+(B_1+B_2H_2)\cdot(1-k_v)\Delta c_1}\cdot\Delta c_1 \tag{6.20}$$

It can be seen from the equation that the condition $c_x = c_0$ can be satisfied only when $H_0 = H_1 = H_2$. The condition $H_1 = H_2$ is satisfied when the interference effect is linear, but the conditions $H_0 = H_1$ and $H_0 = H_2$ are not satisfied regardless of whether the interference effect is linear or nonlinear.

Suppose, however, that the interference effect is linear ($a_0 = a_1 = a_2 = 0$) and is revealed in the same way in the undiluted sample, the diluted sample, and the standard sample ($b_0 = b_1 = b_2 = b$). The condition $H_0 = H_1$ ($= H_2$) is then expressed by the equation:

$$\frac{c_m}{1 + b \cdot c_m} = \frac{k_v c_m}{1 + b \cdot k_v c_m} \tag{6.21}$$

It follows that there may be two situations in which **the sequential method can compensate for the linear interference effect**:

- **when the interferents are present** in the sample **in** such **excess** that a change in their concentration due to dilution of the sample with the standard solution is irrelevant to the nature of their effect on the analyte; in this case, therefore, the interferents exhibit a kind of buffering effect due to the induced interference effect,
- **when the linear interference effect depends**, in the manner expressed by the function (4.10), not on the concentration of interferents alone, but **on the ratio of concentrations of interferents and analyte**, then the equation is valid:

$$\frac{c_m/c_a}{1 + b \cdot c_m/c_a} = \frac{k_v c_m/k_v c_a}{1 + b \cdot k_v c_m/k_v c_a} \tag{6.22}$$

If none of the above situations is the case, then it is to be expected that the analytical results will be subject to systematic errors due to interferers. Thus, it can be concluded that **the sequential method is capable of compensating for linear interference effects, but only to a very limited, well-defined extent.**

6.2.2 Interpolative Variants

In the extensive literature on calibration by the SAM, proposals can be found that involve treating the sample and standard in such a reciprocal manner that the analytical result is determined by interpolation. A classic example is the approach directly referred to as **the interpolative standard addition method (I-SAM)** [17].

The principle of the I-SAM method is similar to some extent to that of the sequential method. In the basic version, the preparative step (Figure 6.15) consists of diluting a sample of volume V_0 and concentration c_0 with two standard solutions of equal volume, V_1, and analyte concentrations Δc_1 and Δc_2, the analyte concentrations in these solutions being greater and less than the analyte concentration in the sample ($\Delta c_1 > c_0 > \Delta c_2$).

The intensities of the signals obtained for all three solutions, Y_0, Y_1, Y_2 ($Y_1 > Y_0 > Y_2$), are represented as shown in Figure 6.16, namely as the relationship between the difference of the signals measured for the sample with the standard solutions added and the signal obtained for the sample alone, ($Y_1 - Y_0$) and ($Y_2 - Y_0$), and the analyte concentrations in the standard solutions.

The difference in the signals measured for the analyte in the sample before and after the addition of the calibration solution should be zero when the added calibration solution contains the analyte at the same concentration as it is in the sample. The measure of the concentration of the analyte in the sample is therefore the analytical result determined by the intersection of the calibration graph with

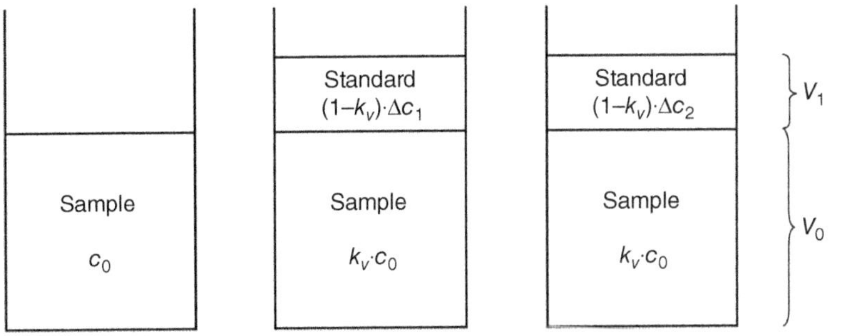

Figure 6.15 Scheme of the basic form of the I-SAM method at the preparative stage of the calibration process.

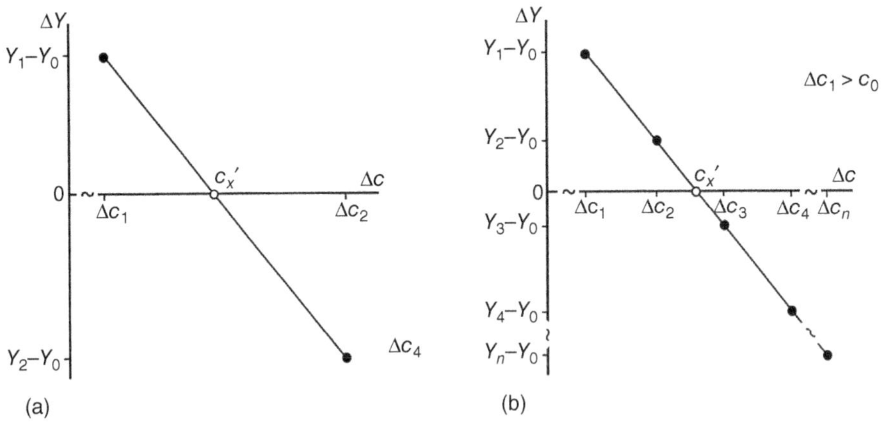

Figure 6.16 Interpretation of the measurement data in the twopoint (a) and multipoint (b) versions of the I-SAM method.

the concentration axis. As shown in Figure 6.15, this result is calculated from the formula:

$$c_x{}' = \frac{\Delta c_1 \cdot (Y_2 - Y_0) - \Delta c_2 \cdot (Y_1 - Y_0)}{Y_2 - Y_1} \tag{6.23}$$

If there are no uncontrolled effects in the sample and in the standard solutions and the real function is linear, this result can be determined from the values Y_0, Y_1, Y_2 of the linear real function taking into account the dilution of the sample with the standard solutions to the extent of k_v:

$$Y_0 = B_1 \cdot c_x \tag{6.24}$$

$$Y_1 = B_1 \cdot [k_v c_x + (1 - k_v)\Delta c_1] \tag{6.25}$$

$$Y_2 = B_1 \cdot [k_v c_x + (1 - k_v)\Delta c_2] \tag{6.26}$$

where k is the dilution degree. From Eqs. (6.54)–(6.56) results that the analytical result is given by Eq. (6.23), that is, $c_x = c_x{}'$. In the I-SAM method, it is not necessary to take into account in the transformation step the degree of dilution of the sample, k_v, with standard solutions.

In the multipoint version of the method the procedure is the same as in the basic version: measurements are carried out on the sample alone and on the sample diluted with several standard solutions with concentrations lower and higher than the analyte concentration in the sample, a multipoint calibration plot is constructed and the apparent concentration of the analyte, c_x, is determined from the formula:

$$c_x = \frac{\Delta c_1 \cdot (\hat{Y}_n - \hat{Y}_0) - \Delta c_n \cdot (\hat{Y}_1 - \hat{Y}_0)}{\hat{Y}_n - \hat{Y}_1} \tag{6.27}$$

where $\hat{Y}_0$, $\hat{Y}_1$, and $\hat{Y}_n$ are signal intensities, determined for the sample alone and for the sample with the addition of a standard of analyte concentration Δc_1 and Δc_n respectively, resulting from the model function fitted (e.g. by the method of least squares) to the measurement points.

The I-SAM method is an attractive analytical tool because, while retaining the nature of the standard addition method (standards are added to the sample), it also allows the concentration of the analyte in the sample to be determined by interpolation. It should be noted, however, that similar to the sequential E-SAM method, potential interferents present in the sample prior to dilution change their concentration as a result of diluting the sample with standard solutions. In this situation, it is again questionable whether this does not take away the primary advantage of the standard addition method, which is the ability to compensate for proportional interference effect.

To demonstrate this, it is necessary to apply, analogously as before, the modeling of the real function by means of the function (4.10). Taking into account formulas (6.23)–(6.26) it is possible to determine the conditions for obtaining an accurate estimate of the analyte concentration in the sample, $c_x = c_0$ when uncontrolled effects are present. If it is assumed that the additive effect is eliminated and the speciation effect is compensated, these conditions are expressed by the formula:

$$c_x = \frac{(1 - k_v) \cdot B_1 + B_2 \cdot (H_0 - k_v \cdot H_1)c_0}{(1 - k_v) \cdot (B_1 + B_2 H_2)} \tag{6.28}$$

It is seen that $c_x = c_0$ only if, as in the case of the sequential method, the condition $H_0 = H_1$ $(= H_2)$.

Thus, **the I-SAM method can lead to accurate analytical results ($c_x = c_0$) only if the interference effect is linear and acts in the same way on the analyte in the undiluted sample and in the diluted sample with the standard, and if the interferents, i.e. if the concentration of interferents is so high that they show a buffering effect or the effect depends on the ratio of interferent and analyte concentrations** (compare Eq. (6.20)).

A typical linearly multiplicative interference effect, for example, is the change in signal measured for magnesium by the FAAS method under the influence of aluminum. In this example, it is shown that after appropriate transformation of the measured results obtained by the I-SAM method, the calibration graph obtained by

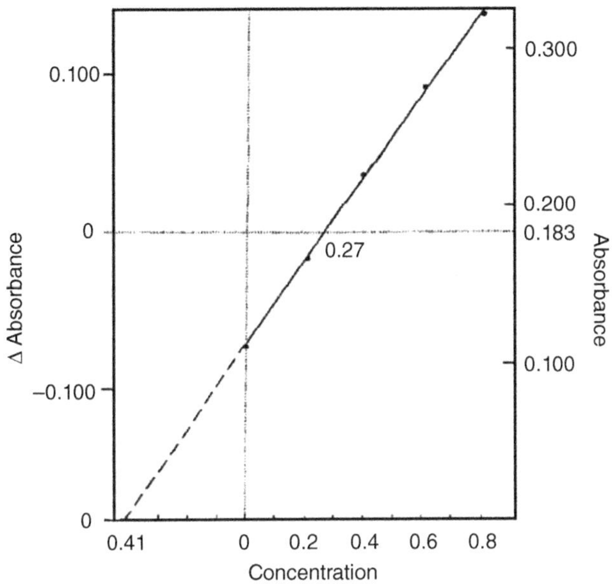

Figure 6.17 FAAS determination of Mg (0.4) in the presence of in the presence of aluminum as interferent (20 mg l^{-1}) by the I-SAM and E-SAM methods using the same calibration graph. Source: Kościelniak [18]/with permission of Elsevier.

Table 6.5 Comparison of the results of the FAAS determination of magnesium (4 mg l^{-1}) in synthetic samples by the E-SAM and I-SAM methods.

Mg concentration, c_0 (mg l^{-1})	Al concentration (mg l^{-1})	Analyte result, c_x (mg l^{-1})	
		E-SAM	I-SAM
0.4	20	0.41 ± 0.04	0.27 ± 0.02
0.4	50	0.42 ± 0.04	0.22 ± 0.02
0.4	100	0.37 ± 0.03	0.19 ± 0.02
0.8	20	0.80 ± 0.06	0.69 ± 0.04
0.8	50	0.75 ± 0.05	0.60 ± 0.03
0.8	100	0.83 ± 0.03	0.52 ± 0.02

Source: Kościelniak [18]/with permission of Elsevier.

this method can be extrapolated to zero signal value according to the E-SAM method, as shown in Figure 6.17 [18]. It can be seen that the analytical results obtained by both methods using the same calibration graph by interpolation and extrapolation are significantly different. That in this analytical system, the E-SAM method leads to more accurate results than the I-SAM method is evidenced by the results summarized in Table 6.5. The advantage of the I-SAM method is manifested in the slight improvement in the precision of the analyte determination compared to that achieved by the extrapolation method.

Another attempt to develop a method for interpolative standard additions is a method called **the sample-to-standard additions method** [19]. As the name suggests, it proposes to reverse the roles of sample and standard compared to previous procedures: the sample is added to the standard, not vice versa.

In the preparative step, an undiluted standard of volume V_0 and concentration c_1 is diluted with a sample solution of volume V_1 and analyte concentration c_0. Assuming that c_0 and Δc_1 are the analyte concentrations of the sample and standard solutions of equal volume V_0, then when the sample is added to the standard, the analyte concentration derived from the sample is $k_v{\cdot}c_0$, and the analyte concentration derived from the standard addition is equal to $k_v{\cdot}\Delta c_1$. This procedure is shown in Figure 6.18.

The measurement results are presented as shown in Figure 6.19.

If the analyte in the sample and in the standard solution is not subject to any uncontrolled effects and the real function is linear, then the linear model function at points Y_0 and Y_1 take values according to the formulas:

$$Y_0 = B_1 \cdot \Delta c_1 \tag{6.29}$$

$$Y_1 = B_1 \cdot [k_v \Delta c_1 + (1 - k_v) c_x] \tag{6.30}$$

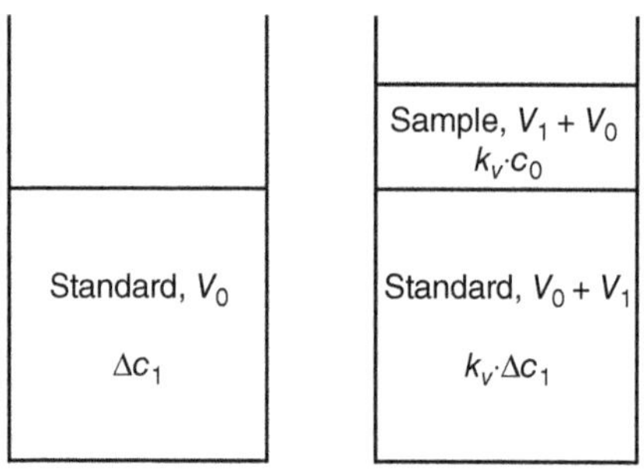

Figure 6.18 Scheme of the basic form of the St-SAM method at the preparative stage of the calibration process.

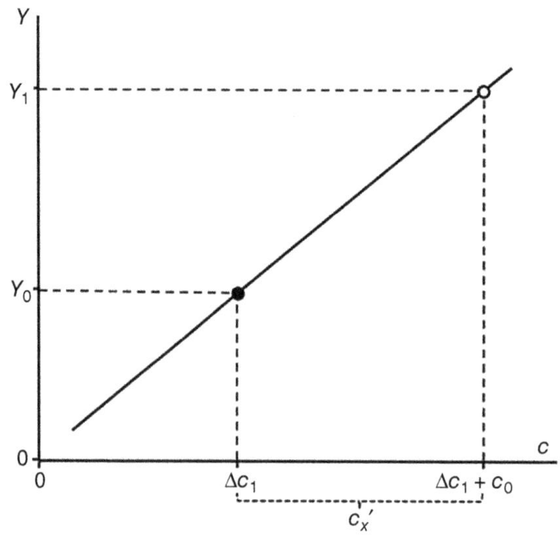

Figure 6.19 Interpretation of the measurement data in the sample-to-standard additions method.

where B_1 is a factor that determines the measurement sensitivity of the analyte. The concentration of the analyte in the sample, c_x, can therefore be determined using the formula:

$$c_x = \frac{Y_1 - k_v \cdot Y_0}{(1 - k_v) \cdot Y_0} \cdot \Delta c_1 \tag{6.31}$$

To obtain the analytical result c_x, which is a measure of the concentration c_0 of the analyte in the sample, from the result c_x' determined experimentally from the graph shown in Figure 6.16, use the formula:

$$c_x = \frac{c_x' + (1 - k_v) \cdot \Delta c_1}{(1 - k_v)} \tag{6.32}$$

Assuming that only multiplicative interference effects are present in the sample, and that other effects are eliminated or compensated for, the equation is obtained:

$$c_x = \frac{B_1 \cdot (1 - k_v)c_0 + B_2 \cdot [H_3 \cdot (1 - k_v)c_0 + H_4 \cdot k_v \Delta c_1]}{(1 - k_v) \cdot B_1} \cdot \Delta c_1 \tag{6.33}$$

where:

$$H_3 = \frac{(1 - k_v)c_m}{1 + a \cdot (1 - k_v)c_0 + b \cdot (1 - k_v)c_m} \tag{6.34}$$

$$H_4 = \frac{(1 - k_v)c_m}{1 + a \cdot k_v c_0 + b \cdot (1 - k_v)c_m} \tag{6.35}$$

If multiplicative effects are linear ($a = 0$), then $H_3 = H_4$, but the effects cannot be offset under any specific conditions.

It follows that **the sample-to-standard additions method**, like the external standard method, **is not resistant to any uncontrolled effects**. From this point of view, its use misses the point: although it is indeed interpolative in nature (see Figure 6.15), it, like the E-SAM method, requires the combination of each sample analyzed with a standard solution and provides results subject to errors due to uncontrolled effects. Of course, if the effects are not present, it can lead to accurate analytical results, but in that case, the external standard method, which is simpler and more efficient, also leads to accurate results.

6.2.3 Indicative Variant

Still another way of mutual treatment of sample and standard and another way of interpretation of measurement results is represented by the method called the **standard addition and indicative dilution method (In-SAM)** [20]. At the preparative stage, it consists first of preparing a solution of the sample and the sample with the addition of a standard with the concentration of the analyte Δc_1 in the manner characteristic of the conventional E-SAM method, that is, keeping the concentrations of the sample components equal. The sample with the standard addition is subjected to measurement Y_1 and then gradually diluted to a degree k_x such that the Y_2 signal for the diluted sample with the addition is equal to the Y_0 signal measured for the sample alone. This simple calibration procedure is shown in Figures 6.20 and 6.21.

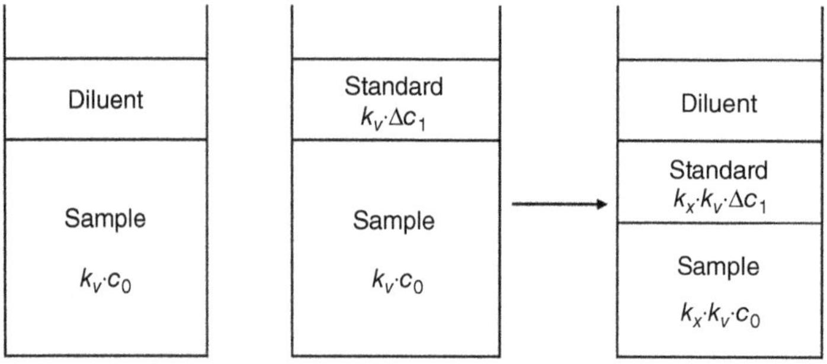

Figure 6.20 The In-SAM method at the preparative stage of the calibration process.

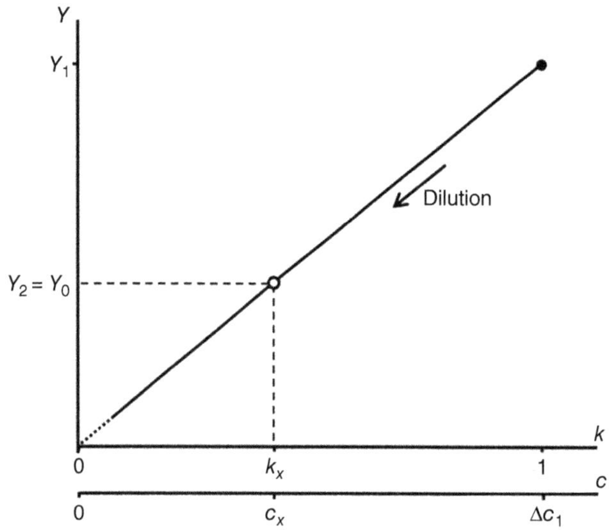

Figure 6.21 The In-SAM method at the measurement and transformation stages: the analytical result, c_x, is indicated by dilution factor k_x, which is achieved by dilution of the sample with standard when the signals Y_0 and Y_2 are the same.

If the Y_0, Y_1, and Y_2 signals are not subject to systematic uncontrolled effects and the real function over these signals is linear, then the model function expresses the dependence of these signals on the concentration of analyte in the sample, c_x, and in the standard solution, Δc_1, by the formulas:

$$Y_0 = B_1 \cdot k_v c_x \tag{6.36}$$

$$Y_1 = B_1 \cdot [k_v c_x + (1 - k_v)\Delta c_1] \tag{6.37}$$

$$Y_2 = B_1 \cdot k_x \cdot [k_v c_x + (1 - k_v)\Delta c_1] \tag{6.38}$$

where B_1 is a factor that determines the measurement sensitivity of the analyte and k_v is degree of the sample dilution caused by addition of diluent and standard (see Fig. 6.20). From the above equations it can be seen that the analytical result is calculated from simple formulas, (6.39) and (6.40), depending on whether the dilution degree k_v is not to be or is to be taken into account:

$$c_x{}' = \frac{k_x}{1 - k_x} \cdot \Delta c_1 \tag{6.39}$$

$$c_x = c_x{}' \cdot \frac{1 - k_v}{k_v} \tag{6.40}$$

The indicative nature of the method, in contrast to the interpolative and extrapolative characters of the other calibration methods discussed so far, is due to the fact that the dilution of the sample with the standard addition is carried out until a single characteristic measurement point is reached, for which $Y_0 = Y_2$. In other words, the prerequisite for obtaining an accurate analytical result is the matching of the model function (calibration graph) to the real function at this one point only. It should be also noted that the form of the calibration graph before and after reaching this point can be practically arbitrary and even unknown to the analyst, and the analytical result can be determined even without knowing the values of any analytical signals. The calibration procedure of the In-SAM method at the approximation and transformation stages is therefore very similar to the well-known and widely used titration procedure, which will be described in more detail in Section 6.2.

It can be proven in the mathematical way applied earlier that when uncontrolled effects are present, the In-SAM method is not robust to either additive interference effect or speciation effect. A linear interference effect is allowed, but only if it reveals itself to the same degree in a sample diluted to the k_v degree and in a sample with the standard addition diluted to the $k_x \cdot k_v$ degree.

Thus, the In-SAM method is similar to the I-SAM method in terms of resistance to interference effects, namely: **it can only lead to accurate analytical results ($c_x = c_0$) if the interference effect is of the linear nature and the interferents exhibit a buffering effect or the effect depends on the interferent to analyte concentrations ratio**. However, the difference is that these conditions apply not to the entire dilution process, but only to the dilution in degree k_x.

The question arises to what extent these rather demanding conditions are met in practice. Of course, the answer can only be given by extensive studies carried out with equal analytical methods. However, some light is shed by the results in Table 6.6 on the determination of alkaline earth metals in the presence of various interferents by the FAAS method [20].

What positively distinguishes the In-SAM method from other SAMs is the greater flexibility of the modeling of the real function as it takes on different forms as the sample is diluted with the standard. This property, achieved by gradually diluting the sample with the standard, can be seen in Figure 6.22 [20]. The possible change in the form of the real form function can be tracked and mapped during dilution over a much wider range of analyte concentrations than in the E-SAM method. This is of great importance when, as a result of a particular interference mechanism (e.g. the formation of difficult calcium and aluminum compounds of different compositions in a sample of increasing dilution), the calibration graph in the E-SAM method is

Table 6.6 Results of the FAAS determination of calcium, magnesium, and strontium in synthetic samples in the presence of different interferents using external standard method (a), E-SAM method (b), and In-SAM method (c) for calibration.

Real concentrations ($\mu g\ ml^{-1}$)		Analyte concentration found, c_x ($\mu g\ ml^{-1}$)			Relative error (%)		
Analyte	Interferent	a	b	c	a	b	c
5 Ca	20 Si	3.23	4.97	4.38	−35.4	−0.6	−12.4
5 Ca	100 Si	2.45	4.92	3.87	−50.9	−1.7	−22.6
0.5 Mg	400 Al	0.36	0.53	0.49	−27.4	+6.4	−2.8
0.5 Mg	1000 Al	0.31	0.52	0.54	−38.8	+4.2	+7.8
5 Ca	5 Al	3.55	3.92	4.79	−28.8	−21.6	−4.1
5 Ca	400 Al	1.67	3.37	4.52	−66.6	−32.5	−9.6
10 Sr	15 Ti	4.70	11.45	7.81	−53.0	+14.5	−21.8
10 Sr	50 Ti	4.15	15.26	6.86	−58.5	+52.6	−31.4

Source: Kościelniak et al. [20], table 2 (p. 1391)/Royal Society of Chemistry.

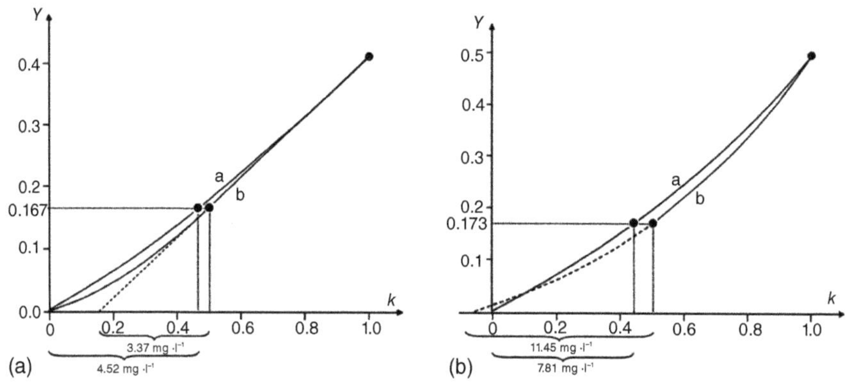

Figure 6.22 Determination of calcium in the presence of aluminum (A) and of strontium in the presence of titanium (B) the In-SAM (a) and E-SAM (b) methods. Source: Kościelniak et al. [20], fig (p. 1391)/with permission of Royal Society of Chemistry.

nonlinear in the extrapolation region, while in the In-SAM method it is linear or nearly linear. A clear improvement in the accuracy of the analytical result, in this case, is shown in Figure 6.22A. A more accurate In-SAM result can also be obtained when the shape of the graph in the E-SAM method is nonlinear and concave (see Figure 6.22B).

The ability of the In-SAM method to accurately determine calcium in the presence of aluminum is particularly important because both elements are present in many types of natural samples. As the results presented in Table 6.7 show, this property of the method can persist even when other interferents (e.g. silicon) are present in the sample showing an effect with a different mechanism. The results obtained with

Table 6.7 Results obtained for FAAS determination of calcium in the iron ore samples (certificated reference materials) of $c_0 = 10\,\mathrm{mg\,l^{-1}}$ using external standard method (a), E-SAM method (b), and In-SAM method (c) for calibration.

Sample code	Main sample components (%)					Calculated analyte amount (%)		
	Ca	Al	Fe	Si	Mg	a	b	c
2.53	0.286	0.37	66.1	1.90	0.17	0.169 ± 0.012	0.223 ± 0.033	0.261 ± 0.020
2.54	2.423	2.97	41.5	6.07	2.21	1.692 ± 0.031	1.782 ± 0.078	2.413 ± 0.026
2.55	0.529	0.51	64.8	2.45	0.89	0.343 ± 0.015	0.423 ± 0.055	0.505 ± 0.019
2.57	0.051	0.71	63.9	1.15	0.02	0.021 ± 0.002	0.017 ± 0.006	0.045 ± 0.005

Source: Kościelniak et al. [20], table 3 (p. 1391)/Royal Society of Chemistry.

this In-SAM method are also characterized by a very good precision, lower than that obtained by the interpolation method, but much higher than that of the determinations made by the extrapolation.

The presented theoretical considerations and performed experiments prove that calibration by the SAM does not have to proceed only by the most common extrapolation route – different versions of the method can be created in various ways at the laboratory stage, in various ways adding standard solutions to the sample and combining this procedure with the dilution process. The above solutions are those found in the literature, but certainly, many more can be developed.

Each of the versions of the SAM presented above has its practical and analytical advantages and limitations. Each gives the possibility of determining the exact concentration of an analyte when uncontrolled effects, especially interference effects, are not revealed in the sample. However, if the analyst is concerned about compensating for interference effects by the SAM, the following guidelines should be followed:

- calibration by the SAM is only able to compensate unconditionally or under certain conditions for proportional interference effects,
- to achieve compensation for this type of interference, each calibration solution subjected to the measurements should contain a sample,
- if the method of adding a standard to a sample with a given concentration of interferents keeps their concentration constant, then a calibration method based on this method allows unconditional, full compensation of interferences,
- if the method of adding the standard to the sample causes a change in the concentration of interferents in one calibration solution in relation to the concentration in another solution, the calibration method based on this method allows interference compensation within a limited range and under strictly defined conditions,
- the possibility of conditional interference compensation is not determined by the nature of the transformation step (interpolative, extrapolative, indicative); nevertheless, unconditional compensation of proportional interference effects can only be achieved by extrapolative transformation.

Therefore, if the analyst, when starting to analyze a given sample, does not have full information (which is most often the case) as to the source and mechanism of the interference effects occurring in this sample, then among all versions of the SAM methods, he should choose the conventional, extrapolation SAM method for calibration, because chemical interferences are most often linearly multiplicative in nature, and this method completely compensates for this type of effect. However, this method should be used with great caution and under strict rules to minimize random and systematic analytical errors resulting from the extrapolation process.

6.3 Titration

Additive calibration methods can include **titration** because it also involves combining a sample with a standard. In contrast to the additive standard addition method, however, the added standard is not the standard of the analyte, but the standard of the substance that reacts chemically with the analyte. Consequently, the measurement and transformation steps of the titration procedure are also quite different than in the SAM.

Titration is one of the oldest methods of quantitative analysis. As early as in 1729 Roku French chemist, C. Geoffroy, in an essay presented to the French Academy and published two years later [21] described an analytical method to determine the strength of vinegar by adding a controlled amount of powdered potassium carbonate to a known amount of vinegar until effervescence ceased. William Lewis (1708–1781), who is also considered one of the early pioneers of titration, for the first time used a color indicator to determine alkali content in American potashes [22]. The first burette has been described by F.-A.-H. Descroizilles in 1794 [23]. In 1824 J.L. Gay-Lussac invented the terms "pipette" and "burette" in paper on the standardization of indigo solutions [24] and then he introduced a verb "titrer," meaning "to determine the concentration of a substance in a given sample"[25]. Several years later K.F. Mohr wrote the first textbook on the titration [26]. Over the years, the method has become increasingly important and is now one of the most recognized and frequently used analytical tools with a wide range of applications.

The importance and significance of titration are best demonstrated by this, it appears as one of the first theoretical and practical analytical topics in all chemistry studies. Indeed, there is no better way to know and feel the essence of chemical analysis than by becoming thoroughly acquainted with the ins and outs of titration and understanding the principles of this analytical process. Titration touches on "real" chemistry, allowing you to get acquainted with different types of chemical reactions and learn such basic concepts as chemical equilibrium, the strength of an acid or a base, a complex compound, precipitation, and coprecipitation, oxidation and reduction, and many others. It also teaches proper analytical procedures based on qualities such as accuracy, precision, meticulousness, purity, and patience. In its modern form, it also allows the student to become familiar with various types of instrumental measurement methods – their principle of operation, construction, and measurement specifics.

Unfortunately, the topic of titration is under no circumstances presented and taught in the context of analytical calibration. However, titration, commonly regarded as a "classical" analytical method, is not only subject to calibration but is in fact an empirical calibration method with its own specificity [27]. The author will try to convince of the validity of viewing titration in this particular way, without going too much into other aspects of this process, which are well developed and described in other books.

The titration process consists of the successive, controlled addition of a known, precisely defined quantity of a substance, the so-called titrant, to the sample without making it up to a constant volume. This is shown in Figure 6.23. As stated earlier, the titrant added to the sample does not contain an analyte, but another chemical compound that is capable of reacting with the analyte present in the sample according to a known and well-defined stoichiometry. If the initial amount of titrant resulting from the reaction stoichiometry is in some excess relative to the amount of analyte in the sample, then as the titration proceeds a state is reached in which the amount of standard added is chemically equivalent to the amount of analyte. This state corresponds to the so-called equivalence point of the titration.

From a calibration point of view, the change in analyte and titrant concentration that occurs as a result of a chemical reaction depending on, for example, the volume of titrant added in the sample is a calibration real function, and the titration equivalent point corresponds to the actual analyte concentration present in the sample. In practice, the course of a chemical reaction is followed with the aid of a suitable measuring instrument, producing a so-called titration curve, which in fact is a kind of calibration graph. and the analytical result, c_x, is determined when the so-called end point (EP) corresponding to the theoretical equivalent point is reached. Examples of titration curves with the most common forms are shown in Figure 6.24.

The shape of the titration curve (model function) depends primarily on the type of chemical reaction taking place and includes the chemical state of the sample before and after the end point is reached. By their very nature, therefore, titration curves are not similar to the simple, linear graphs found in other calibration methods, usually not clearly distinguishable in any of their parts. However, the amount of analyte in the sample is not determined by the form of the entire model function, but only by

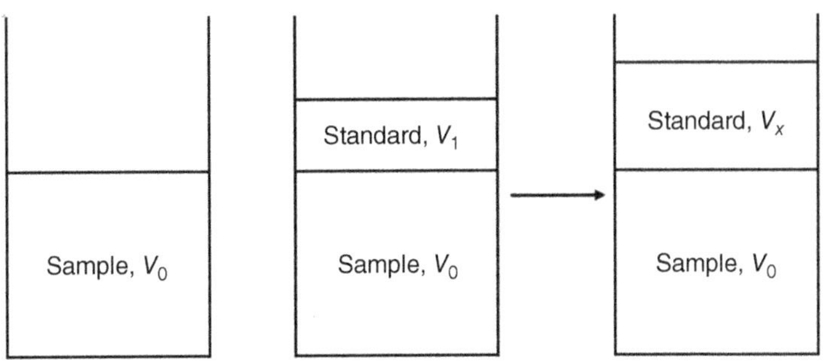

Figure 6.23 Scheme of the basic form of the titration method at the preparative stage of the calibration process.

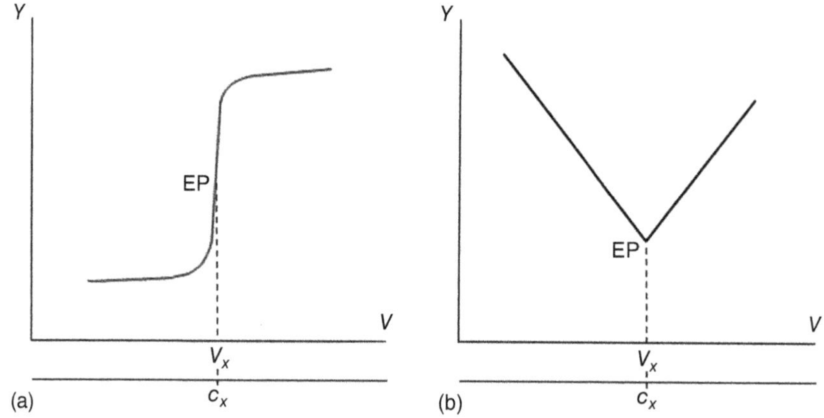

Figure 6.24 Titration curves of sigmoidal (a) and section (b) shapes occurring in the titration method: the analyte concentration, c_x, is determined on the basis of the titrant volume, V_v, corresponding to the end point (EP) of titration.

the position of its characteristic point on the graph, indicating the volume of titrant, V_x, needed to reach that point. This shows that titration has an indicative character (in contrast to the interpolative and extrapolative character of most other calibration methods)

Because the transformation of the analytical signal to the analytical result takes place via the indicative way, the titration can be carried out classically, i.e. without the use of a measuring instrument. Although it is not then possible to construct a titration curve, the end point can be determined visually, the corresponding volume of titrant can be determined and the analytical result calculated.

This result is determined by the following principle. If in a given chemical reaction an equilibrium is reached between an analyte of mole number a and a titrant of mole number b, the value of c_x (mol l^{-1}), is determined from the volume V_x by the formula:

$$c_x = \frac{V_x}{V_0 + V_x} \cdot \frac{b}{a} \cdot c_t \tag{6.41}$$

where c_t is the concentration of the titrant and V_0 is the initial volume of the sample to be measured. Often the volume of titrant added is so small compared to the initial volume of the sample that the volume increase of the solution due to titration is neglected. In many cases, Eq. (6.41) requires one to take into account additional quantities (e.g. dissociation constant, solubility product) related to the specificity of the chemical reaction being used.

Thus, as can be seen, the titration procedure consists of the steps typical of a calibration procedure: preparative, measurement, and transformation. As in any calibration method, accurate determination of the analytical result requires accurate representation of the real function (the course of the chemical reaction) by the model function (the titration curve). Because of the indicative nature of the titration, this mapping should be particularly accurate at the titration end point.

Titration is among the calibration methods that provide the greatest precision in analytical results. This is largely due to the relatively high general knowledge of the theory and technique of titration. which, as mentioned, is given much attention in chemistry curricula. The titrant receives special attention. Many scientific and didactic positions have been written about the proper selection of a chemical compound for the role of titrant, which meets the requirements of the primary standard (i.e. of the highest metrological quality), its preparation, and the so-called setting (i.e. in essence, determining the exact concentration in solution). Similarly, the titration technique itself (selection and calibration of measuring vessels, ensuring their cleanliness, the problem of correct volume measurement of liquids) has been the subject of separate theoretical and practical studies. Following the recommended rules makes it possible to minimize accidental human errors associated with the preparation of the titration process to a very large extent.

Another factor that has a great influence on the very good precision of titrimetric determinations is the fact that this method is relatively simple. Although the chemical reactions leading to the determination of an analyte are often multistep and require from the analyst considerable chemical knowledge and laboratory skills, these processes – after careful preparation of the reagents and equipment – do not cause significant errors due to their very good reproducibility. The titration process usually does not require additional instrumental sample processing steps, hence the sources of accidental measurement errors are few and the errors themselves are relatively small.

The natural properties of any chemical reaction mean that the observed end of the reaction never perfectly coincides with the actual end of the process. For this reason, reactions forming the basis of titrations should be fast and end at the theoretically predicted moment (in particular, they should not be accompanied by other accompanying reactions).

Typical preparative and measurement errors (mainly manifested in the classical version of titration) also include: "droplet error" – consisting in exceeding the end point of the titration as a result of adding the smallest volume of titrant to the unit sample, "run-off error" caused by the phenomenon of adhesion of the solution on the walls of the vessels, or "reading error" consisting in incorrect determination of the volume of the solution due to the parallax effect.

At the transformation stage, the precision and accuracy of the titration method are largely determined by the slope of the titration curve to the volume axis in the vicinity of the end point. The parameter that characterizes the sigmoid curve in this respect is the so-called "titration jump" (TJ) (see Figure 6.25), i.e. the difference in the value of the analytical signal before and after reaching the end point, caused by the addition of a small amount of titrant (e.g. a drop). The smaller the slope of the curve, the greater the titration spike, i.e. the measurement tolerance of the end point with a certain precision increases. The value of this parameter increases with increasing analyte and titrant concentration, and also depends on the values of the equilibrium constants of the chemical reactions. The possibility of changing it under the conditions of a given titration is therefore very limited.

The possibility of fully mechanizing and automating the entire titration process is also of great importance for achieving precise results. Laboratory operations such as measuring out portions of very small volumes of liquid and adding and mixing them

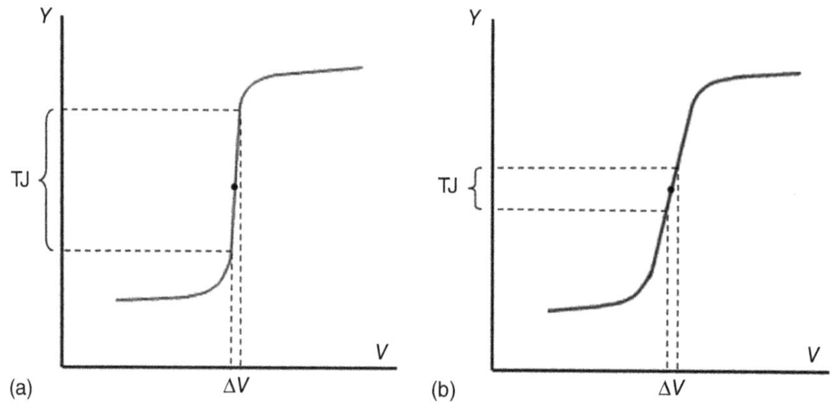

Figure 6.25 Sigmoidal calibration graphs of small (a) and large (b) titration jump (TJ).

with another liquid, or even, for example, gradually reducing the volume of a portion of titrant as the end point is approached, can easily be instrumented and carried out in this form with great precision and accuracy. The instruments for titration, the so-called titrators, are becoming technically more and more perfect and reliable, and they ensure that analyses are carried out not only with small errors but also with high speed, low reagent consumption, and minimal chemical waste production. If we add to this the relatively low cost of this apparatus, it is no wonder that titration is the analytical (and calibration) method, which among other methods, is most frequently used in chemical laboratories in the form of automation.

A critical step in the titration procedure is the determination of the end point and the assignment of the corresponding titrant volume to this end point (see Figure 6.24). This problem is minor when the titration curve has the shape shown in Figure 6.24b, whereas the sigmoidal curve illustrated in Figure 6.24a can present more difficulties due to its rather complicated, nonlinear shape. However, many ways have been developed to graphically and numerically determine the end point in such cases. Figure 6.26 shows the three most commonly used graphical methods, where the end point of the titration is taken to be the inflection point of the sigmoidal curve.

Titrations generally have very good accuracy, although of course, they are not free from many uncontrolled effects. Like random errors, systematic errors can be minimized by careful and strict adherence to the recommended rules of procedure at each stage of the calibration procedure. This is facilitated by the well-tested rules of titration analysis developed during its long development. They are all the more valuable because, despite the passage of years, they remain relevant, since the essence of titration is in the chemical processes and not in the instrumental aspects. Thanks to the possibility of using many different kinds of chemical reactions in titration and, in recent decades, also various types of measurement methods, the number of reliable titration procedures developed are enormous. This allows for a fairly free choice of the appropriate analytical procedure for the determination of a given analyte in the sample under study in terms of obtaining results of the expected accuracy.

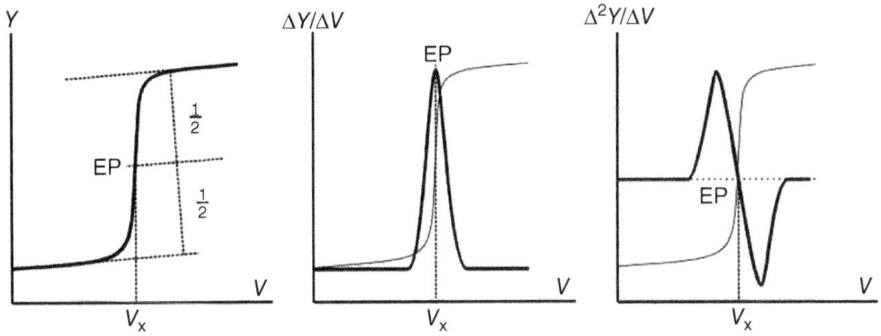

Figure 6.26 Graphical methods for determining the end point (EP) of titration on the sigmoidal titration curve: secant method (a), first derivative method (b), and second derivative method (c).

Apart from preparative effects, which are quite easy to avoid in titration, the most serious source of systematic errors are the effects connected with the chemical reactions taking place. It is assumed that, from the point of view of the accuracy of the analytical result, the reaction underlying the titration should, first of all, proceed quantitatively (stoichiometrically) and with the participation of chemical compounds that are stable under titration conditions. The problem, however, lies in the fact that no reaction in such a complex environment as the analytical sample ever takes place perfectly in accordance with theoretical predictions. Any deviation of the course of the reaction in the sample from that assumed causes a change in the intensity of the signal corresponding to the actual amount of analyte in the sample. These effects are of course sometimes so small that they can be neglected, but in many cases, they can result in a significant decrease in the accuracy of the analyte determination.

The effects associated with a chemical reaction are mainly interference effects of various types and nature. Most often they result from the fact that the selected reaction does not occur only with the selected analyte, but also with other components of the sample. Many examples of this type of interference can be given because they occur regardless of the type of reaction used. Such well-developed and frequently used methods as complexometric titration with ethylenediaminetetraacetic acid (EDTA) (e.g. metals from the first three groups of the periodic table take part in the reaction with magnesium), precipitation Mohr method (e.g. carbonates and phosphates interfere in the determination of chlorides) or redoxometric titration with potassium permanganate (e.g. iron ions interfere in the determination of vanadium) are not free from such interferences.

The influence of substances accompanying the analyte in the sample can also be caused by their activities, such as a shift in the equilibrium of the reaction (e.g. due to the common ion effect), a change in the concentration product of the precipitate solubility (e.g. due to the salt effect), or a change in the reaction rate (e.g. molybdate ions interfere in this way in the iodometric determination of hydrogen peroxide).

In titration, the typical speciation effect does not occur because, in principle, the analyte standard is not used in the process. However, this effect can manifest itself in other ways than through a difference in the chemical form of the analyte in the sample and in the standard. Namely, the analyte may be present in the sample at least partly in a different chemical form than the predicted form and in this different form not enter the selected chemical reaction. It is obvious that in such a situation only a part of the analyte will be determined in the sample, i.e. a systematic error will be made.

It should be remembered that each titration analysis should be approached individually, with full knowledge of its theoretical basis, and in particular with very careful recognition of the physicochemical properties of the analytical system under study and the resulting specificity of the chemical reaction. The equilibrium state of a chemical reaction in any titration may depend not only on the type of this reaction, but also on the type of analyte, titrant, reaction environment, temperature, etc. All these factors affect the form of the real function and, in particular, value of the function that corresponds to the equivalent point. Proceeding routinely, without recognizing the theoretical ins and outs of titration, can result in this point being sought on the titration curve not always where it actually is, which can lead to an inaccurate determination of the analyte. Figure 6.27 shows sigmoid titration curves made under similar experimental conditions from the same type of reaction, characterized by typical and untypical positions of the equivalent point, using selected examples.

In titration analysis, a whole set of approaches has been developed to either increase the applicability of the titration or to minimize errors. Among the basic ones is the use – besides the technique of direct determination of the analyte by means of added titrant (or vice versa) – of various other titration techniques.

Thus, the indirect technique consists of selecting a third substance that reacts stoichiometrically with the analyte and forms a new compound with it, which is then titrated with the titrant. This technique is used when a commonly used and proven in other titrations titrant does not react directly with the chosen analyte in such a way as to provide results of sufficient precision and accuracy. If the titrant reacts slowly with the analyte or an excess of titrant is required to reach the end point, the reverse titration is used. A known amount of titrant in excess of the analyte is then added to the sample, and the amount of titrant remaining after the reaction is titrated with another standard solution.

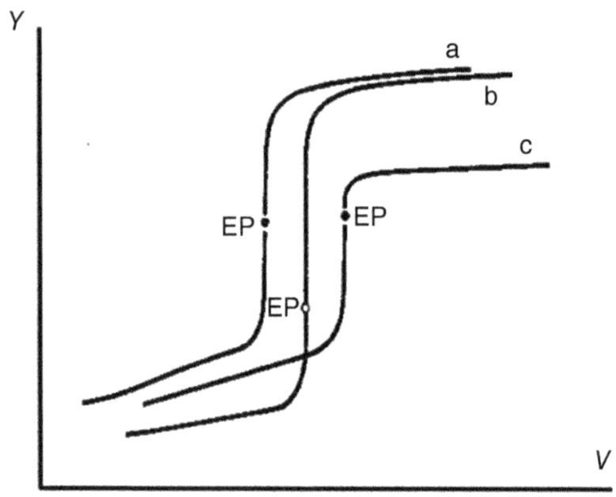

Figure 6.27 Curves obtained during titration of Fe^{2+} with Ce^{4+} (a), As^{+3} with Ce^{+4} (b), and Fe^{+2} with $Cr_2O_7^{2-}$ (c) in similar chemical condition; EP – equivalent point. Source: Minczewski and Marczenko [28].

Other unconventional techniques are more specific to a particular type of chemical reaction. For example, when the complex compound formed in the reaction of the titrant with the analyte is not sufficiently stable, the so-called basal titration is used. This involves adding another element to the sample of the complex compound, chosen so that this element can be quantitatively displaced from the complex by the analyte and titrated in place of the analyte. In the so-called dead-end titration used in amperometric analysis, the electrochemical system is built so that current stops flowing through the system at the equivalent point. This technique helps in increasing the precision of analytical results.

One of the major problems of acid-base titration is the large errors made when weak acids and bases are determined in aqueous solutions. It is also usually not possible to titrate organic or inorganic compounds insoluble or sparingly soluble in water directly under these conditions. In such cases, titration in a nonaqueous medium is used, which is chosen so that the acid or base being determined is sufficiently strong in it.

Another very simple way to improve the precision and accuracy of results in titration analysis is to titrate with an auxiliary solution, prepared separately, and brought to a state that best matches the end point of the titration currently being performed. The titration analysis of the sample is then performed until the same state (e.g. same color) as the witness is reached.

However, titration end point indicators have the greatest impact on increasing the analytical capability of the titration. These are chemical compounds (usually synthetic dyes) that exhibit a change in their physicochemical properties, usually resulting in a change in the color of the solution, under conditions defined by the end point of a given titration. Indicators play a key role in classical titration, where they enable the visual determination of an end point in all those cases where the titrated solutions are not colored or where the color change is not sufficiently pronounced. Despite the instrumental development of titration analysis, its importance remains very high if only because the most widely used apparatus for end point detection is the visible light absorption spectrophotometer (VIS).

For example, in the precipitation titration, the color change of an indicator can be related to a specific change in adsorption of a dye on the surface of the precipitate being formed, and in the complexometric titration – to the formation of a specific colored complex with the analyte. The most widely used are indicators of hydrogen ion concentration (pH) in the sample solution, used mainly in alkacimetric titration. Examples of such indicators, which change their color in different pH ranges, are shown in Table 6.8.

An example of instrumental support of a titration technique to improve its utility and analytical capability is a procedure called **Tracer Monitored Titration (TMT)** first presented in 2006 [29]. Its principle is that the dilution process of a sample during titration is tracked by means of an additional substance, the so-called tracer, which by no means enters into the chemical reactions underlying the titration in question. During the titration, the intensities of the signals from the analyte (or titrant) and the tracer are measured in parallel, and the degree of dilution of the sample when the end point of the titration is reached is calculated from the ratio of

Table 6.8 A list of common laboratory pH indicators.

Indicator	pH range	Indicator	pH range
Thymol blue	1.2 (red)–2.8 (yellow)	Chlorophenol red	5.2 (yellow)–6.8 (red)
Bromophenol blue	3.0 (yellow)–4.6 (blue)	Bromothymol blue	6.0 (yellow)–7.6 (blue)
Congo red	3.0 (blue)–5.0 (red)	Phenol red	6.8 (yellow)–8.2 (red)
Methyl orange	3.0 (red)–6.3 (yellow)	Naphtholphthalein	7.3 (colorless)–8.7 (blue)
Alizarin red S	4.0 (red)–5.6 (yellow)	Phenolphthalein	8.0 (colorless)–10 (pink)
Dichloro-fluorescein	4.0 (colorless)–6.6 (green)	Cresolphthalein	8.2 (colorless)–9.8 (red)
Methyl red	4.2 (pink)–6.2 (yellow)	Thymolthalein	8.8 (colorless)–10.5 (blue)
Bromocresol purple	5.2 (yellow)–6.6 (purple)	Indigo carmine	11.6 (blue)–14.0 (yellow)

the signal obtained for the tracer at that point to the signal measured for it before the dilution.

The TMT procedure relieves the analyst of the need to measure the volume of titrant needed to reach the end of the titration. Its usefulness in terms of its ability to improve the precision and accuracy of analytical results has not yet been fully demonstrated. It seems that in certain specific cases it can contribute to a reduction of random and systematic errors in titrations, but it is certainly not universal in this respect. The basic condition for the positive action of a tracer is a linear change in the analytical signal within the limits encompassing its initial and final concentrations. The compound should therefore not only be chemically inert with respect to the analyte and the titrant, but also not subject to interference effects and not to be an interferent in the sample environment.

Regardless of various procedural modifications and instrumental improvements, titration thus has all the characteristics of an empirical calibration method. Although there is a specific nomenclature associated with the process, which has been formed over the years, this nomenclature, as shown, essentially overlaps with that used in analytical calibration. Titration has traditionally been treated as some type of analytical method associated with a specific measurement method (e.g. spectrophotometric titration method). However, this is not justified because it is only a way of arriving at an analytical result, which may accompany a given group of analytical methods (e.g. spectrophotometric methods) alternatively to other ways of this type (most frequently to the external standard method) recommended for use in this group.

6.4 Isotope Dilution Method

All calibration methods of quantitative analysis described so far have concerned the determination of elements or compounds without going into their isotopic composition. The differences between the properties of the isotopes of an analyte in the sample are so small that in the usual chemical procedures practically no changes in the isotope abundance are produced. Furthermore, the usual chemical measurements do not differentiate between a compound having a normal isotope distribution among its elements and this one in which one or more of elements have an "abnormal" isotope distribution.

The ability of measuring signals from individual isotopes emerged with the development first of radiometric methods and then of mass spectrometry. Both of them have naturally inspired the creation of a specific calibration approach called **the isotopic dilution analysis (IDA)**. In general, **the IDA method consists in changing the natural isotopic composition of an analyte in a sample by labeling the sample, i.e. by addition of another isotope or other isotopic composition to the sample**. Spike dilutes the sample and hence the name of the method. Depending on whether the added standard contains radioactive or stable isotopes, their amount in the sample is determined by measuring the radioactivity or number of ion counts, respectively.

The precursor to isotope analysis was György Hevesy, Hungarian radiochemist, codiscoverer of hafnium, and Nobel Prize laureate (1943). In 1932 Hevesy and Hobbie were the first to report the use of a radioactive isotope to solve an analytical problem [30] and then, in 1934, Hevesy and Hofer introduced the IDA procedure by determining water in the human body using deuterium-enriched water (heavy water) [31]. The name "isotope dilution" was used for the first time by D. Rittenberg and G. L. Foster [32] in application to the determination of amino acids and fatty acids.

After II world war research on the nuclear energy for the civil purposes required enriched stable isotopes. In 1946 the U.S. Atomic Energy Commission made available separated stable isotopes of many elements so that the IDA became a practical tool for chemical analysis. As adapted to mass spectrometry IDA was initially developed in 1950 by J.H. Reynolds [33] and applied to elemental analysis using thermal ionization mass spectrometry. In 1991 K. Okamoto for the first time combined the IDA procedure was combined with inductively coupled plasma mass spectrometry [34]. Nowadays, it enjoys steady recognition under the name "isotope dilution mass spectrometry" (IDMS) and it is widely used particularly with the use of chromatography, capillary electrophoresis, and ICP, all coupled with mass spectrometry.

It should be noted that IDA, like titration, is not widely recognized as a calibration method. Nevertheless, it should be treated as such because it contains all the attributes that support it: both radiometric and spectrometric procedures require the use of a chemical standard, the making of measurements, and the transformation of the measured results into an analytical result. Therefore, its placement among other calibration methods is completely justified. Despite its name, it should be classified as an additive calibration method because a key part of its preparative procedure is the addition of isotopic standards to the sample.

IDA is not a universal method as it requires the use of special measurement methods for the detection of analyte isotopes. With mass spectrometry becoming more widely used as a detection system combined with a variety of analytical methods, its use in quantitative analysis is becoming increasingly versatile. In addition, IDA is worthy of general interest because of its specificity. Its special features arise primarily from the unique opportunity to use in a standard an analyte with the same chemical properties as the native analyte in a sample while being able to differentiate the two forms using separate analytical signals. As a result, the different IDA variants are characterized by their own ways of mapping the real function and transforming the measurement signals that are not found in "traditional" calibration methods.

6.4.1 Radiometric Isotope Dilution

The IDA method in its radiometric (classical) version is based on the linear dependence of the radioactive activity (analytical signal Y) measured for a specific material containing a radioactive isotope of mass m:

$$Y = a \cdot m \tag{6.42}$$

where a is the specific activity. Because of this type of relationship, in methods of quantitative isotopic analysis, the amount of an analyte is expressed by its mass and not by its concentration. For the same reason, dilution of a solution containing a given mass of analyte does not reduce the analytical signal and does not require a corresponding correction to be made to the calculation of the analytical result.

Several useful versions of radiometric IDA procedures have been developed. In case of **the simple isotopic dilution method** (Simple-IDA), a standard in the form of a well-defined amount, m, of a radioactive isotope of that analyte with a known specific activity, a_0, is added to a sample with an unknown amount of the analyte, m_0. This is shown schematically in Figure 6.28.

The activity of the isotope in the standard before and after adding it to the sample is the same, so:

$$a_0 \cdot m_\Delta = a_x \cdot (m_x + m_\Delta) \tag{6.43}$$

where a_x is the specific activity of the sample with added standard. Thus,

$$m_x = \left(\frac{a_0}{a_x} - 1\right) \cdot m_\Delta \tag{6.44}$$

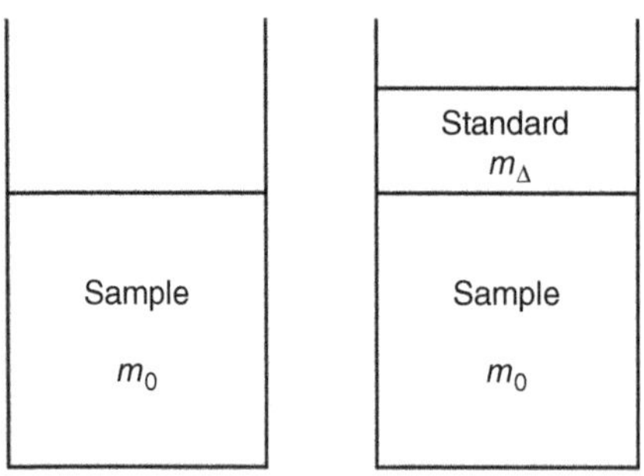

Figure 6.28 Scheme of the Simple-IDA method at the preparative calibration stage before the separation process.

If the mass of the radioactive isotope is much less than the mass of the natural isotope of the analyte in the sample, then Eq. (6.3) can be simplified to:

$$m_x = \frac{Y_x}{a_x} \tag{6.45}$$

where Y_x is the signal measured for the standard-subsidized sample. To account for the unknown value of specific activity, a_x, in formula (6.45), any amount of analyte, m_1, is separated from the isotope-subsidized sample and this amount is determined using a "traditional" measurement method (e.g. spectrophotometric). This takes advantage of the fact that the specific activity of the separated sample fraction is the same as the specific activity of the sample before separation. The analytical result can therefore be calculated from the formula:

$$m_x = \frac{Y_x}{Y_1} \cdot m_1 \tag{6.46}$$

where Y_1 is the measured activity of the separated sample fraction.

A graphical description of this procedure is shown in Figure 6.29. From a calibration point of view, it is interesting to note that the calibration graph is extrapolated to the signal obtained for the dotted sample, and the analytical result is obtained by interpolating this signal to the extrapolated part of the graph. Thus, it can be said that the transformation step of the method is extrapolation–interpolation.

The graph shows that the error of the analytical result depends mainly on the precision and accuracy of the determination of the analyte in the separated portion of the sample. Thus, it is very important to choose a suitable method for this purpose. The mass of the separated sample portion should be relatively large compared to the original sample mass, which may create practical difficulties. On the other hand, the method does not require quantitative separation of the analyte and thus can be used for determination of the analyte in samples containing components with similar properties to the analyte (i.e. posing a problem in quantitative separation).

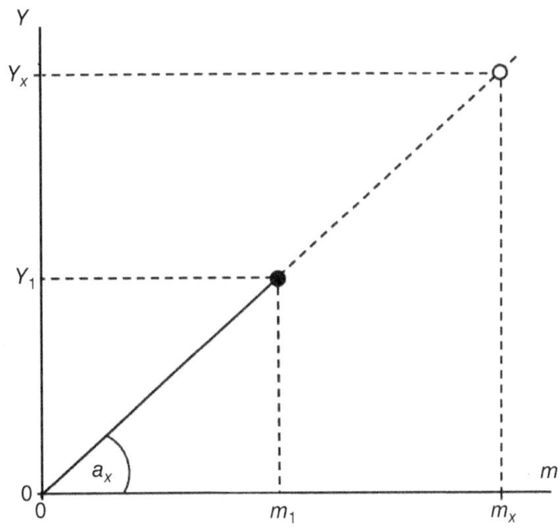

Figure 6.29 The principle of the simple isotope dilution method: a_x – specific activity of a radiotracer after addition to the sample, Y_x, Y_1 – signals obtained for the sample spiked with tracer and before and after separation of the analyte in mass m_1, m_x – unknown mass of the analyte in the sample (analytical result).

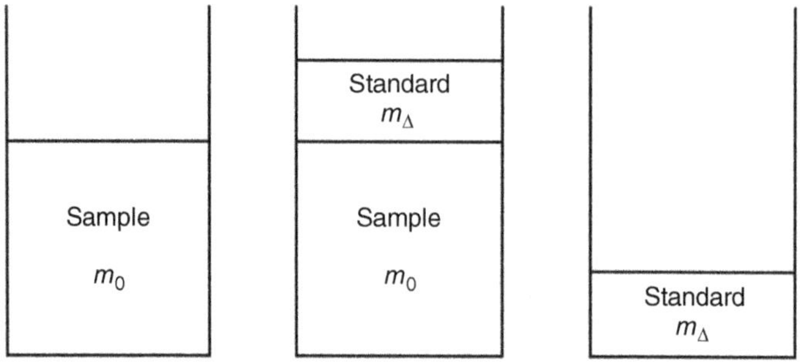

Figure 6.30 Scheme of the Subst-IDA method at the preparative calibration stage before the separation process.

The **sub-stoichiometric isotopic dilution procedure** (Subst-IDA) requires an additional step to the Somple-IDA procedure. It involves preparing the radioactive isotope of the analyte itself (Figure 6.30), isolating a specific portion of it with mass m_2, and measuring its Y_2 activity.

It can then be assumed that:

$$\frac{Y_x}{m_\Delta} = \frac{Y_2}{m_2} \tag{6.47}$$

From Eqs. (6.44) and (6.47) the relation follows:

$$m_x = \left(\frac{Y_2 \cdot m_1}{Y_1 \cdot m_2} - 1\right) \cdot m_\Delta \tag{6.48}$$

If the mass of the separated fractions of the sample with the isotope, m_1, is equal to the mass of the fraction of the isotope alone, m_2, then Eq. (6.48) simplifies to the form:

$$m_x = \left(\frac{Y_2}{Y_1} - 1\right) \cdot m_\Delta \tag{6.49}$$

The separation of equal amounts of the substance to be determined from the isotope solution and the dotted sample is carried out by adding equal, sub-stoichiometric amounts of reagent (e.g. precipitant, extractant) and separating the reaction product from the excess reagent. Hence the name of this variant of the method.

Formula (6.49) allows the determination of an analyte in a sample without the need to know the specific activities of the radioactive isotope before and after its addition to the sample and without the need to measure the Y_x signal for the sample solution with isotope addition. Furthermore, it avoids the exact determination of the mass, $m_1 = m_2$, of the separated portions of the sample with isotope addition and of the isotope itself. This is particularly important when the mass of the sample analyzed is very small or the analyte content of the sample is traced.

A graphical interpretation of the calibration formula (6.50) is shown in Figure 6.31. As can be seen, in this case, the analytical result is obtained by interpolation.

The Subst-IDA method generally has a very good precision of determination, especially when the analyte is present in the sample in a relatively high concentration.

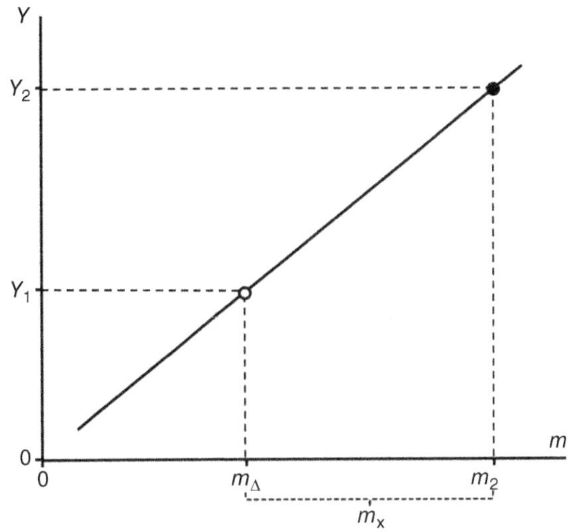

Figure 6.31 The principle of the substoichiometric isotope dilution method: Y_1, Y_2 – signals obtained for the sample spiked with tracer and for the tracer, respectively, both after separation in the same amounts, m_Δ – mass of isotope added to the sample, m_x – unknown mass of the analyte in the sample (analytical result).

The precision of the determination is also usually very good provided that the masses m_1 and m_2 of the separated sample fractions are identical, i.e. the efficiency of the process of isolation of the analyte from the sample with the addition of the radioactive isotope and of the isotope itself is exactly the same. This requirement is not always fulfilled in practice, especially when the sample and standard have very different chemical compositions.

In the case of a sample with a rich matrix composition, one has to expect – in both the Simple-IDA and Subst-IDA methods – the possibility of an interference effect of foreign sample components in the separation process. If this type of effect is not present, and the separation procedure is simple, the reliability of both methods is well represented by the results shown in Table 6.9.

Table 6.9 Results of the determination of silver in the synthetic samples by the Simple IDA and Subst-IDA methods based on the precipitation reaction with chloride ions.

	Results obtained			
	Simple-IDA		**Subst-IDA**	
Result expected c_0 (mg)	**c_x (mg)**	**Error (%)**	**c_x (mg)**	**Error (%)**
110.5	111.6 ± 1.2	+1.0	111.3 ± 1.1	+0.7
55.2	—	—	54.8 ± 0.9	−0.7
33.1	—	—	32.4 ± 0.3	−2.1
11.1	11.4 ± 0.2	+2.7	11.1 ± 0.8	0.0
5.6	—	—	5.9 ± 0.8	+5.3

Source: Ikeda and Noguchi [35], table 2 (p. 110)/Springer Nature.

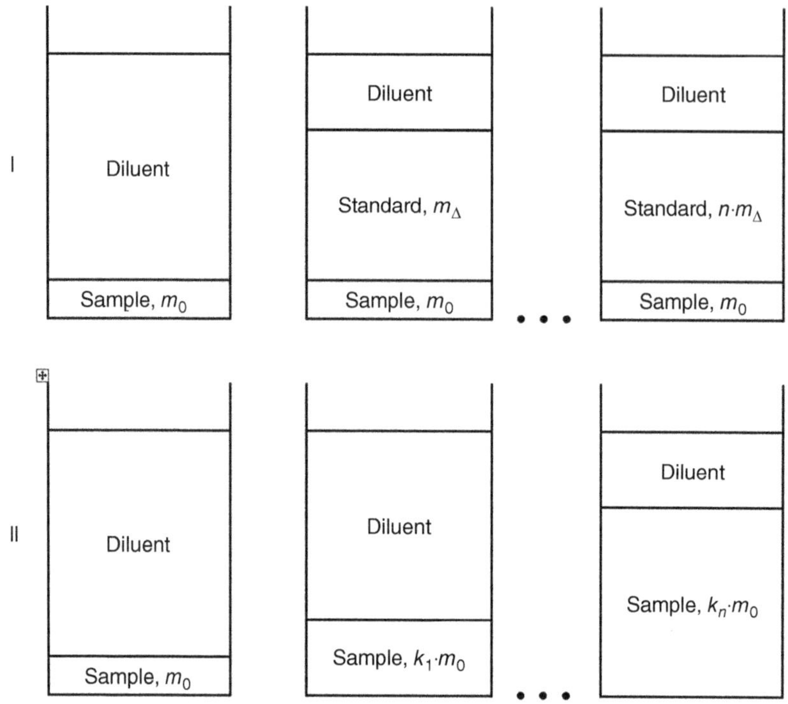

Figure 6.32 Scheme of the SSE-IDA method at the preparative calibration stage before the separation process.

The need for identical analyte separation efficiencies from sample and standard is avoided in **the sub-superequivalence isotope dilution method (SSE-IDA).** The procedure is relatively complex as it involves preparing two series of $n + 1$ calibration solutions (see Figure 6.32). In the first series (I), all solutions contain the same amount, m_x, of the analyte labeled with a constant amount of radioactive isotope and then they are spiked with regularly increasing amount $i{\cdot}m_\Delta$ ($i = 0, 1, \ldots, n$) of the stable analyte so that the specific activities are changed in accordance to $Y_0/(m_x + i{\cdot}m_\Delta)$. In the second series (II) each solution contains k_i times the amount m_x of the analyte and radioisotope used in the first series and no carrier is added, so their specific activity is the same and equal to Y_0/m_0. Finally, all solutions, I and II, are brought to the same volume by addition of the solvent.

Among the calibration solutions I and II shown in Figure 6.32 one can find a hypothetical jth pair of solutions in which the analyte concentration is equal, i.e.:

$$m_x + j \cdot m_\Delta = k_j \cdot m_x \tag{6.50}$$

Hence, the analytical result can theoretically be determined from the formula:

$$m_x = \frac{j \cdot m_\Delta}{k_j - 1} \tag{6.51}$$

How to determine this result in practice? This is the next step of the calibration procedure. Namely, to all solutions, I and II equal substoichiometric amount of a reagent is added. After completion of the reaction with analyte, the products of amount $m_{x+i\Delta}'$ and m_{ix}' are isolated from solutions I and II, respectively, and the corresponding activities Y_{x+iy} and Y_{ix} are measured. As the total activity of the analyte is not changed due to the separation, the specific activity in both series is defined by:

$$\frac{Y_0}{m_x + i \cdot m_\Delta} = \frac{Y_{x+i\Delta}}{m_{x+i\Delta}'} \tag{6.52}$$

$$\frac{Y_0}{m_x} = \frac{Y_{ix}}{m_{ix}'} \tag{6.53}$$

Hence:

$$\frac{Y_{ix}}{Y_{x+i\Delta}} = \frac{i \cdot m_\Delta}{m_x} \cdot \frac{m_{ix}'}{m_{x+i\Delta}'} + \frac{m_{ix}'}{m_{x+i\Delta}'} \tag{6.54}$$

If the analyte concentrations in two solutions are the same (see Eq. (6.50)), the degree of reaction and separation should be the same from thermodynamic point of view. So, the amounts m_{jx}' and $m_{x+j\Delta}'$ isolated from jth pair of solutions should be the same, $m_{jx}' = m_{x+j\Delta}'$, and Eq. (6.54) takes the form:

$$\frac{Y_{jx}}{Y_{x+j\Delta}} = \frac{j \cdot m_\Delta}{m_x} + 1 \tag{6.55}$$

Thus, it follows from Eqs. (6.10) and (6.14) that

$$\frac{Y_{jx}}{Y_{x+j\Delta}} = k_j \tag{6.56}$$

The analytical result, m_x, can be then obtained graphically as shown in Figure 6.33, i.e. by plotting Y_{ix}/Y_{x+iy} against $i{\cdot}m_\Delta$, finding the abscissa value $j{\cdot}m_\Delta$ corresponding to $Y_{ix}/Y_{x+i\Delta} = k_j$ and finally applying Eq. (6.51).

A good example of the usefulness of the SSE-IDA method is the determination of a trace amount of Sb^{3+} in the presence of arsenic ions by means of a redox reaction to Sb^{5+} with potassium dichromate [36]. As seen in Figure 6.34, the amount of Sb^{5+} separated increases nonlinearly with increasing concentration of Sb^{3+} at constant concentrations of As^{3+}. This confirms that As^{3+} interferes by competing with Sb^{3+} for oxidation with $K_2Cr_2O_7$.

Under conditions presented in Figure 6.34 the Subst-IDA method fails while SSE-IDA gives results with very good accuracy, as evidenced by the analytical results obtained [34] and shown in Table 6.10. Even more accurate results were obtained using the modified SSE-IDA method (SSE-RA), in which the sample in series I of the calibration solutions (see Figure 6.32) is not a natural but a synthetic sample with a known, strictly determined amount of analyte [37].

Thus, as proved, the classic, radioisotope IDA method can be used in equal variants adapted to specific current circumstances and analytical expectations (precision, accuracy, limit of detection). The most important problem connected with this type of analysis is the question of safety of work with radioactive isotopes. A general limitation of IDA procedures is the availability of a suitable tracer to act as a standard. Most important is its purity, half-life, and type of radiation emitted. The half-life must be long enough that sufficient activity is available during analysis for good

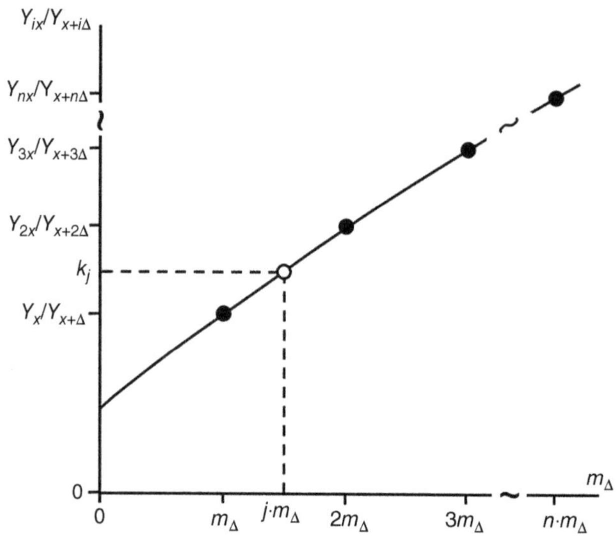

Figure 6.33 The principle of the SSE-IDA method: $Y_{x+i\Delta}$, Y_{ix} – signals obtained for the reaction products isolated from solutions I and II (see Figure 6.28), respectively, m_Δ – amount of the analyte added to solutions I, k_j – multiple of the initial analyte concentration in jth solution of series II equal to total analyte concentration in jth solution of series I.

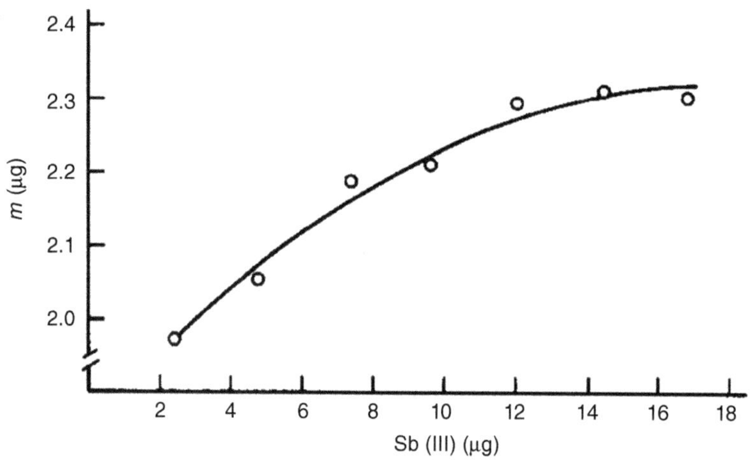

Figure 6.34 Influence of As^{3+} ions on Sb^{5+} ions (m) separated from the sample after oxidation of increasing concentrations of Sb^{3+} ions. Source: Adapted from Ikeda and Noguchi [35].

counting statistics. However, half-lives that are too long can be problematic due to low specific activity and storage and disposal issues. The type of radiation is important primarily with respect to ease of measurement.

A more substantive weakness of the method is that to determine the specific activity of the separated fraction of the sample, the amount of analyte must be determined

Table 6.10 Results of the determination of Sb^{3+} ions in a synthetic sample using the Subst-IDA, SSE-IDA, and SSE-RA methods under the same conditions as related to Figure 6.30.

Result expected	Subst-IDA			SSE-IDA		SSE-RA	
c_0 (μg)	c_x (μg)	Error (%)	k_j	c_x (μg)	Error (%)	c_x (μg)	Error (%)
2.41	2.82	+17.0	2	2.32	−3.7	2.41	0.0
			3	2.40	−0.4	2.38	−1.2
			4	2.47	+2.5	2.41	0.0
			5	2.44	+1.2	2.41	0.0
			6	2.47	+2.5	2.42	+0.4

Source: Adapted from [36, 37].

by a second measurement technique (weighing, titration, instrumental). This limits the sensitivity of the method to the sensitivity of that technique. However, with the current state of detection techniques, this problem is of decreasing importance.

6.4.2 Isotope Dilution Mass Spectrometry

As already mentioned, in the last time the analytical interest is more widely directed to the **isotope dilution mass spectrometry** (IDMS) method than to the classical IDA method. IDMS is realized using stable or almost stable isotopes, listed in Table 6.11, and mass spectrometry as the measurement method. Since it requires at least two isotopes of the analyte (one in the sample, and another one in the standard) it cannot be used for the determination of those elements that naturally occur as single stable isotopes. This is most serious natural limitation of IDMS. The positive feature of IDMS, on the other hand, is that in contrast to radiochemical IDA a single measurement method can be used to measure the signal intensities of the stable isotopes in the sample and in the standard. Of course, the ability to avoid the dangers of working with radioactive isotopes has to be taken into account.

The calibration procedure of the basic IDMS variant includes the following preparative stage. The sample containing the analyte in the form of isotopes A and B in unknown concentration c_0 and with percentage amounts (abundances) of A_x and B_x is spiked with the standard containing the analyte in known concentration c_Δ in the form of the same isotopes A and B, but with different abundances of A_Δ and B_Δ. This stage of the calibration procedure is schematically shown in Figure 6.35. The number of solutions (liquid or solid) containing the sample with the standard added at different weights is arbitrary.

The isotope amount ratio, $R_{x\Delta}$, in the sample spiked with the standard is then given by the formula:

$$R_{x\Delta} = \frac{A_\Delta \cdot n_\Delta + A_x \cdot n_x}{B_\Delta \cdot n_\Delta + B_x \cdot n_x} \tag{6.57}$$

Table 6.11 Stable or very long-lived isotopes existing in nature.

Number of isotopes	Element
1	Be, F, Na, Al, P, Sc, Mn, Co, As, Y, Nb, Rh, I, Cs, Pr, Tb, Ho, Tm, Au, Bi, Th
2	H, He, Li, B, C, N, Cl, V, Cu, Ga, Br, Rb, Ag, In, Sb, La, Eu, Lu, Ta, Re, Ir, Tl
3	O, Ne, Mg, Si, Ar, K, U
4	S, Cr, Fe, Sr, Ce, Pb
5	Ti, Ni, Zn, Ge, Zr, W
6	Ca, Se, Kr, Pd, Er, Hf, Pt
7	Mo, Ru, Ba, Nd, Sm, Gd, Dy, Yb, Os, Hg
8	Cd, Te
9	Xe
10	Sn

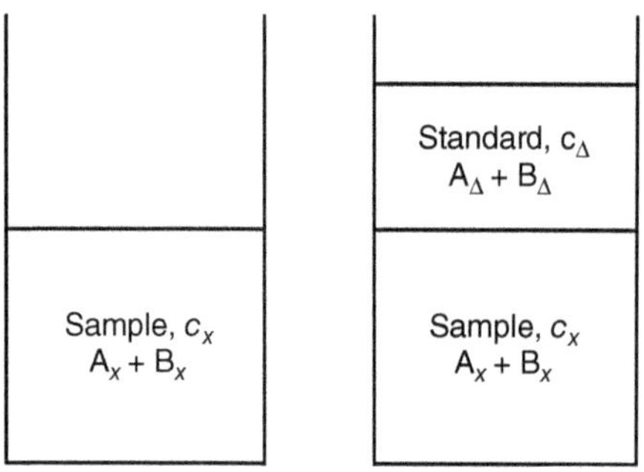

Figure 6.35 Scheme of the IDMS method at the preparative calibration stage before the separation process (for details see text).

where n_x and n_Δ are the number of moles of the analyte in the sample and the standard, respectively. From Eq. (6.1) follows that:

$$c_x = c_\Delta \cdot R_m \cdot \frac{A_\Delta - R_{x\Delta} \cdot B_\Delta}{R_{x\Delta} \cdot B_x - A_x} \tag{6.58}$$

where c_x and c_Δ are the mass concentrations of the analyte in the sample and standard, respectively, and R_m is the ratio of the sample and standard masses. Equation (6.58) can also be expressed as:

$$c_x = c_\Delta \cdot R_m \cdot \frac{R_\Delta - R_{x\Delta} \cdot \sum_{i=1}^{n} R_{xi}}{R_{x\Delta} - R_x \cdot \sum_{i=1}^{n} R_{\Delta i}} \tag{6.59}$$

where $R_x = A_x/B_x$, and $R_\Delta = A_\Delta/B_\Delta$, while $\sum_{i=1}^{n} R_{xi}$ and $\sum_{i=1}^{n} R_{\Delta i}$ are the sums of the amount ratios of every analyte isotope present in the sample or in the standard to B_x or B_y, respectively.

In the case of determination of inorganic components the isotopically enriched analogs are used. Such a standard is not an array of individual labeled molecules, but rather a mixture of individual isotopes that contribute directly to the enrichment of the array, but in different proportions to the sample. For instance, for the determination of analytes containing mercury atoms in the molecule, for which the "natural" primary isotope is ^{200}Hg, the standards enriched in other, remaining mercury isotopes are used. An example of the isotopic composition of a mercury standard is shown in Table 6.12.

The determination of organic compounds takes advantage of the fact that the main elements present in this type of compounds, i.e. hydrogen, carbon, and nitrogen, exist in the form of two stable isotopes. As standards, isotopically labeled analogs are used, which contain isotopes ^{2}H, ^{13}C, or ^{15}N. Thus, for the determination of organic compounds by the IDMS method simplified form of Eq. (6.59) is used:

$$c_x = c_y \cdot R_{\mathrm{m}} \cdot \frac{R_\Delta - R_{x\Delta} \cdot (R_x + 1)}{R_{x\Delta} - R_x \cdot (R_\Delta + 1)} \tag{6.60}$$

In special cases, the model (6.60) can be greatly simplified. If the amount of isotope B in the sample is very small, then:

$$c_x = c_\Delta \cdot R_{\mathrm{m}} \cdot \frac{R_\Delta - R_{x\Delta}}{R_\Delta + 1} \tag{6.61}$$

and if, in addition, the amount of isotope A in the standard is very small, the formula is obtained:

$$c_x = c_\Delta \cdot R_{\mathrm{m}} \cdot R_{x\Delta} \tag{6.62}$$

Equation (6.59) is the basic evaluation model function in IDMS calibration. Despite its rather complicated mathematical form, its practical use is very simple. Values m_Δ, R_Δ and $R_{\Delta i}$ are usually given in the certificate of the standard used, and

Table 6.12 Calculation results needed to obtain the analytical result by the IDMS method on the example of the mercury determination.

	Isotopic abundance (%)			
Isotope	**Sample**	**Standard**	R_{xi}	$R_{\Delta i}$
^{196}Hg	0.15	0.11	0.009	0.002
^{198}Hg	9.97	2.93	0.591	0.044
^{199}Hg (B)	$B_x = 16.87$	$B_\Delta = 65.98$	1.000	1.000
^{200}Hg (A)	$A_x = 23.10$	$A_\Delta = 18.15$	$R_x = 1.369$	$R_\Delta = 0.275$
^{201}Hg	1318	3.96	0.781	0.060
^{202}Hg	29.86	7.43	1.770	0.113
^{204}Hg	6.87	1.44	0.407	0.021
Sum	100.00	100.00	$\sum_{i=1}^{7} R_{xi} = 5.927$	$\sum_{i=1}^{7} R_{\Delta i} = 1.515$

the isotopic ratios R_x and R_{xi} – as invariant values, characteristic of a given isotope of a given analyte – are available in the information provided by IUPAC. From the values of R_x and R_{xi}, the sums of $\sum_{i=1}^{n} R_{xi}$ and $\sum_{i=1}^{n} R_{\Delta i}$ are calculated. This calculation step of the calibration procedure is partially shown in Table 6.5 using the example of the mercury standard.

The values that must be determined experimentally are R_m and R_x. The mass ratio R_m is determined gravimetrically, and the isotopic ratio R_x of the selected isotope pair A and B in the sample and standard mixture is determined by measurement with a mass spectrometer. As a result, on the basis of the calibration model (6.59), it can be said that the IDMS method has an indicative character: under certain experimental conditions (for a certain mass ratio R_m and a certain standard composition), the ratio of signals obtained for selected isotopes present in the sample with the standard, R_x, unambiguously indicates the analytical result c_m.

However, the IDMS method is characterized by many specific analytical effects that are potential sources of random and systematic errors. One of the most important problems is the achievement of isotopic equilibrium in the sample and standard mixture after the weighing step. The goal is to achieve a state in which the isotopic ratio R_x is stable over time, for then in each portion of the mixture taken for analysis this value is the same.

At the preparative stage, care must also be taken to select an isotopically enriched analog of appropriate isotopic composition. It has been shown that the most favorable conditions are provided when the ratio of the mass concentrations of the selected isotopes A and B after addition of the standard to the sample is approximately equal to 1. To achieve such a status, different experimental approaches under the common name "the signal matching techniques" are used. For inorganic IDMS the error propagation plots can be also applied to calculate the optimum analyte-spike isotope amount ratio [38]. However, it is sometimes very difficult or even infeasible to achieve optimal conditions, especially when the analyte is present in the sample in very high concentration, which would require unacceptably high amounts of the standard. The possibility of moving away from the optimum isotopic ratio whilst retaining the key benefits of the approximate matching technique is discussed in [39].

Among instrumental effects, a phenomenon characteristic of mass spectrometry called mass discrimination is a major problem. This effect consists in the fact that heavier isotopes are transported more efficiently through the measuring system than lighter isotopes, which causes the value of R_{xy} ratio to be shifted towards the heavier ion of a given isotope pair. To compensate for this effect, a correction factor (key factor) is introduced, which is determined by taking into account the measured and known isotopic ratio in the standard.

Another effect associated with the mass spectrometer is a detection uncontrolled effect called "detector dead" . When ions are detected using electron multiplier in pulse counting mode during and after an electronic pulse the detector is "dead," i.e. is not able to detect any ions. If dead time is not taken into account there will be an apparent reduction in the number of pulses at high count rates.

Additional instrumental sources of random and systematic errors are obviously related to the apparatus with which the sample is prepared for measurement and introduced into the mass spectrometer. Thus, ICP suffers from the quenching effect resulting in a decrease in signal intensity over time, as well as the phenomenon of gradual deposition of precipitate on the sampler cone. In the case of separation methods, errors can result from all the sources affecting the efficiency of sample introduction into the chromatographic column or the capillary of the electrophoretic system. Multiplication of potential error sources in the complex analytical systems (so-called hyphenated systems) is a challenge facing not only the IDMS method but also other calibration methods.

It is assumed that analyses with MS detection are generally free of multiplicative interference effects. In some cases, the sample components may influence the chemical processes in plasma causing effects of such character. Undoubtedly, however, much more problematic are the effects involving the presence in the sample of isotopes of the same mass number as the analyte (eg. ^{204}Hg and ^{204}Pb) or on the formation in the plasma of adducts with the same m/z value as the analyte has (np. ^{38}Ar^{40}Ar, ^{38}Ar^{40}Ca, ^{41}K^{37}Cl on ^{78}Se). Regardless of the nature of these spectral effects, they have an additive character.

In some exceptional cases, simple isobaric interference effects can be mathematically canceled out. For example, the influence of ^{204}Hg on ^{204}Pb is effectively corrected by measuring the signal at ^{202}Hg ($Y_{\mathrm{Hg},x}$) and using an the following equation to obtain the true signal for ^{204}Pb ($Y_{\mathrm{Pb},0}$):

$$Y_{\mathrm{Pb},0} = Y_{\mathrm{Pb},x} - Y_{\mathrm{Hg},x} \cdot \frac{a_{^{204}\mathrm{Hg}}}{a_{^{202}\mathrm{Hg}}} \tag{6.63}$$

where $Y_{\mathrm{Pb},x}$ is the signal measured for ^{204}Pb, and $a_{^{204}\mathrm{Hg}}$, $a_{^{202}\mathrm{Hg}}$ are the atom fractions of isotopes ^{204}Hg and ^{202}Hg [40]. If the interference is caused by a polyatomic ion it can sometimes be removed instrumentally, e.g. using mass spectrometer equipped with sector field (SF-MS) or dynamic reaction cell (DRC) with appropriately selected reaction gas.

So why, with so many different dangers from analytical effects, is the IDMS method increasingly valued and used by analysts? This is undoubtedly influenced by several of its important features. First of all, according to the principle of the method, standards are added to the sample before it is processed according to the specified analytical procedure and thus they have a chance to undergo to the same extent (provided the state of isotopic equilibrium is reached) all the transformations that the sample itself undergoes. As it was emphasized before, from the calibration point of view, such a procedure is the most correct and recommended one, because it favors the compensation of preparative effects in the sample and in the standard (that is, the exact representation of the calibration real function by the model function). Consequently, the IDMS method is able to lead to accurate analytical results even when the analytical procedure is complex, multistep, and time-consuming.

A specific feature of the IDMS method is also that it uses a special type of standards which, although they have the same chemical properties as the native analyte in the sample, their amount when added to the sample can be determined by a separate, individual analytical signal. Furthermore, the analytical result is calculated not from the absolute value of the analytical signal, but from the ratio of the signals measured for two different isotopic forms of the same analyte. The standard added to the sample, therefore, acts to some extent as an internal standard – the signals measured for the isotopes of the standard undergo random measurement fluctuations to the same extent as the isotopes of the native analyte, resulting in compensation for random effects. This applies mainly to instrumental and preparative effects, although in some cases different isotopes of the same analyte also undergo small interference effects of similar magnitude and direction. All this leads to analytical results with increased precision.

A certain light on the analytical value of the IDMS method is given by comparing the results obtained by this method and the external standard method (ESM) under identical optimized experimental conditions free of significant analytical effects. These are shown in Table 6.13 and concern the determination of Pt, Pd, and Ag by mass spectrometry in two different synthetic multielement lead samples [41].

It is clearly seen that for every element in both samples the IDMS results are more accurate than those obtained by the ESM method. The highest relative errors were −13.3% for ESM and only +3.4% for IDMS. For 4 and 7 out of 10 cases, no statistically significant bias (at a confidence level of 95%), was established between the reference value and the concentration obtained by ESM and IDMS, respectively. At a confidence level of 99%, no bias is statistically significant for IDMS, while in the case of ESM the difference still is significant in 3 out of the 10 cases. The results obtained by IDMS are also shown to be excellent in terms of precision and much more precise than those obtained by ESM.

The IDMS method offers the possibility of determining of sub-trace amounts of analytes if only an instrument providing very good measurement sensitivity is used for the analysis. In Table 6.14 such spectacular results are presented achieved for ^{226}Ra determined in geological samples with the use of thermal ionization mass

Table 6.13 Results of the ICP-MS analysis lead samples using ESM and IDMS methods for calibration.

		Concentrations (μg g^{-1})		
Sample	**Analyte**	**Expected, c_0**	**Obtained, c_x, by ESM**	**Obtained, c_x, by IDMS**
EM5	Pt	98.0 ± 2.0	95.5 ± 0.6	98.9 ± 0.2
	Pd	99.0 ± 2.0	95.2 ± 1.9	98.4 ± 0.2
	Ag	108.0 ± 2.2	101.3 ± 0.5	109.9 ± 0.2
KS3	Pt	9.60 ± 0.20	10.16 ± 0.57	9.93 ± 0.10
	Pd	10.50 ± 0.21	9.1 ± 2.0	10.26 ± 0.16
	Ag	52.9 ± 1.1	50.57 ± 0.33	54.49 ± 0.16

Source: Adapted from Compernolle et al. [41].

Table 6.14 Results of the ICP-MS determination of ^{226}Ra in femtogram amounts in geological sample using IDMS for calibration.

Sample	Replicate analysis	Mass concentration obtained, c_x (fg·g^{-1})
Mount Lassen volcanic rock	1	$(1.063 \pm 0.010) \cdot 10^3$
	2	$(1.068 \pm 0.011) \cdot 10^3$
Midocean basalt RD65-6	1	69.1 ± 0.8
	2	67.8 ± 0.9
Midocean basalt A1374-2B	1	14.4 ± 0.2
	2	13.9 ± 0.2

Source: Adapted from Volpe et al. [42].

spectrometry working in the positive ion mode (PTI-MS) [42]. In accordance with the procedure developed, a small sample (100–500 mg) after appropriate preparation is subjected to the measurements of the ^{226}Ra/^{228}Ra signal ratio. By doing so, the analyte can be determined in concentrations reaching the fg·g^{-1} level with precision better than 1.5%. The results are also characterized by excellent reproducibility. As the authors point out, the ability to measure with high precision radium isotopic ratios and concentrations in small samples containing subpicogram amounts of radium is critical to determine the chronology of young volcanic rocks.

There are also quite many examples in the literature of successful application of IDMS to ICP-MS direct analysis of solid samples using laser ablation technique. For example, the procedure has been developed for simultaneous determination of chlorine, bromine, and iodine in rocks, sediments, and similar samples [43]. A sample in the pulverized form and of well-defined masses was spiked with about each of a ^{37}Cl–, ^{79}Br–, and ^{129}I enriched spike solution and mixed up, dried and pressed to pellets of 20 mm in diameter. Although a fractionation between analyte and spike could be observed for isotope-diluted samples during the ablation process the results agreed well with reference values (see Table 6.15). Thus, the possibility has been proven for time-effective and accurate halogen determinations in many matrices and can be used for certification of chlorine, bromine, and iodine in a geological reference materials.

As already mentioned, taking into account the accuracy of analytical results the crucial problem is to estimate the amount of analyte that was possibly lost during sample processing prior to measurement. For this purpose, the so-called recovery factor of the analyte from the sample is determined and the use of isotopic standards greatly facilitates this procedure. This is particularly important for organic samples where the analyte may be present in different chemical forms and each of these forms may be lost at different rates. In such a situation, studies can be performed using the instrumental system HPLC/ICP-MS, which allows separation of individual forms of the analyte (using HPLC) and then their determination by IDMS together with evaluation of recovery (using ICP-MS).

Table 6.15 Results of halogen determinations by LA-ICP-IDMS in sediment (SRM 1646, SRM 2704) and rock (granite [GS-N], disthene [DT-N], bauxite [BX-N]) samples using IDMS for calibration.

	Concentrations (μg g^{-1})					
	Chlorine		Bromine		Iodine	
Sample	Indicative, c_0	Obtained, c_x	Indicative, c_0	Obtained, c_x	Indicative, c_0	Obtained, c_x
SRM 1646	14 200 ± 800	16 100 ± 1900	115 ± 10	137 ± 15	33 ± 2	31.0 ± 1.0
SRM 2704	116 ± 28	132 ± 18	6.0 ± 0.4	5.1 ± 0.7	2.1 ± 0.3	2.6 ± 0.2
GS-N	442 ± 3.3	407 ± 57	2.3 ± 0.1	2.56 ± 0.37	0.031 ± 0.003	<0.1
DT-N	27.0 ± 0.5	30 ± 10	0.058 ± 0.012	<1.7	0.063 ± 0.002	<0.1
BX-N	27.7 ± 0.3	30 ± 4	0.983 ± 0.014	<1.7	2.52 ± 0.05	3.0 ± 0.2

Source: Adapted from Boulyga and Heumann [43].

Table 6.16 Results of application of the IDMS method to the HPLC-ICP/MS determination of total selenium and to estimation of the recovery coefficients of its inorganic (Se) and organic (SeMet) forms in the reference yeast material (SEAS 6) and commercial yeast-based supplements after acid microwave digestion and different enzymatic extraction procedures (using protease (A), driselase/protease (B) and in vitro enzymolysis (C)).

	Total selenium content		Recovery of Se (%)			Recovery of SeMet (%)		
Sample	Reference value, c_0	Obtained value, c_x	A	B	C	A	B	C
SEAS 6	1374 ± 101	1288 ± 31	87 ± 6	92 ± 4	—	68 ± 2	76 ± 2	—
Arkovital	100	117 ± 7	—	105 ± 10	97 ± 6	54 ± 4	60 ± 4	37 ± 3
Verdalia	100	108 ± 2	—	108 ± 10	103 ± 2	58 ± 9	63 ± 6	26 ± 2
ACE	200	202 ± 1	—	59 ± 2	87 ± 3	51 ± 1	49 ± 4	26 ± 1

Source: Adapted from Reyes et al. [44].

The example of such studies has been described in relation to the determination of total selenium, total solubilized selenium, and the selenomethionine (SeMet) in yeast and yeast-based nutritional supplements [44]. The samples were spiked with both ^{77}Se-enriched SeMet and after acid microwave digestion, they were subjected to different enzymatic extraction procedures. After separation of individual species, the analytical results were calculated based on the ^{78}Se/^{77}Se and ^{80}Se/^{77}Se isotope ratios. Results show (see Table 6.16) that all enzymatic procedures tested were able to solubilize total selenium quantitatively from the solid. However, the recovery for SeMet was seriously affected by the enzymatic procedure employed. After accounting for the recovery of both forms of the analyte, the total amount of selenium contained in the samples agreed very well with the certified value.

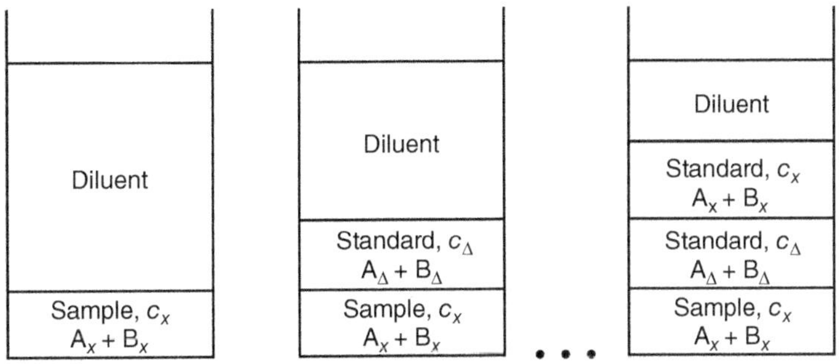

Figure 6.36 Scheme of the preparative stage of the IDMS combined with the standard addition method.

6.4.2.1 Modified Procedures

The growing interest in the IDMS method is also reflected in various proposals to modify this method and to adapt it for proper operation in particularly difficult analytical situations. From the calibration point of view, the most interesting proposal is a **combination of the IDMS method with the SAM method**. It was developed and tested on the determination of silicon in aqueous tetramethylammonium hydroxide (TMAH), sulfur in biodiesel fuel, and iron bound to transferrin in human serum [45].

The method requires the preparation of a series of calibration solutions, with constant masses of sample and isotopic standard and with increasing masses of added reference material. This material contains the same analyte with the same isotopic composition as the analyte (or, at least, very close to it). The preparative procedure is schematically shown in Figure 6.36.

The advantage of this method is based on the fact that the measured intensity ratios are calculated without the need for mass bias correction. Furthermore, the authors expected the method to be useful when a matrix effect is present against an analyte present in the sample in trace amounts ($ng\,g^{-1}$). However, it seems that a more convincing proof of these advantages of the method still requires its application to the analyses of many other analytical systems.

Recently, the concept of the **pseudo isotope dilution (PID)** method has been developed and verified experimentally on the example of the europium determination by ICP-MS [46]. Due to the similarities in chemical properties, in natural samples europium (^{151}Eu and ^{153}Eu) is usually coexisting with barium and then suffers spectral interferences from barium oxides and hydroxides However, the composition of major barium isotopes (^{135}Ba, 6.59% and ^{137}Ba, 11.23%) contributing to the spectral interferences with europium isotopes is significantly different from that of the later (^{151}Eu, 47.8% and ^{153}Eu 52.2%) permitting the treatment of barium interferents as a pseudo isotopic spike of europium. The adequate equation for calculating the analyte concentration in a sample has been derived. The analyses of the synthetic samples and natural water samples showed that PID is able to be valuable tool effectively correcting spectral interferences and considerably improving accuracy of the analytical results.

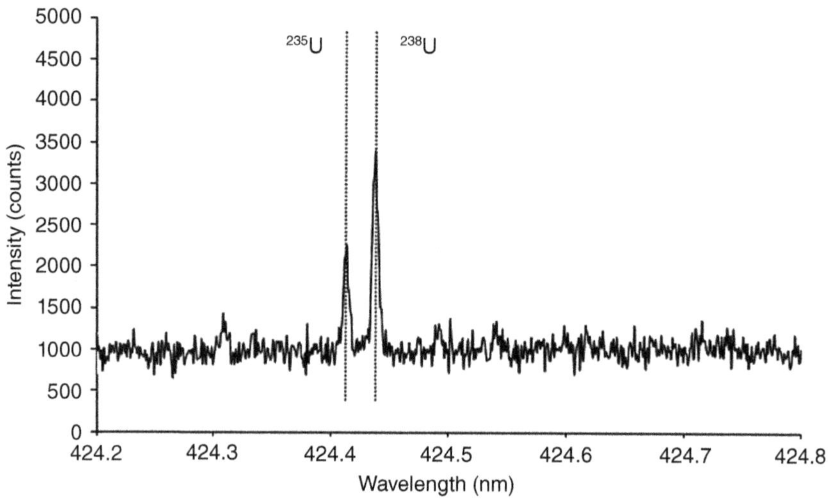

Figure 6.37 Spectrum obtained in the 424 nm region for uranium isotopic reference material SRM U-750 (5 mg l^{-1}). Source: Vigneau et al. [47], fig 3 (p. 2354)/with permission of Springer Nature.

To avoid some of the interferences characteristic of mass spectrometry, attempts are made to adapt the ICP-AES technique to calibration by IDMS. This is in principle possible provided that an AES spectrometer with sufficiently high resolution is used. Such a procedure has been tested under the name **IDEAS** for the determination of uranium isotopes ^{235}U and ^{238}U [47]. Figure 6.37 shows that the isotopic shift in the atomic emission lines for both isotopes is large enough to be used for IDMS purposes. The analytical results obtained were quite similar in accuracy to those reported by ICP-MS, although inferior in precision, which is a consequence of the higher accuracy of measurements offered by the ICP-MS technique than by ICP-OES. However, the IDAES method may be more useful for uranium measurements, for instance, when significant amounts of lead or plutonium are present in the sample because in ICP/MS the isotopes of lead and plutonium form isobaric interferences.

It is worth mentioning that the mass spectrometers dedicated to the measurement of isotopic ratios are manufactured. This technique is called **isotope ratio mass spectrometry (IRMS)**. These instruments contain several ion detectors arranged in such a way that multiple ions can be measured simultaneously. These spectrometers are traditionally used to measure natural variations in the isotopic ratios of elements. They are also suitable for measuring isotopic ratios in geochemical applications and in the certification of isotopically enriched elements or labeled molecules.

In summary, the IDA method, which is actually represented by different variants of a different preparative, measurement, and transformation stages, should be treated, also in the nomenclatural sense, as an individual and separate calibration method with different detailed procedures. Furthermore, IDMS in the IDA version most closely meets the theoretical requirements for “ideal” calibration methods, and thus provides higher accuracy and precision compared to other calibration strategies. This is the reason why IDMS is widely used in applications where high accuracy and confidence are required (for example, it is widely used for certification of reference materials). Due to its analytical advantages, it is ranked – along with gravimetry, titration, and coulometry – among the methods with the highest metrological value, i.e. among the basic measurement methods.

References

1 Foster, J.S., Langstroth, G.O., and McRae, D.R. (1935). Quantitative spectrographic analysis of biological material. I – A method for the determination of lead in cerebrospinal fluid. *Proceedings of the Royal Society of London. Series A: Mathematical and Physical Sciences* 153: 141–152.

2 Hohn, H. (1937). *Chemischen Analysen mit dem Polarographen*. Berlin: Springer Verlag.

3 Varasova, E.N. (1937). The polarograhic analysis of minerals I. *Zavodskaya Laboratoriya* 6: 804–807.

4 Lingane, J.J. and Kerlinger, H. (1941). Polarographic determination of nickel and cobalt. *Industrial and Engineering Chemistry, Analytical Edition* 13: 77–80.

5 Chow, T.J. and Thompson, T.G. (1955). Flame photometric determination of calcium in sea water and marine organisms. *Analytical Chemistry* 27 (6): 910–913.

6 Kemula, W. (1966). A simple apparatus for the determination of trace amounts of different ions using the hanging drop-mercury electrode. *Microchemical Journal* 11: 54–61.

7 Kościelniak, P. (1999). Non-linear calibration by the standard addition method. *Chemometrics and Intelligent Laboratory Systems* 47 (2): 275–287.

8 Stafiński, M., Wieczorek, M., and Kościelniak, P. (2013). Influence of the species effect on trueness of analytical results estimated by the recovery test when determining selenium by HG-AFS. *Talanta* 117 (12): 64–69.

9 Hosking, W., Oliver, K.R., and Sturman, B.T. (1979). Errors in the atomic absorption determination of calcium by the standard addition method. *Analytical Chemistry* 51 (2): 307–310.

10 Czakow, J. and Minczewski, J. (1960). Spectrographic analysis by the powder shifting method. III. Addition and division mathematical method (in Polish). *Chemia Analityczna (Warsaw)* 7: 721–730.

11 Pietzka, G. and Chun, H. (1959). Flammenphotometrie I (in German). *Angewandte Chemie* 71: 276–283.

12 Fukuschima, S., Yukawa, K., Shigemoto, M. et al. (1958). Another new application of standard-addition technique in flame spectrophotometry. *Microchimica Acta* 46: 553–570.

13 Andrade, J.M., Terán-Baamonde, I., Soto-Ferreiro, R.M. et al. (2013). Interpolation in the standard additions method. *Analytica Chimica Acta* 780: 13–19.

14 Li, Y., Yu, L.-J., Li, M.-T. et al. (2006). A new approach to the standard addition method for the analysis of F, Al and K content in green tea. *Microchimica Acta* 153 (1): 109–114.

15 Zenkevich, I.G. and Morozova, T.E. (2014). Areas of application and characteristics of quantitative chromatographic analysis by the consecutive standard addition method. *Journal of Analytical Chemistry* 69 (4): 327–335.

16 Brown, R.J.C., Roberts, M.R., and Milton, M.J.T. (2007). Systematic error arising from 'sequential' standard addition calibrations: quantification and correction. *Analytica Chimica Acta* 587 (1): 158–163.

17 Tyson, F.J. (1981). Low cost continuous flow analysis. Flow-injection techniques in atomic-absorption spectrometry. *Analytical Proceedings* 18: 542–545.

18 Kościelniak, P. (1996). A critical look at the interpolative standard addition method. *Analusis* 24 (1): 24–27.

19 Israel, Y. and Barnes, R.M. (1989). Flow injection sample – to standard addition method. Spectrophotometric determination of hydrochloric acid and orthophosphate. *Analyst* 114 (7): 843–848.

20 Kościelniak, P., Kozak, J., and Wieczorek, M. (2011). Calibration by the standard addition and indicative dilution method in flame atomic absorption spectrometry. *Journal of Analytical Atomic Spectrometry* 26 (7): 1387–1392.

21 Geoffroy, C.J. (1731). Examen du Vinaigre Concentré pour la Gelée (in French). *Mémoires de l'Académie Royale des Sciences* 1729: 68–93.

22 Lewis, W. (1767)). *Experiments and Observations on American Potashes. With an Easy Method of Determining Their Respective Qualities.* London: Royal Society of Arts.

23 Descroizilles, F.A.H. (1795). Description et usages du berthollimêtre, ou instrumentd'épreuves pour l'acide muriatique oxigené liquide, pour l'indigo et pour l'oxide de manganese; avec des observations sur l'art de graver le verre par le gaz acide fluorique (in French). *Journal des Arts et Manufactures* 1: 256–276.

24 Gay-Lussac, J.L. (1824). Instruction sur l'essai du chlorure de chaux (in French). *Annales de chimie et de physique.* 2nd series 26: 162–175.

25 Gay-Lussac, J.L. (1828). Essai des potasses du commerce (in French). *Annales de chimie et de physique.* 2nd series 39: 337–368.

26 Mohr, K.F. and Friedrich, K. (1855). *Lehrbuch der Chemisch-Analytischen Titrirmethode.* Braunschweig: Friederich Vieweg und Sohn.

27 Kościelniak, P. (2022). Unified principles of univariate analytical calibration. *TrAC Trends in Analytical Chemistry* 149: 116547.

28 Minczewski, J. and Marczenko, Z. (1985). *Chemia Analityczna. Część II.* (in Polish). Warsaw: PWN.

29 Martz, T.R., Dickson, A.G., and DeGrandpre, M.D. (2006). Tracer monitored titrations: measurement of total alkalinity. *Analytical Chemistry* 78 (6): 1817–1826.

30 Hevesy, G. and Hobbie, B. (1932). Über die Ermittlung des Bleigehaltes von Gesteinen (in German). *Fresenius' Zeitschrift für Analytische Chemie* 88: 1–6.

31 Hevesy, G. and Hofer, E. (1934). Elimination of water from the human body. *Nature* 134: 879–879.

32 Rittenberg, D. and Foster, G.L. (1940). A new procedure for quantitative analysis by isotope dilution, with application to the determination of amino acids and fatty acids. *Journal of Biological Chemistry* 133 (3): 737–744.

33 Reynolds, J.H. (1952). A mass spectrometric investigation of branching in Cu^{64}, Br^{80}, Br^{82} and I^{128}. *Physical Review* 79 (5): 789–795.

34 Okamoto, K. (1991). Preparation and certification of rice flour unpolished reference material. *Science of the Total Environment* 107: 29–44.

35 Ikeda, N. and Noguchi, K. (1979). Substoichiometric method in the simple radiometric analysis. *Journal of Radioanalytical Chemistry* 53 (1–3): 107–113.

36 Yoshioka, H. and Kambara, T. (1984). Comparison of the sub- and super-equivalence and substoichiometric methods of isotope dilution analysis for the determination of a trace amount of antimony. *Talanta* 31 (7): 509–513.

37 Yoshioka, H., Hasegawa, T.H., and Kambara, T. (1987). Redox sub-superequivalence method of radiometric analysis. Determination of trace amount of antimony. *Journal of Radioanalytical and Nuclear Chemistry* 117 (1): 47–59.

38 De Bièvre, P.J. and Debus, G.H. (1965). Precision mass spectrometric isotope dilution analysis. *Nuclear Instruments and Methods* 32 (2): 224–228.

39 Hearn, R., Evans, P., and Sargen, M. (2005). Development of a high ratio isotope dilution mass spectrometry calibration technique for chemical metrology. *Journal of Analytical Atomic Spectrometry* 20 (10): 1019–1023.

40 Evans, E.H. and Clough, R. (2005). Isotope dilution analysis. In: *Encyclopedia of Analytical Science*, 2e (ed. P. Worsfold, A. Townshend and C. Poole), 545–553. Elsevier.

41 Compernolle, S., Wambeke, D., De Raedt, I. et al. (2012). Evaluation of a combination of isotope dilution and single standard addition as an alternative calibration method for the determination of precious metals in lead fire assay buttons by laser ablation-inductively coupled plasma-mass spectrometry. *Spectrochimica Acta B* 67: 50–56.

42 Volpe, A.M., Olivares, J.A., and Murrell, M.T. (1991). Determination of radium isotope ratios and abundances in geologic samples by thermal ionization mass spectrometry. *Analytical Chemistry* 63 (9): 913–916.

43 Boulyga, S.F. and Heumann, K.G. (2005). Direct determination of halogens in powdered geological and environmental samples using isotope dilution laser ablation ICP-MS. *International Journal of Mass Spectrometry* 242 (2-3): 291–296.

44 Reyes, L.H., Marchante-Gayón, J.M., García Alonso, J.I. et al. (2006). Application of isotope dilution analysis for the evaluation of extraction conditions in the determination of total selenium and selenomethionine in yeast-based nutritional supplements. *Journal of Agricultural and Food Chemistry* 54 (5): 1557–1563.

45 Brauckmann, C., Pramann, A., Rienitz, O. et al. (2021). Combining isotope dilution and standard addition – elemental analysis in complex samples. *Molecules* 26 (9): 2649.

46 Zhu, Y. and Itoh, A. (2021). Pseudo isotope dilution (PID) as an approach for correcting barium-related spectral interferences on the measurement of europium

by inductively coupled plasma mass spectrometry (ICP-MS). *Analytica Chimica Acta* 1180: 338854.

47 Vigneau, O., Arnal, N., and Feline, N. (2016). Isotopic dilution analysis using ICP/AES: application to uranium samples and comparison to MC ICP/MS measurements. *Journal of Radioanalytical and Nuclear Chemistry* 307 (3): 2347–2357.

7
Calibration in Nonequilibrium Conditions

In the dilution method, the analyte concentration is determined from the apparent concentrations calculated during dilution of the sample and standard, but the dilution process is performed gradually reaching steady states with well-defined values of the dilution factor. In the standard addition and indicative dilution method, the sample with the standard can be diluted continuously, but the value of the analytical result is determined by the signal measured for the sample itself at a steady state. Titration also involves a continuous change of reaction state between the analyte and the titrant, but the result is calculated when a certain chemical equilibrium state is reached. Because of the instability of some isotopes, some radiochemical methods are classified as nonequilibrium methods, but isotopic dilution methods presented in chapter 6 are not so. Thus, all calibration methods discussed earlier can be considered as being performed in physical and chemical equilibrium.

Typical analytical methods operating in nonequilibrium conditions include **flow injection analysis (FIA)** and **kinetic analysis**. It is, therefore, of interest how analytical calibration can be performed in these cases, and whether and to what extent the procedures of the learned calibration methods are modified under conditions of changing analytical signal over time.

7.1 Flow Injection Analysis

Flow analysis involves the use of instrumental systems by which a sample is fully prepared for measurement as it flows from the point of collection to the point of detection of the analytical signal. The concept was first introduced by L.T. Skeggs Jr. in 1956 [1] and quickly found acceptance among analysts. Over the years, many types of flow systems have been constructed operating on different principles and using a variety of sample processing methods [2]. Nowadays, flow analysis is a very significant and rapidly developing area of analytical chemistry, arousing interest both from the scientific side and from the practical use in analytical laboratories.

Flow injection analysis (FIA) was introduced into analytical practice in 1975 by J. Růžička and E.H. Hansen [3]. It is based on the technique of sample introduction into flow systems by dosing (injection) in a strictly specified volume (of the order of several to several hundred microliters) into another solution (carrier) by means

Calibration in Analytical Science: Methods and Procedures, First Edition. Paweł Kościelniak.

of a specially designed injection valve. Depending on the required analytical procedure, the sample in this form can be combined with other solutions, as well as subjected to chemical reaction and physical transformations (mixing, dilution, concentration, etc.) to give it the appropriate physicochemical form, optimal for a given measurement method. A schematic of the basic FIA system is shown in Figure 7.1.

The sample undergoes a gradual dispersion in the carrier stream from the moment of injection. As a result of this phenomenon, the central part of the sample segment with the highest local concentration of analyte (and other sample components) is formed and an asymmetric gradient of analyte concentration on both sides of the central part is created. Reaching the measuring system in this form, the sample induces a transient signal in the shape of an asymmetric peak of maximum height corresponding to the maximum analyte concentration in the segment. This is illustrated in Figure 7.2.

Instrumentation of sample processing provides the opportunity to automate the entire process, which promotes improved precision of analytical results. In addition,

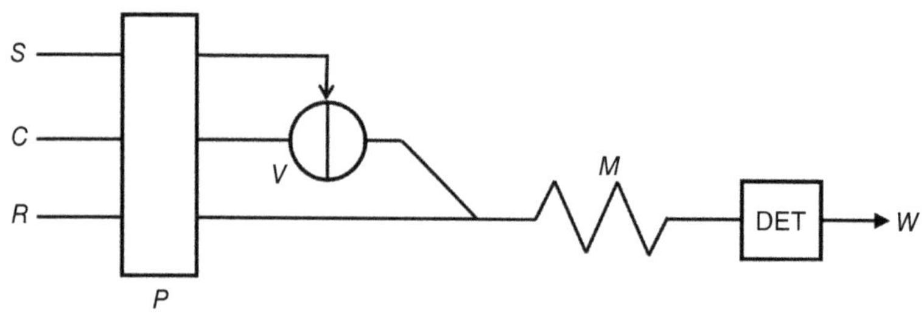

Figure 7.1 Basic flow injection system: solid lines – tubes (with internal diameter <1 mm), *S*, *C*, *R* – sample, carrier, and reagent solutions, respectively, *P* – pump, *V* – injection valve, *M* – mixing coil, DET – detection system, *W* – waste.

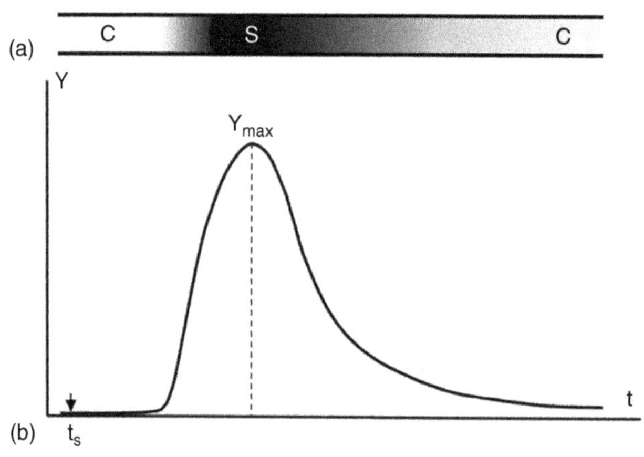

Figure 7.2 Injected and dispersed sample zone (a) and transient unsymmetrical injection peak (b) of a shape corresponding to the local analyte concentrations in the sample zone: *S*, *C* – sample and carrier, respectively, Y_{max} – maximum height of the peak, t_s – moment of the sample injection.

compared to the traditional procedure, the analytical process is faster and occurs with less use of reagents and less production of used solutions. Transport of solutions in closed tubes avoids contact with hazardous reagents and protects these solutions from contact with the environment. Thus, FIA provides significant analytical and economic benefits, and meets the principles of "green chemistry."

The calibration process is usually performed in flow-through systems in the traditional manner: the calibration solutions are prepared in advance according to the principle suitable for the chosen calibration method and introduced one after the other into the system through the sample line (see Figure 7.1). For each solution, the maximum peak height formed is measured, from this, a calibration graph is constructed and the concentration of the analyte in the sample is determined. This procedure maintains all the abovementioned advantages of FIA. An additional advantage of this procedure is that all calibration solutions have the chance to be subjected to the same treatments and prepared for measurement equally. From a calibration point of view, this is, as is well known, a very important action, since it favors the compensation of uncontrolled effects and the obtaining of analytical results with increased precision and accuracy.

The flow-based mode of analysis, however, provides the opportunity to perform calibrations according to such procedures that are difficult or even impossible to perform in a traditional manner [4, 5]. In many cases, these modifications contribute to improving specific features of a given calibration method. In general, modifications to the calibration process can take place using two types of techniques:

- **manipulation techniques,** which involves controlled, deliberate changes to the design of the underlying flow system (see Figure 7.1), and
- **gradient technique,** based on the use of the rich measurement information provided by the characteristic signal measured in the FIA technique, i.e. the asymmetric injection peak (see Figure 7.2).

Both of these routes, skillfully applied by the analyst based on his or her knowledge, experience, and imagination, often yield surprisingly ingenious results. Several approaches are described in what follows that demonstrate how some of the calibration methods discussed so far can be implemented by both techniques in injected flow analysis.

7.1.1 Manipulation Techniques

Appropriately designed modules installed in flow systems primarily allow the preparation of a series of standards using a single, basic standard solution. These solutions can be used to perform calibrations with different calibration methods mostly in their multi-standard versions. **The calibration is based on the modification of the flow system to operate on the principle of either a controlled dilution of a single standard solution or a controlled change in its dispersion**.

An example of a calibration module of the first type is the closed circulation system shown in Figure 7.3 [6]. A standard solution with a well-defined analyte concentration and initial volume fills the loop α and is directed back to the vessel from which it

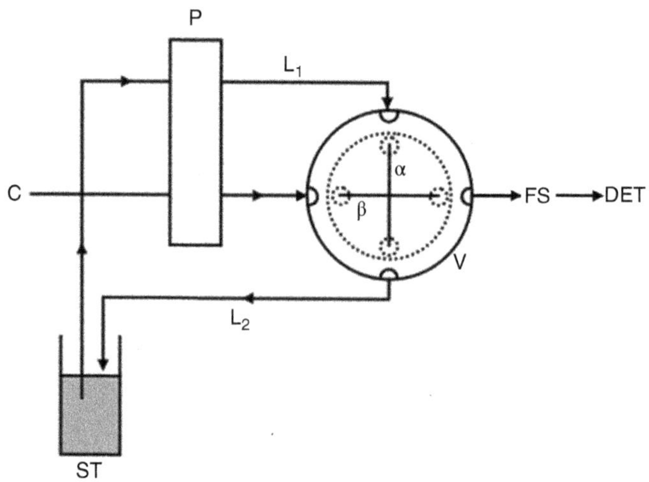

Figure 7.3 Flow injection module for preparation of a set of standard solutions: L_1, L_2 – tubes, ST, *C* – standard and carrier, respectively, *P* – pump, *V* – injection valve with stator (solid line) and rotor (dashed line), *M* – mixing coil, FS – flow system, DET – detection system, α, β – injection loops.

is drawn. When the valve rotor is rotated 90°, the standard segment is injected from the loop α and directed into the flow system, and at the same time, the carrier enters the standard vessel from the loop β and causes dilution. After successive position changes, new standard solutions are created with decreasing analyte concentrations.

A series of calibration solutions can also be made from a single solution using a full rotary valve (FRV), whose rotor can rotate 360° relative to the stator and assume eight positions relative to the rotor. This can be used for calibration purposes as a selection valve as shown in Figure 7.4 [7]. At a given valve position, the primary standard solution, st, and the diluent solution, *D*, combine and mix in a common single line

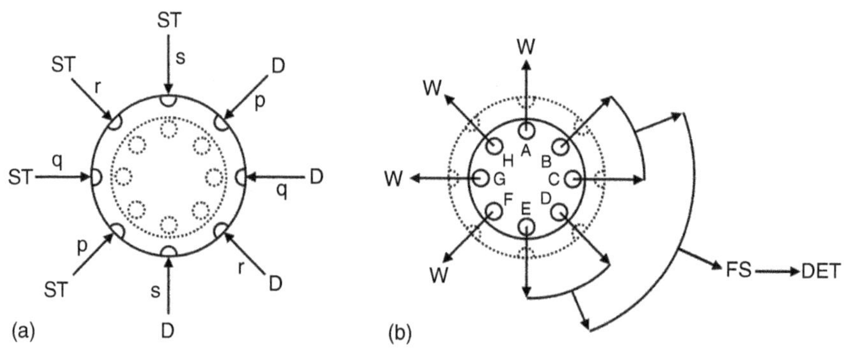

Figure 7.4 Fully rotary valve with stator (a) and rotor (b) used for preparation of a set of standard solutions: ST, *D* – standard and diluent, respectively, *W* – waste, FS – flow system, DET – detection system, *p*, *q*, *r*, *s* – flow rates, A, …, F – channels. Source: Kościelniak et al. [7], fig 1 (p. 112)/with permission of John Wiley & Sons.

Table 7.1 Flow rates, $p, \ldots, s$, of the streams generating the standard solutions in individual positions, A, …, H, of the FRV and the analyte concentrations in these standards.

Rotor position	Diluent rate	Standard solution rate	Total rate of the solution generated	Analyte concentration (normalized units)
A	$p+q+r+s$	0	$p+q+r+s$	0
B	$q+r+s$	p	$p+q+r+s$	$p/(p+q+r+s)$
C	$r+s$	$p+q$	$p+q+r+s$	$(p+q)/(p+q+r+s)$
D	s	$p+q+r$	$p+q+r+s$	$(p+q+r)/(p+q+r+s)$
E	0	$p+q+r+s$	$p+q+r+s$	1
F	p	$q+r+s$	$p+q+r+s$	$(q+r+s)/(p+q+r+s)$
G	$p+q$	$r+s$	$p+q+r+s$	$(r+s)/(p+q+r+s)$
H	$p+q+r$	s	$p+q+r+s$	$s/(p+q+r+s)$

Source: Kościelniak et al. [7], table 1 (p. 114)/John Wiley & Sons.

at a specified ratio. Changing the valve position creates the ability to accurately and precisely generate eight standard solutions with different analyte concentrations.

Furthermore, the streams of standard solution and diluent can flow into each other at different velocities (p, q, r, s). The analyte concentrations in the standards and consequently the distribution of measurement points on the calibration graph then depend on the values of these velocities, as shown in Table 7.1.

An additional advantage of the device is that the analyte concentrations in the generated solutions are to some extent immune to its erroneous operation. This is because accidental or even systematic changes in the flow ratio of the streams cause symmetrical deviations of the measurement points with respect to the calibration graph, as shown in Figure 7.5.

In both of the above described, a solution of the sample to be analyzed is fed into the flow system from a separate vessel, the analytical signal is measured for it, and the analytical result is determined according to the external standard method. Another possibility is to treat the sample in the same way as the basic standard solution, i.e. to subject it to dilution in several or more of the same degrees as the standards were diluted. Given the known dilution values of the sample and standard, the measurement data obtained in this case allow calibration according to the multipoint dilution ESM method or the dilution method.

To perform calibrations by other methods, the generated standards must be combined with solutions of appropriate substances. Once a standard has been injected into a carrier, its segment can be combined with another solution flowing either as a continuous stream (**zone-to-stream technique**) or as a separate segment injected with a second valve into the same or another carrier (**merging zone technique**). As shown schematically in Figure 7.6, in both cases after a sufficiently long time a single solution of a mixture of non-reacting or reacting starting solutions is formed, which in this form undergoes further processing in the flow system or is directed directly to the measuring apparatus.

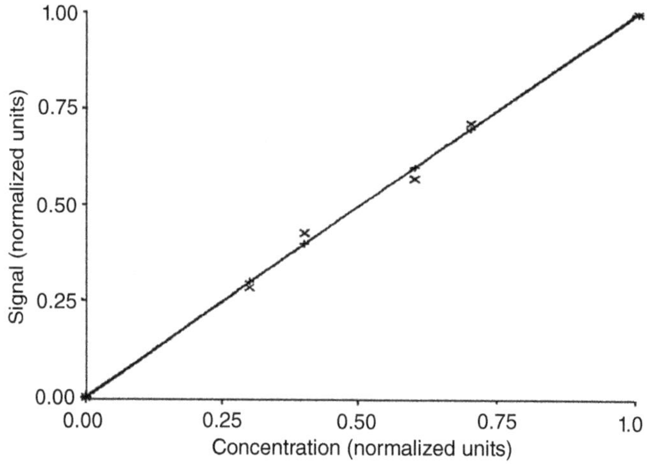

Figure 7.5 Calibration graph obtained in the external standard method using FRV in configuration q–p–q–q (see Figure 7.4 and Table 7.1) for $p/q = 0.333$; two sets of measurement points are produced in the system working perfectly (+) and working with the flow rates changed up to $p/q = 0.5(x)$. Source: Kościelniak et al. [7], fig 6 (p. 118)/with permission of John Wiley & Sons.

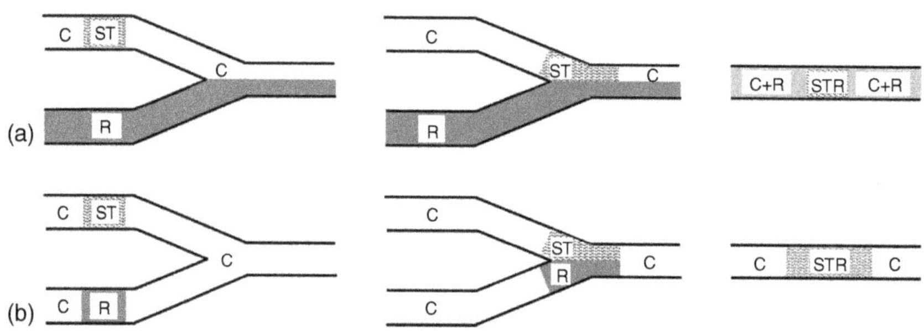

Figure 7.6 Two possible ways of merging of the standard solution (st) injected to carrier (C) with reagent (R): zone-to-stream (a) and merging zone (b) techniques.

By combining the analyte standard with the corresponding analyte-reactive compound or internal standard via these routes, the calibration can be performed by the indirect or internal standard method, respectively. Of course, to complete the calibration procedure, the sample must then also be combined with the respective substances in the same way.

If a sample is introduced in place of the reactant using one of the techniques shown in Figure 7.6, the standard series will be combined with the sample, which of course enables calibration by the standard additive method. If the calibration is to be carried out according to the extrapolative standard addition method (E-SAM), which allows compensation for linearly multiplicative interference effects, then initially the sample stream or segment should be combined with a segment of the diluent itself (of the same capacity as the standard segments). The absence of this diluent and

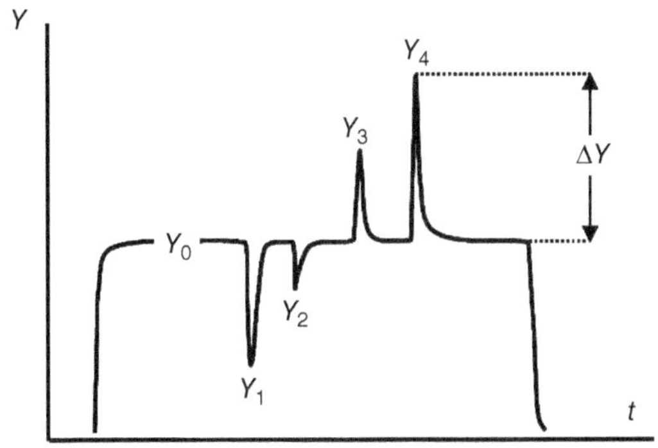

Figure 7.7 Measurement image typical of the calibration by the interpolative standard addition method basing on the reverse flow technique: Y_0 – steady-state signal measured for a sample, $Y_1, \ldots, Y_4$ – peaks measured for the standard solutions.

the combination of standard solutions with nonzero analyte concentrations with the sample leads to calibration by the interpolative standard addition method (In-SAM).

The measurement results obtained when a series of standard solutions are introduced into the continuous sample stream are shown in Figure 7.7 [8]. Standards with higher analyte concentrations than the sample concentration appear as positive peaks on the continuous sample signal, while standards with lower concentrations appear as negative peaks. After determining the differences between the intensities measured for the sample and for the standards, the appropriate calibration graph is constructed (see Figure 6.15) and the concentration of analyte in the sample is determined by interpolation. Because each standard dilutes the sample, the method, as already demonstrated, has limited ability to compensate for interference effects.

A unique feature of FIA is the phenomenon of dispersion of the injected solution in the carrier solution. An example of using this phenomenon for calibration purposes is the preparation of standard solutions with the so-called **sampling zone technique** [9]. Its principle is explained in Figure 7.8.

A segment of standard solution introduced into the system through the first valve (V1) is directed to the second valve (V2) undergoing dispersion along the way. Upon reaching valve V2, the segment fills a loop of less than its volume. The rotation of the valve, therefore, injects only a specific segment of this segment and diverts it to the rest of the system. Due to the heterogeneous distribution of the analyte caused by the dispersion phenomenon, the local maximum analyte concentration in a given

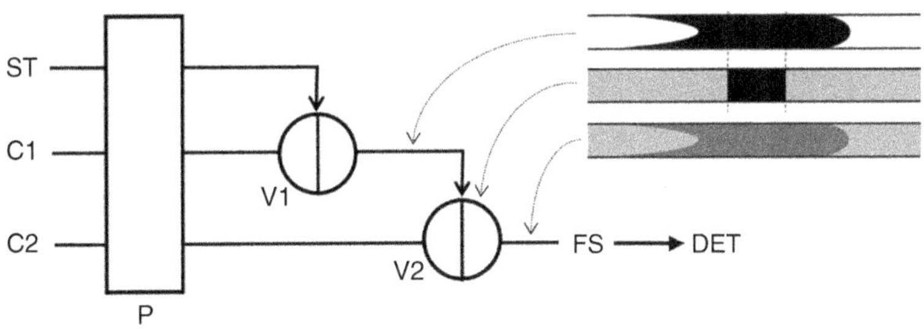

Figure 7.8 Preparation of a set of standard solutions using sampling zone technique: ST – standard, *C*1, *C*2 – carriers, *V*1, *V*2 – injection valves, *W* – waste, *P* – pump, FS – flow system, DET – detection system.

fragment of the injected segment depends on the time that elapses between one injection and the other. By regulating this time-lapse with the flow rate of the carrier C1, several segments of the standard can be obtained with different maximum analyte concentrations and different, corresponding to these concentrations, intensities Y_i of the analytical signals.

It should be noted, however, that in this type of system the exact values of analyte concentration in the segments remain unknown. However, the results obtained can be interpreted according to the parametric external standard method (ESM) method (see Figure 5.9). The Y_i signals are then assigned to the known c_{st} concentration of the analyte in the standard solution injected into the system and a series of two-point calibration graphs are constructed, then the signals measured for a sample injected under the same conditions as the standard are related to these graphs and the analytical result determined. The parameter characterizing the calibration conditions is in this case the so-called dispersion coefficient calculated from the formula:

$$D_i = \frac{Y_{max}}{Y_i} \tag{7.1}$$

where Y_{max} is the signal intensity measured for the standard solution in concentration c_{st}.

It is worth noting that the value $1/D_i$ can be thought of as the dilution degree, k_i, of the ith segmented sample relative to the continuous stream sample. The same measurement results that are interpreted according to the parametric ESM method can also be used for calibration by dilution ESM method.

Another very ingenious flow calibration module using the dispersion phenomenon is the system shown in Figure 7.9 [10]. After injecting a standard with an analyte concentration of c_{st}, its segment is directed to a location where it splits into three segments (**splitting zone technique**; Figure 7.9a). These flow further through the tubes of different lengths so selected that when the segments are combined, the peaks recorded for them partially overlap with each other. The segments' different ways of the measuring system also cause them to disperse to a different extent, which results in the overlapping peaks having different heights (Figure 7.9b). Thus, three maximum points and two minimum points can be distinguished in the measurement image, which can be assigned different values for the dispersion coefficient, D_i, as an experimental parameter. Thus, if a sample is injected into the system next, it also undergoes dispersion in degrees D_i, which allows the determination of the analyte concentration in the sample according to the principles of the parametric ESM method or the dilution ESM method.

To further differentiate the signals recorded for a single standard and sample solution, an FRV valve can be used as an injection valve (see Figure 7.10) [11]. It is equipped with four central injection loops of different capacities and an additional side loop that connects two of the central loops. At successive valve positions, the standard and sample solutions are injected one after the other and are separated by the media injected from the side loop. After rotating the rotor 180°, the standard and sample are injected from all four central loops, with each second solution flowing in tandem with the previous one reaching the measurement system dispersing slightly

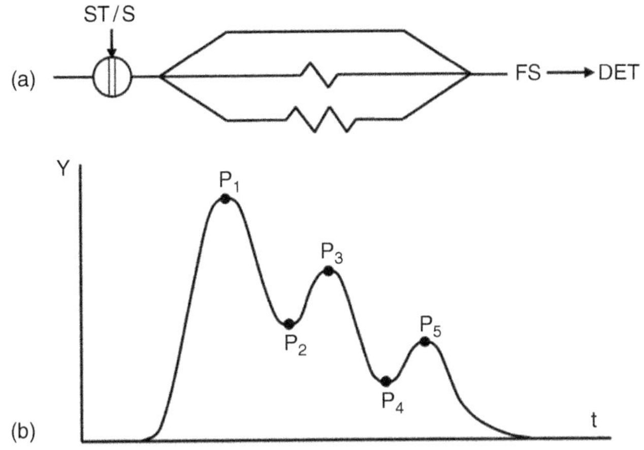

Figure 7.9 Flow system for generation of a set of standard solutions using splitting zone technique (a) and typical signal produced (b): *S*, *ST* – sample and standard, respectively, FS – flow system, DET – detection system, $P_1, \ldots, P_5$ – local extremum measurement points used for preparation of a set of two-point calibration graphs (see Figure 5.9).

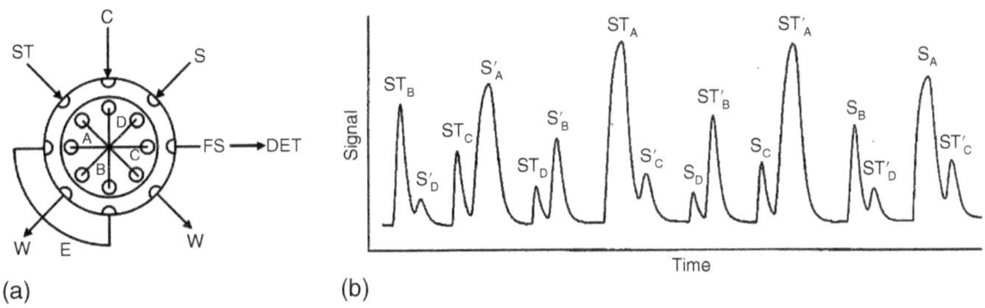

Figure 7.10 Fully rotary valve (a) of configuration allowing eight peaks to be obtained (b) for the sample ($S_A, \ldots, S_D, S'_A, \ldots, S'_D$) and for the standard ($ST_A, \ldots, ST_D, ST'_A, \ldots, ST'_D$) of heights differentiated by different loop volumes and different dispersion degree; *S*, *ST*, *C* – sample, standard, and carrier, respectively, A, …, *E* – loops, *W* – waste, FS – flow system, DET – detection system. Source: Kościelniak and Janiszewska [11], fig 14 & 15 (p. 52–53)/with permission of John Wiley & Sons.

more. After a further rotation of 180° (i.e. returning to the starting position), the sample and standard are injected from the same loops as before, but in pairs of inverted order. This results in eight peaks for the sample and eight peaks for the standard with different (but corresponding) heights, as illustrated in Figure 7.10b.

The amount of useful calibration information obtained in this way is so large that the calculated values of the eight apparent concentrations of the analyte can be presented as a function of $1/D_i$ as measures of the degree of dilution of standard and sample, and the analytical result is then determined according to the dilution method. A factor favoring good precision of the analytical result in both methods is the alternate and identical treatment of sample and standard in the preparative and measurement stages.

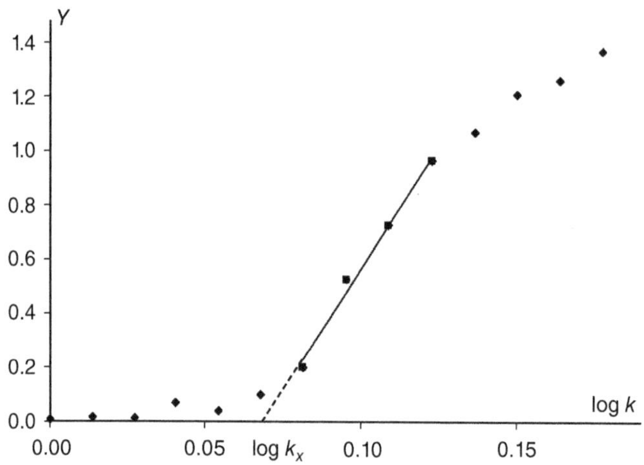

Figure 7.11 Titration curve obtained by the merging zones technique in case of titration of HCl (0.220 mol l^{-1}) with the use of NaOH (0.100 mol l^{-1}), Y – integrated absorbance, k – dilution factor, k_x – value corresponding to the titration end point. Source: Wójtowicz et al. [12], fig 2 (p. 1008)/with permission of Elsevier.

In flow analysis, many systems have been developed to accommodate titration-based calibration. Most of them use manipulative techniques in a way that mimics the traditional titration technique. For example, the procedure described in [12] relies on titration of a sample that is gradually diluted in a strictly controlled way in the flow injection system developed. On each step of sample dilution, equal volumes of the sample and titrant solutions are simultaneously injected into two carrier streams and the zones are merged with each other. Then, they are mixed completely in the mixing chamber, merged with a stream of indicator. The resulting titration curve is shown in Figure 7.11.

In the literature, one can also find examples of using FIA to implement the IDMS method [13], but the flow system is then only used to give the sample the form of an injected segment and thus achieve better analytical performance of the IDMS method.

In all calibration solutions based on manipulation techniques, a key role from the point of view of accuracy and precision of analytical results is played by the correct and reproducible operation of the applied calibration modules, i.e. those elements of the flow system that determine the correct and reproducible preparation of calibration solutions. These modules should be as simple as possible in terms of design and strictly controlled by the analyst.

7.1.2 Gradient Technique

A peak recorded in FIA analysis is most often treated as in the examples described above – as a single analytical signal whose value is determined by measuring the peak maximum point or the area under the peak. However, **a peak can also be interpreted as a set of points with varying values of signal intensity,**

corresponding to a gradient of analyte concentration in a segment with a certain degree of dispersion. A particular property of the signals recorded in flow analysis is that, under fixed instrumental conditions, a given peak measurement point appearing after time t after solution injection provides a well-defined and highly reproducible measure of the local analyte concentration at that point in the segment that reaches the detector after time t.

Thus, one of the oldest gradient approaches to calibration consists of recording a peak under conditions that provide a well-defined and visible gradient of measurement points for a standard solution of c_{st} concentration and assigning to selected points on the descending portion of the peak values of the time elapsed between injection and the appearance of these points [14]. In a separate experiment, the values of local dispersion coefficients are determined for these points. From this, the corresponding analyte concentration values in the standard segment are calculated and a calibration graph is constructed. This procedure is shown in Figure 7.12.

The concentration of the analyte in the sample is determined by interpolation according to the ESM method from the value of the signal intensity recorded at one of the selected moments after sample injection and the value of the dispersion coefficient corresponding to this moment. It is worth noting that the predetermined dispersion coefficients can be used to determine the analyte concentration in many samples provided that the experimental conditions are strictly kept constant.

Among the gradient procedures developed for calibration with the standard addition method, the procedure called gradient scanning standard addition method (GSSAM) [15]. In this case, another feature of the FIA is exploited: pairs of measurement points of equal value can be found on the increasing and decreasing parts of the peak, which correspond to equal dispersion of the injected solution.

In the GSSAM method, in a preliminary step, these points, selected after times, t_i and t_j (Figure 7.13a), are determined on the peak recorded for the standard solution, and after first determining the value of the constant signal for this solution, these points are assigned a local dispersion coefficient, D_{ij}. At the actual stage of the

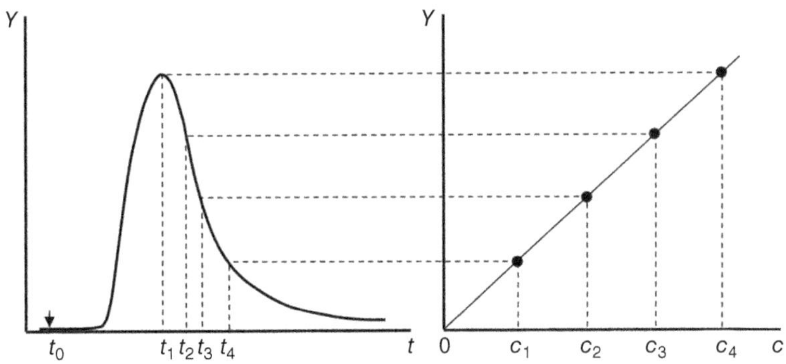

Figure 7.12 Basic gradient approach to calibration by the external standard method: individual points of the injection peak of well-defined delay times, t_1, …, t_4, are transformed to the local analyte concentrations in the standard segment, c_1, …, c_4, on the basis of dispersion coefficients initially determined.

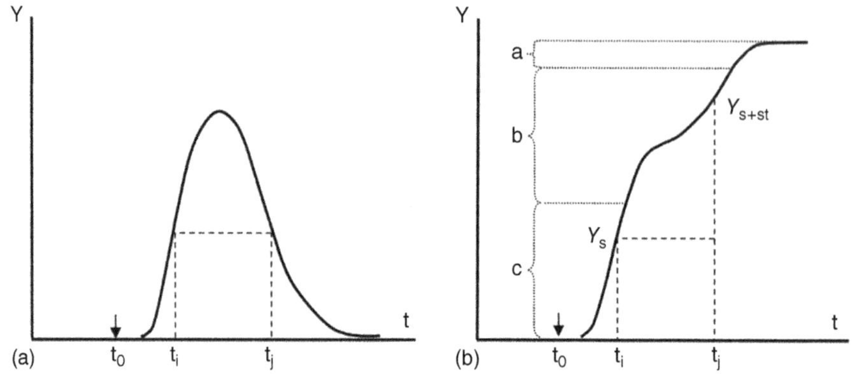

Figure 7.13 Gradient scanning standard addition method: two delay times, t_i an t_j, established using the standard peak (a) serve to find signals Y_S and Y_{s+st} on the carrier-sample-standard profile (b) enabling calculation of the analytical result; a, b, c – the sample, sample + standard and standard parts of the signal profile, respectively.

calibration procedure, the sample is injected through a valve with a configuration such that the sample segment is located between the carrier solution and the standard solution, As a result of the dispersion taking place, the sample gradually mixes with the standard solution and consequently, upon reaching the measuring system, produces the complex measurement image shown in Figure 7.13b. For the determination of the analyte concentration in the sample the signals Y_s and Y_{s+st} appearing in this image after times t_i and t_j, corresponding to the zone of the sample alone and the sample combined with the standard are taken into account. The analytical result is then determined from the formula:

$$c_x = \frac{Y_s}{Y_{s+st} - Y_s} \cdot D_{ij} \cdot c_{st} \tag{7.2}$$

An interesting question is whether the GSSAM procedure provides such a preparation of the sample and the sample with the standard that preserves a constant concentration of the sample components in both solutions. Appearing after the times t_i and t_j, the measuring points Y_s and Y_{s+st} correspond to such places of the complex segment, which have equal degree of dispersion. Thus, if a sample and a sample with a standard arrive at the measurement system at these times, the concentrations of the sample components should be equal, and the Y_{s+st} signal is greater than the Y_s signal only because of the additional amount of analyte coming from the added standard. The accuracy of the GSSAM method and its full utility in terms of its ability to eliminate multiplicative interference effects thus depends largely on the accuracy of the selection of points t_i and t_j relative to the sample and sample-with-standard zones.

The number of pairs of t_i and t_j values meeting these requirements is theoretically large, allowing multiple estimations of analyte concentration in a single sample. However, one must additionally take into account the fact that each pair corresponds to a different ratio of analyte concentrations in the standard and in the sample and therefore some of them may be practically useless from the point of view of precision of analytical results. A once predetermined series of these values allows the analysis of any number of samples under experimental conditions with parameters maintained at constant optimum levels.

Other gradient methods take advantage of the phenomenon shown in Figure 7.14 whereby, during flow, the carrier gradually penetrates more and more into a segment of that solution causing dilution. The measuring points forming the falling parts of the injection peak on either side of the maximum point can therefore be interpreted as the response of the measuring instrument to the decreasing concentration of analyte in the sample or standard segment due to the progressive dilution of these solutions.

Such interpretation of measurement results in the FIA analysis is the basis, among others, of the procedure described in the literature under the original name gradient ratio calibration method (GRCM) [16]. In this method, the number of pairs of measurement points are determined on two peaks recorded for a standard and a sample after respectively equal time since both solutions were injected into the carrier. This is illustrated in Figure 7.15a. From each ith pair of signal intensities, $Y_{\text{st},i}$ and $Y_{\text{s},i}$, corresponding to these points, the apparent analyte concentrations, c_i, in the sample at different dilutions are calculated from the formula:

$$c_i' = \frac{Y_{\text{s},i}}{Y_{\text{st},i}} \cdot c_{\text{st}} \tag{7.3}$$

where c_{st} is the analyte concentration in the injected standard solution. The graphical interpretation of Eq. (7.3) is shown in Figure 7.15. The determined values of apparent concentrations are presented as a function of the dilution degree, k_i determined as the ratio of a given signal, $Y_{\text{st},i}$, obtained for a standard after time t_i to the signal corresponding to the maximum peak point of that standard. The GRCM method is thus essentially an injection-transfer version of the dilution method.

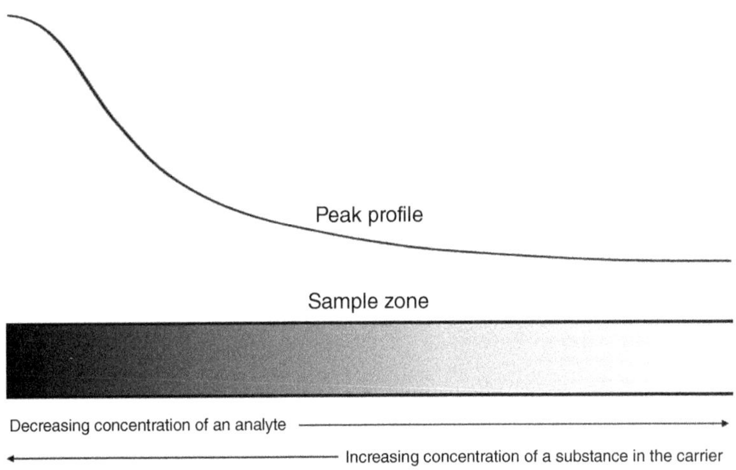

Figure 7.14 The process of spontaneous dilution of a sample segment with a carrier solution and the resulting shape of the declining part of the injection peak.

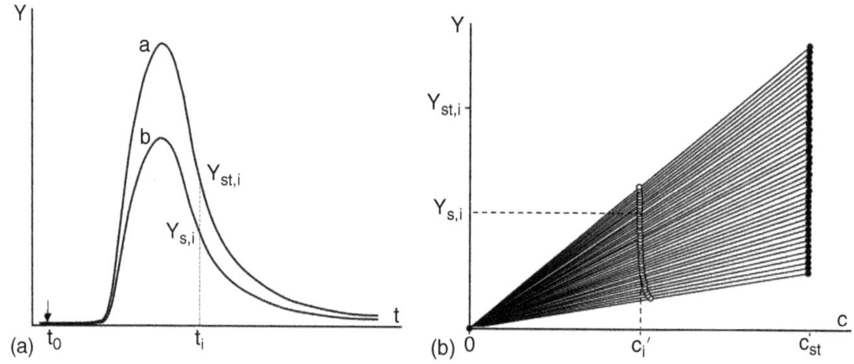

Figure 7.15 Gradient ratio calibration method: pairs of the signals obtained for the sample, $Y_{s,i}$, and standard, $Y_{st,i}$, after well-defined delay time t_i (a) are used for construction of many two-point calibration graphs and calculation of apparent analyte concentrations, c_i', in the sample segment (b); a, b – sample and standard peaks, respectively.

The gradient analysis provides the opportunity to determine a much larger number of apparent concentration values than in traditional analysis and also allows for higher sample dilution with satisfactory measurement precision. As a consequence, the dependence of these concentrations on the degree of dilution of the sample and more accurate analytical results can be obtained. This procedure also has the advantage of avoiding the need for preliminary experiments to determine local values of the dispersion coefficients. On the other hand, the final accuracy of these results is largely determined by the accuracy of such computer matching of the peaks recorded for the sample and standard segments that the pairs of measurement points taken into account in the calculation of apparent concentrations correspond exactly to the equal dilution of the two segments by the carrier solution.

If strong interference effects are present, the GRCM method provides an easy way to eliminate them gradually by using a special reagent. When discussing the dilution method it was mentioned that the action of the reagent is very effective when its solution is used to dilute the sample and standard. Thus, in the GRCM method, the reagent plays this role when it is present in the carrier solution diluting the calibration solution as shown in Figure 7.14 [17]. The effect of the special reagent introduced into the FIA system in this way is shown in Figure 7.16.

Comparing these results with those obtained by the traditional dilution method (see Figure 5.16), it can be seen that the same interferents are eliminated under the same conditions with equal efficiency, but the GRCM method creates the conditions for determining the analytical result with greater accuracy. Equally spectacular results are obtained when interference effects are compensated in the GRCM method by interferents present in the carrier solution [18].

The original way to use the gradient technique for calibration is the so-called gradient titration described as early as 1977 [19]. In this method, the sample solution is usually injected into a titrant, which acts as a carrier. During the flow, the titrant reacts with the analyte, simultaneously penetrating deeper and deeper into the sample segment on either side and gradually increasing its concentration relative to the

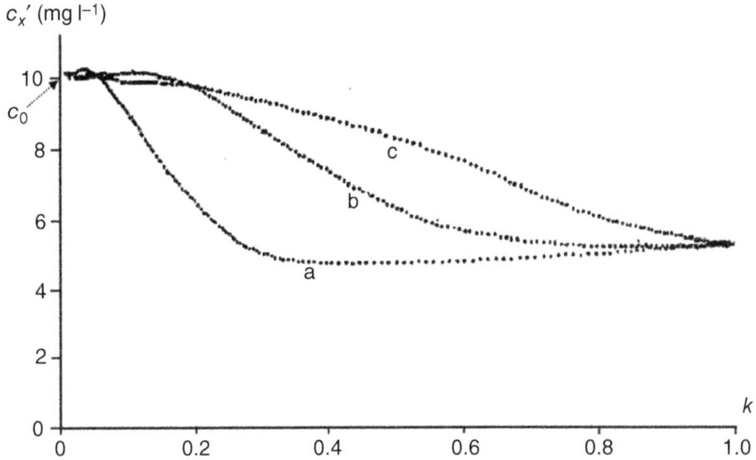

Figure 7.16 Calibration by the GRCM method: results obtained for FAAS determination of calcium in the synthetic sample contained 10 mg l^{-1} with 80 mg l^{-1} Al and diluted with lanthanum solution of concentrations 20 (a), 50 (b), and 100 (c) mg l^{-1} La: concentration of Ca in the standard solution: 10 mg l^{-1}; maximum dilution: $k = 0.005$. Source: Adapted from Kościelniak [18].

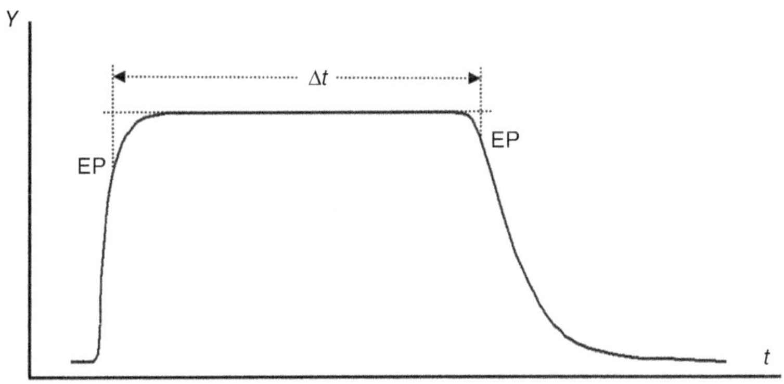

Figure 7.17 Characteristic cut injection peak in gradient titration: EP – end point, Δt – peak width as a measure of analyte concentration (the example relates to acid–base titration with spectrophotometric detection, Y – transmittance).

analyte concentration (see Figure 7.14). The moment when the entire amount of analyte has been reacted, the end point (EP) of the titration is manifested by a sharp change in the analytical signal recorded on both sides of the injection peak, as shown in Figure 7.17.

It can be shown that the measure of the analyte concentration in the sample is the width of the peak between the titration end points. To maximize the sensitivity of the determination, i.e. to create conditions for registration of peaks of the greatest possible width, flow systems intended for titration are therefore constructed in such a way as to ensure the greatest possible dispersion of the sample (by installing, e.g. a mixing chamber in the system).

Using the described procedure it is practically impossible to determine the volume of titrant or sample consumed until the end point is reached. It is therefore not possible to transform the measurement signal (peak width) into the concentration of the analyte in the sample. The only way to obtain analytical results with satisfactory precision and accuracy in gradient titration is to use analyte standards and to perform calibration by the external standard method. In addition to measuring the peak width for the sample, this parameter should be measured for a series of standards treated in the flow system as the sample, and then the analyte concentration in the sample is determined from the constructed calibration graph. This additional step significantly distinguishes gradient titration from classical titration, all the more so because in this flow mode of titration the analytical result can be determined without knowledge of the titrant concentration.

The gradient calibration method described above can be adapted in various ways to two-component analysis. It has been shown that in spectrophotometry, speciation analysis can be performed by gradient titration using the different stability of complexes formed by the two chemical forms of the analyte present in a sample with a specific titrant [20]. For example, both Fe^{2+} and Fe^{3+} ions form with sulfosalicylic acid and 1,10-phenanthroline stable complexes in acid environment (pH ≈ 3), but when EDTA (titrant) is injected into the sample, it replaces sulfosalicylic acid forming with Fe^{3+} more stable colorless complex. Interestingly, the concentrations of both analytes can be then estimated on the basis of a single peak registered at wavelength 530 nm (see Figure 7.18) as the peak width is a measure of Fe^{3+}, whereas the peak high corresponds to Fe^{2+}. Obviously, the method requires additional calibration by the ESM method (in accordance with FI titration rule) taking into account the calibration graphs made for both analytes.

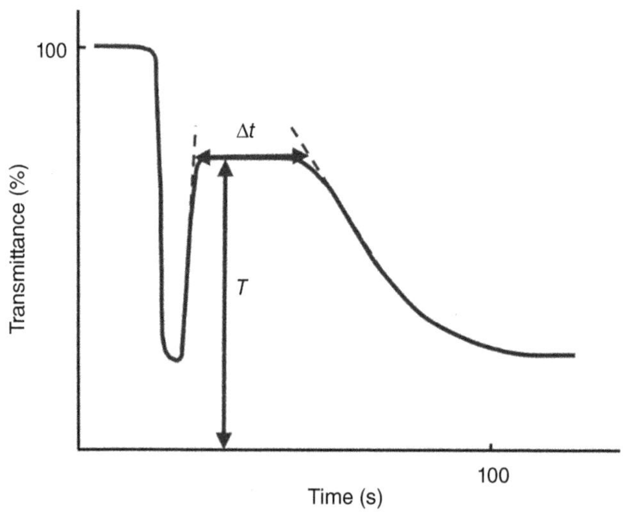

Figure 7.18 The signal obtained in two-component single-peak speciation analysis based on the gradient titration: height, T, and width, Δt, of the signal correspond to Fe^{2+} and Fe^{3+} concentrations, respectively. Source: Kozak et al. [20], fig 2 (p. 10)/with permission of Elsevier.

There have been many discussions as to whether such a procedure can in fact be considered as "titration," not only because of the above deviations from classical titration. The basic issue was that, in contrast to the classical approach, the gradient titration involves two simultaneous processes, namely chemical reaction and mass flow of the reagents, and, as a consequence, the reagents react in nonstoichiometric proportions. Moreover, it was revealed that cut-off peaks can be obtained without a reagent in the flow system, and therefore a chemical reaction is not in fact needed to obtain a calibration graph. On the other hand, it was proved experimentally that all traditional titration techniques, i.e. acid–base, complexometric, redox, and precipitation techniques, can be performed in the gradient mode. In such a situation even the authors of this concept admitted that "… the similarity of FIA and classical bath titrations is useful to recognize because such recognition turns our attention to the wealth of chemistries exploited by classical titrations that are accessible to FIA adaptation" [21].

Continuing this discussion, it was shown in [22] that in FIA analysis, in principle, any chemical reaction from which the determination of a given analyte is performed using indirect method calibration can also be used to perform the determination of that analyte in the manner characteristic of gradient titration. The difference between the two approaches comes down to technical rather than substantive aspects: when treating a given calibration method as an indirect method, standards and samples are injected into the reagent from a loop of relatively small capacity and the measurement results are read off on the basis of the height of typical injection peaks formed under these conditions, whereas the same analytical result can be obtained by peak width measurements if only the injection loop has the appropriate, much larger capacity.

The discussion of gradient titration may be concluded by saying that however gradient titration is treated and named, it cannot be denied that in flow analysis it is based on an original procedure that has all the characteristics of a calibration procedure.

The examples presented above obviously cover only a part of all the calibration solutions that have been developed so far in flow analysis. However, these examples already show that, in principle, any calibration method implemented in the traditional way can be adapted to FIA analysis while retaining all its characteristic features. This not only reduces the time and cost of analysis but in particular allows for rapid calibration with different methods in their multi-pattern versions, which is conducive to obtaining more precise analytical results than in traditional analysis. The automation of the calibration procedure provided by FIA analysis also favors the minimization of uncontrolled effects, which finds its consequences in the improvement of assay accuracy.

However, regardless of the method used, calibration modules should be installed upstream of those parts of the flow system that are responsible for processing the sample and standards to properly prepare them for measurement. This provides an opportunity to treat all calibration solutions equally, which is known to at least partially compensate for various uncontrolled effects that may occur during the analysis. It is not always possible or advantageous to build such a system, but on the other hand, if it is possible, introducing standards and samples via different routes into the flow system should be treated as a kind of negligence and a calibration error.

It should be made clear, however, that although the flow mode improves and facilitates the performance of calibrations by various methods, it is in no way, even with the help of the most modern and automatic instrumental system, able to change the principles and nature of calibration methods. Every calibration procedure developed in flow analysis can be classified as one of the calibration methods known in traditional analysis (although in some cases, as gradient titration shows, this is not quite obvious and may be debatable). However, the manipulative and gradient techniques at the disposal of flow analysis release the imagination and inventiveness of the analyst, undoubtedly contributing to the development of analytical calibration from the technical, laboratory, and interpretive side.

As stated earlier, flow analysis is a constantly evolving branch of chemical analysis. Flow systems of different designs and operating on different principles, such as multi-pump systems, multi-commutated systems, lab-on-valve systems, and others, appear in the literature. There are also continuous efforts toward miniaturization of these systems. All these instrumental solutions can be associated with a variety of calibration procedures, either as outlined above or quite different. However, FIA analysis has become somewhat exemplary in this respect, exemplifying – mainly in the 1980s and 1990s – the wide possibilities that, in terms of modifying analytical calibration methods, can exist in flow analysis.

7.2 Kinetic Analysis

Kinetic analysis involves the determination of an analyte based on a chemical reaction involving it. The nonequilibrium nature of kinetic analysis is due to the fact that as the reaction proceeds; the initial amount of analyte in the sample may nonlinearly (exponentially) increase or decrease to a maximum or zero value, respectively, as shown in Figure 7.19.

Historically, kinetic procedures for quantitative analysis have their origins in intensive research since the nineteenth century into new characteristic reactions and their application in qualitative analysis for the detection of inorganic and organic substances. The rate of a chemical reaction was first applied to quantitative analysis as early as in 1876 by Guyard who reported a determination of vanadium based on its catalytic effect on the oxidation of aniline by chlorate [23].

Systematic research on client analytical methods was initiated before World War II. In 1934, Sandell and Kolthoff reported the iodide determination based on the reaction of arsenic with cerium [24], and a little later, Krumholz and Watzek described a procedure for the determination of gold with gold as a catalyst in the reaction of silver with sulfate ions [25]. In 1941, Poluektov first applied the method of standard additions to the kinetic analysis for the determination of rhenium [26], and the concept of using the reaction rate for titration was submitted by Yatsimirskii and Fedorova in 1962 [27], calling this type of analysis catalimetric titration.

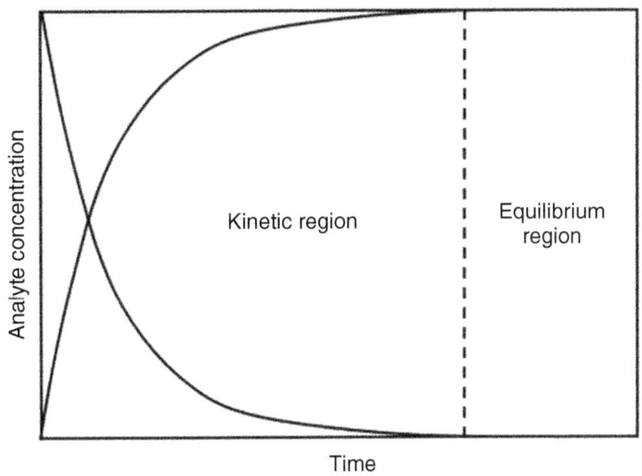

Figure 7.19 Changes in analyte concentration in a chemical reaction.

A particularly intensive development of kinetic methods in application to quantitative analysis occurred, both from the theoretical and experimental sides, in the 1950s and 1960s. Based on these studies, the first monographs were written [28, 29]. Further development of analytical kinetic methods was provided by flow injection techniques, which developed intensively in the 1980s and 1990s [30]. In 1995, a study published by IUPAC appeared [31], which reviewed methods of catalytic analysis and presented their characteristics and classification. Nowadays, kinetic analysis is a constant but – in comparison with other fields of analytical chemistry – relatively infrequent topic of scientific papers.

It is quite symptomatic that this analytical area is also quite rarely included in the academic textbooks on instrumental analysis. It should be emphasized, however, that many excellently developed kinetic methods have found their permanent application primarily in clinical and biochemical analyses.

The basis of calibration methods in kinetic analysis is the measurement of the rate of an indicator reaction, that is, a reaction whose rate depends on the concentration of the analyte. The rate of a reaction is defined as the change in concentration of the reactant per unit time. If the reaction follows the general equation:

$$\alpha + \beta = \gamma + \delta \tag{7.4}$$

then the reaction rate can be determined either from the loss of initial concentration of substrates (α, β) or products (γ, δ). In addition to "ordinary" kinetic reactions, there are also catalytic reactions:

$$\alpha + \beta \overset{\kappa}{=} \gamma + \delta \tag{7.5}$$

whose rate depends on the initial concentration of catalyst κ. In both cases, characteristic and analytically indifferent is also the presence of the induction period of the reaction, i.e. the time during which, despite the initiation of the reaction, it does not proceed with the observed rate.

The role of the analyte can be either a substrate or product of the reaction (7.4) or a catalyst in the reaction (7.5). Furthermore, the rate can be determined not only by changes in the amount of analyte in the solution under study but also by increase or decrease in the concentration of the reagent involved in the reaction along with the analyte. In other words, in kinetic analysis, the analyte can be determined directly or indirectly, which has to do with the appropriate qualification of the procedure used for a specific calibration method.

The first calibration step is therefore the determination of the reaction rates for the calibration solutions (sample and standards). For the measurement step, at least two standard solutions with different initial analyte concentrations, c_1, c_2, and a sample with an unknown analyte concentration c_x, must be prepared in an identical manner. These solutions are subjected to the specified chemical reaction under identical conditions and the analyte signal, Y, is measured at well-defined time intervals. The determinations are carried out by an appropriately chosen measurement method, e.g. spectrophotometry, potentiometry, etc. The results obtained can be presented as in Figures 7.20 and 7.21, i.e. as a dependence of the change in analyte signal, Δ (in relation to the initial concentration) vs. time.

There are various ways of determining the reaction rate corresponding to each solution from the data obtained. When the number of measurements made is large enough for an accurate determination of the relationship under study, the so-called **tangent method** is used. If the reaction is fast and its rate can be approximated by a linear function in the initial phase, the values of the tangents of the angles of the recorded straight lines are considered (Figure 7.20a). If, on the other hand, the reaction takes a relatively long time to complete, then the tangent of the angle of the tangent to the initial portion of the recorded curve is a measure of the rate of reaction of the analyte in a given solution (Figure 7.20b).

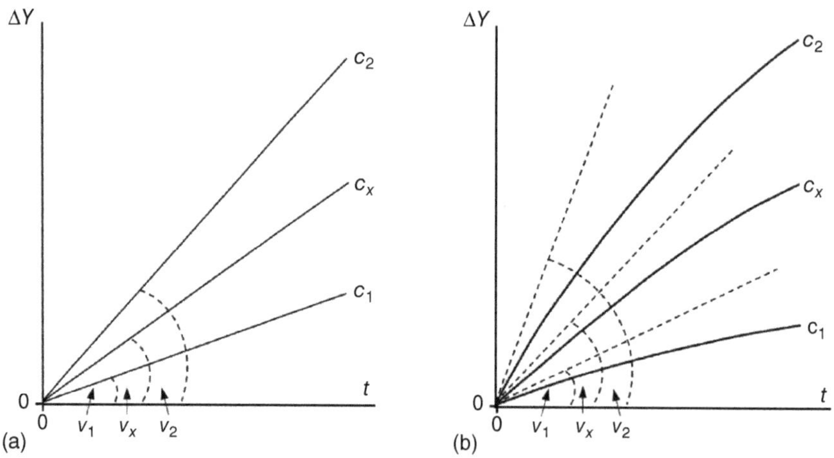

Figure 7.20 Determination of reaction rates, v_1, v_2, and v_x, for the analyte initial concentrations c_1, c_2, and c_x in the calibration solutions by the tangent method on the basis of linear (a) and nonlinear (b) ΔY vs. t dependences.

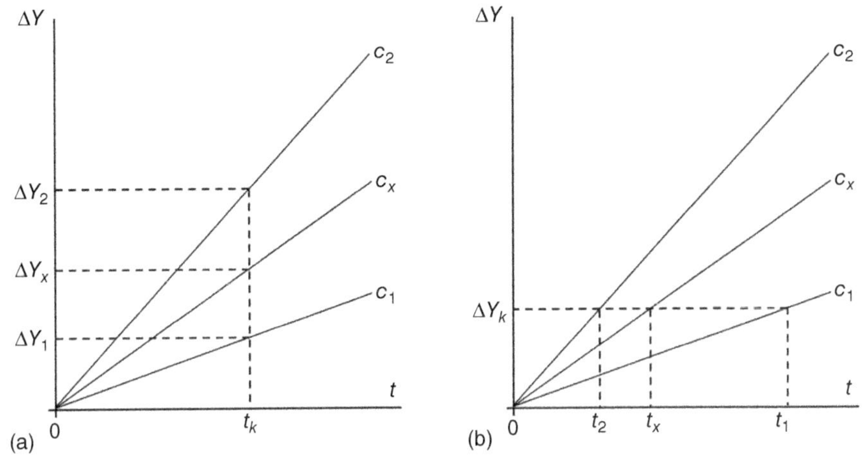

Figure 7.21 Determination of the reaction rates for the analyte initial concentrations c_1, c_2, and c_x in the calibration solutions by the fixed time method (a) and the fixed concentration method (b).

Another method is the so-called **fixed time method**. This consists in running the reaction for a well-defined time t_k and after this time the change in signal for the analyte in all the solutions tested is determined. The measure of the reaction rate is the Y/t ratio. This procedure is shown in Figure 7.21a. The **fixed concentration method**, on the other hand, uses the assumption of a linear dependence of the analyte concentration on the inverse of the time required to reach a certain analyte concentration. Thus, one proceeds as shown in Figure 7.21b: one runs the reaction in each solution up to some fixed change in concentration corresponding to the change in signal Y_k, one determines for each solution the values of time t_1, t_2, and t_x needed to reach this state, and the inverses of these values are taken as measures of the reaction rate of each solution.

The subsequent calibration procedure is based on the assumption that the reaction rate measured for an analyte of initial concentration c is well defined for that concentration and varies proportionally with its change:

$$v = K \cdot c \tag{7.6}$$

Equation (7.2) is a model calibration function that relates the reaction rate (in the role of analytical signal) to the concentration of analyte in the sample being analyzed.

Thus, from the determined reaction rates for the standard solutions, a calibration graph is constructed in the coordinate system: reaction rate vs. analyte concentration, as shown in Figure 7.22a. The measured reaction rate v_x for an analyte of unknown initial concentration c_x in the sample is then referred to the calibration graph and the value of this concentration is determined by interpolation.

This type of procedure, therefore, leads to calibration by the ESM. If the reaction rate for the calibration solutions is determined indirectly, i.e. from measurements

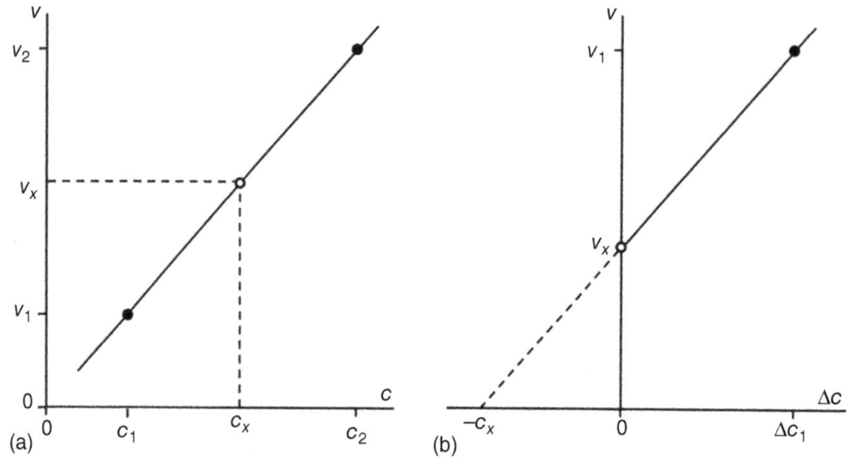

Figure 7.22 Basic calibration methods in kinetic analysis: reaction rates, v, measured for standards prepared in concentrations c_1 and c_2 separately from the sample (a) or added in concentration c_1 to the sample (b) allow for estimation of the analyte concentration, c_x, in the sample in interpolative and extrapolative ways, respectively.

made for the reagent involved in the indicator reaction, the calibration is carried out according to the indirect method.

The procedure described can be modified so that the reaction rate is determined according to one of the methods mentioned above for a sample containing the analyte at a concentration c_x and for the same sample with the addition of a standard with a concentration c_1 of the analyte. If the concentration c_x of the analyte in the sample remains constant after the addition of the standard, measuring the reaction rate of the analyte in the sample, v_x, and in the sample with the addition of the standard, v_1, allows this concentration to be extrapolated as shown in Figure 7.22b. In this way, the analyte is determined according to the E-SAM.

It should be noted that a kind of new calibration method arises when the reaction rates for the sample solutions and the sample with standard addition are determined indirectly from measurements made for the reagent involved in the indicator reaction. This procedure can be called the extrapolative indirect method (as opposed to the conventional indirect method, which is interpolative in nature).

Thus, a characteristic of kinetic calibration methods is that the real function (7.6) is not determined directly with a specific measuring instrument, but indirectly by measuring the change in analytical signals Y. One source of error may be that the measured change in analytical signal over time is not directly proportional to the change in analyte concentration, and this quantity is an appropriate measure of the rate of the reaction taking place. Therefore, in many cases, the measurement of the signal is replaced by the determination of the analyte concentration in the calibration solution after a specified reaction time. This procedure is of course more tedious and longer but often gives more accurate analytical results.

Certainly, both steps – measurement and determination of the reaction rate – contribute to the errors made in kinetic analysis. The measurement method largely determines the measurement sensitivity and the limit of quantification of the analyte. It can also, of course, contribute its characteristic uncontrolled effects, in particular, instrumental and interference effects. It has also been shown that irrespective of the measurement method used, the steady-state concentration method, although simpler to implement than the tangent method, leads to the determination of reaction rates with less accuracy. However, as numerous experimental results show, this overlap between kinetic and measurement factors is not as great an analytical limitation of kinetic methods as one might think on purely theoretical grounds.

Numerous reactions that find application in kinetic methods of quantitative analysis are characterized by a certain induction period. In particular, these are multistep reactions. In the first, slow step, there is a "preparation" of the reactants for the next steps, which may consist of catalyst formation in the case of autocatalytic reactions, catalyst activation, or the formation of crystallization centers in precipitation reactions, etc. It has been found that the induction period, t_i, is inversely proportional to the concentration of the reactants, so in particular also to the concentration of the analyte:

$$c = \frac{K}{t_i} \tag{7.7}$$

where K is a constant quantity for a given reaction. Equation (7.7) can therefore be used for calibration purposes by measuring the induction period for calibration solutions and determining the analyte concentration by interpolation from a calibration graph created in the $1/t$ vs. c coordinate system. The ESM method implemented in this way is illustrated in Figure 7.23a.

Still another kinetic calibration method is based on the fact that the analyte concentration, c_t, determined after time t of the reaction taking place is related to the initial concentration of that analyte, c, by the following formula:

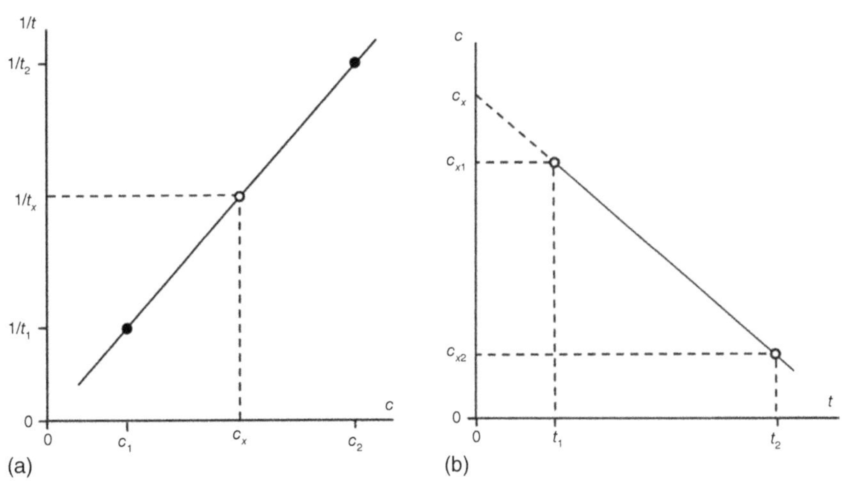

Figure 7.23 Calibration by the external standard method with the use of induction periods of a chemical reaction (a) and calibration by the kinetic version of the dilution method (b).

$$\frac{c_t}{c} = e^{-k \cdot t} \tag{7.8}$$

From this follows the equation:

$$\ln c_t = \ln c - k \cdot t \tag{7.9}$$

Thus, the graph of natural logarithm of analyte concentration vs. time should be a straight line with the directional coefficient $-k$ (i.e. characteristic for the given reaction taking place under specified conditions) and with the point of intersection with the axis of ordinates having the value ln$c$.

In practice, calibration is carried out in such a way that the tested sample is subjected to a chemical reaction and after strictly defined time intervals, t_1, t_2, the analyte is determined in it by means of the conventional ESM method, determining the values of c_{x1} and c_{x2}. From these values, the initial concentration of analyte in the sample, c_x, is calculated as shown in Figure 7.23b, which follows from Eq. (7.9). This method is an alternative to the previously described kinetic calibration approaches and can be used primarily when the chemical reaction is too fast to allow the analyte to be determined before it begins.

It should be noted that in the transformation step of the calibration procedure, this approach is analogous to the dilution method. This becomes evident when the chemical reaction and dilution processes are treated as analogous sources of changes in analyte concentration in the sample. The concentrations determined during the chemical course can be called apparent concentrations, which under zero time boundary conditions (in analogy to zero dilution step) indicate the exact concentration of the analyte in the sample in the state prior to the chemical reaction (in analogy to the state prior to the dilution process). An additional justification for treating this approach as a dilution method is the transformation step of the calibration procedure, which involves an extrapolative route to determine the analytical result.

Calibration in kinetic analysis can also be carried out by titration, which in this case is called **catalimetric titration**. It is based on the effect of a substance (anion, ligand, or metal) on a reaction catalyzed by a metal or anion and is typically used to determine either an inhibitor or a catalyst. It involves the use of a reaction in which a catalytic titrant added to the sample reacts rapidly and stoichiometrically with the analyte (inhibitor). This basic reaction is accompanied by an indicator reaction, which under the given conditions can only occur at a noticeable rate if an excess of titrant is present in the system. The end point is detected by a sudden increase or decrease in the rate of the indicator reaction. This is because the first drop of titrant in excess is not used for the primary reaction, but acts as a catalyst or inhibitor for the indicator reaction. Therefore, a very small excess of titrant is sufficient to catalyze large amounts of the indicator reagent mixture. Such catalytic end points are therefore clearly sensitive and very easily detectable. It is assumed that the amount of catalyst needed to reach the end point is directly proportional to the amount of inhibitor present in the sample.

As an example, in Figure 7.24 a titration curve is shown that was obtained when Au(III) present in some pharmaceuticals was titrated with a solution of potassium

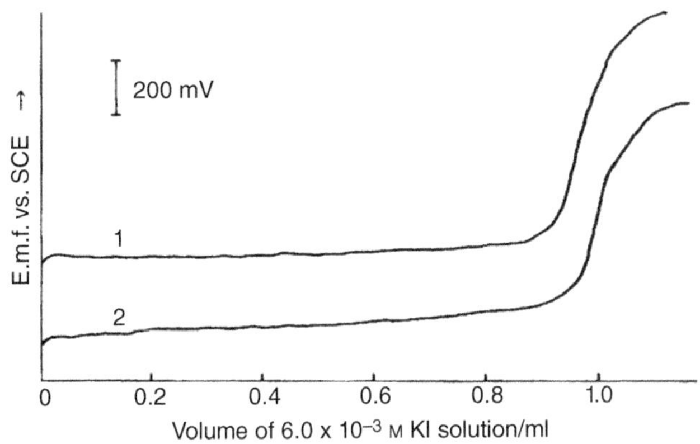

Figure 7.24 Typical catalytic titration curves obtained in the determination of 2.079 mg of gold sodium thiomalate in Tauredon (1) and of 2.154 mg of gold thioglucose in Solganol B oleosum. Source: Abramovic et al. [32], fig 3 (p. 81)/with permission of Royal Society of Chemistry.

Table 7.2 Results of the Au determination in Tauredon ampoules by catalimetric and potentiometric titrations.

	Catalimetric titration			Potentiometric titration		
Sample	c_x **(mg)**	**RSD (%)**	**RE (%)**	c_x **(mg)**	**RSD (%)**	**RE (%)**
1	26.12	1.08	3.44	26.22	0.45	3.86
2	26.32	0.27	4.24	25.92	0.42	2.66
3	26.33	0.25	4.27	26.62	0.83	5.43
4	26.10	0.58	3.37	26.01	0.59	3.03
5	26.19	0.99	3.72	25.85	0.45	2.38

Source: Adapted from Abramovic et al. [32].

iodide in the presence of components of the indicator reaction Ce(IV)–As(III) [32]. The analytical obtained are shown in Table 7.2 (the RE error was determined by comparison with the results reported by the official pharmacopoeial methods). A very good performance of the catalytic titration is seen, comparable in terms of both precision and accuracy with that achieved in potentiometric titration.

Kinetic methods have many advantages that demonstrate their very high analytical utility. Calibration procedures are simple and fast, and do not usually require complex sample and standard processing before they can be used for reaction rate measurements. Standards are prepared for measurement and subjected to measurement in the same manner as samples, which is known to compensate for analytical effects and minimize analyte determination errors. Furthermore, the preparative step of the calibration procedure can be easily miniaturized and automated, which further promotes improved precision and accuracy of analytical results. Owing to their specificity, kinetic methods, although inherently associated with the use of different types of chemical reagents, thus follow in many aspects the rules of green analytical chemistry.

Kinetic analyses can be based on many chemical reactions of different types, e.g. complexation, acid–base, oxidation–reduction, and other reactions, making them suitable for the determination of many biological and biochemical substances as well as inorganic cations and anions. As early as the 1960s, about 200 kinetic procedures adapted to the determination of about 30 elements were developed [27]. Catalytic reactions, including those in which various types of enzymes are the catalysts, are particularly widely used. In catalytic methods, the role of the analyte is generally played by the catalyst, and the rate of reaction is measured from changes in the concentration of one of the reagents of the catalyzed reaction. The calibration process is therefore indirect in these cases.

Kinetic methods can be used for the determination of analytes at both relatively high and low concentrations. In the former case, non-catalyzed first-order reactions are generally used, and in the latter case, reactions involving a catalyst as analyte. Catalytic reactions are characterized by extremely high sensitivity, comparable to that of activation analysis, which usually translates into the possibility of performing analyses with very low limits of quantification. Already in the early years of analytical kinetic studies, some elements were determined at concentrations of the order of 10^{-5}–10^{-6} μg ml^{-1} [27].

From a practical point of view, it is important that the rate of the indicator reaction is fixed within certain limits since reactions that are too fast or too slow are inconvenient for analytical purposes. It is assumed that the optimum time for measuring the reaction rate is 10–15 minutes. The duration of catalytic reactions can be regulated to some extent by the addition to the catalyst of a certain suitable substance (so-called promoter or inhibitor) increasing or decreasing the rate of the indicator reaction accordingly.

Kinetic determinations are, of course, subject to a variety of uncontrolled analytical effects (UASs), although because of the simplicity of the analytical procedure and the predominance of chemical reaction, these effects are not as dangerous to the precision and accuracy of analytical results as in many other methods. Of the preparative effects, the greatest danger comes from uncontrolled changes in the temperature at which the reaction occurs. The temperature effect manifests itself by a change in the rate constant of the reaction, and is therefore proportional in nature. During measurements, it is necessary to carefully thermostat the reaction vessels and, of course, to maintain exactly the same temperature throughout the course of the reaction in the sample and standards.

From the point of view of the influence of various factors on the reaction rate, it is very important that the reagents used are as pure as possible. Errors also depend on whether the sample and standards have the same pH and ionic strength, whether the reaction vessels do not differ in surface type and in the presence of adsorbed substances, or whether they contain even very small amounts of solids (e.g. dust).

Interference effects are mainly based on the participation of interferents in a given reaction used for the determination of a specific analyte. Their occurrence, therefore, depends on the degree of selectivity of the chosen reaction. Interferences can be of proportional, additive, or mixed nature. Proportional interference effects can of course be eliminated by performing calibrations by the standard addition method.

A specific additive effect in catalytic analyses is the overlapping of the natural rate of the basic reaction with that which depends on the concentration of catalyst as analyte; in such cases, therefore, the value of the rate without the presence of catalyst must be taken into account.

In kinetic analyses, the type and effect of interferents on the rate of a chemical reaction can often be predicted and appropriate measures are taken to eliminate or compensate for them. In some cases, it is also possible to mathematically evaluate and correct for this influence. Nevertheless, interference effects remain a significant problem and must always be taken into account. Figure 7.25 provides an example of how important it is to do preliminary studies to get a good idea of the type and magnitude of chemical interference [33].

Today, kinetic methods are often performed in dedicated flow analysis systems. FIA is naturally suited to the implementation of kinetic assays based on different types of calibration procedures since the rate and time of reaction can be easily controlled by setting the flow rate of the solutions and the path length of the reaction mixture to the detector. Rapid and efficient mixing of samples and reagents occurs in the FIA system, without the risk of contamination of reacting solutions with the environment. Sample and standards are automatically treated in the same way in all respects. Analyses are fast, making it possible to carry out determinations based on reactions that are too fast for use in stationary analysis conditions.

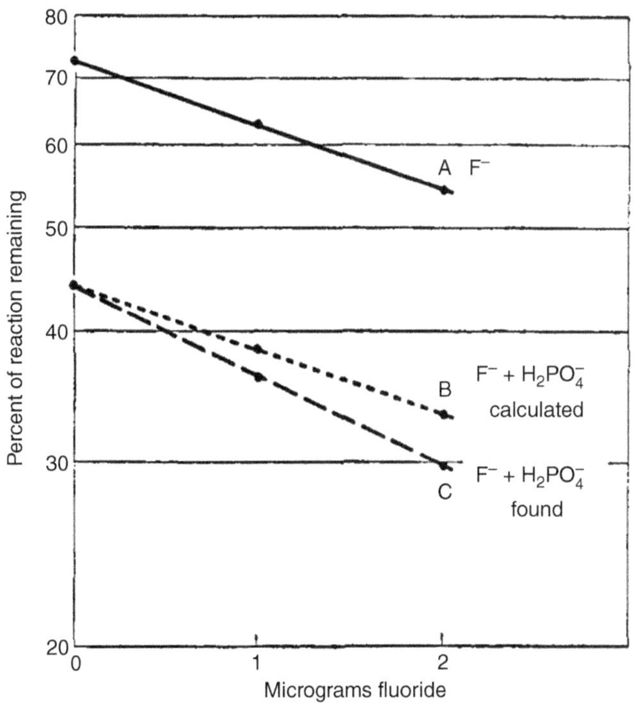

Figure 7.25 Interference between phosphate and fluoride in cerous–ceric electron transfer reaction: line A – no effect, line B – calculated (expected) additive effect, line C – observed, additive, and proportional effect. Source: Armstrong and Singer [33], fig 3 (p. 1048)/with permission of American Chemical Society.

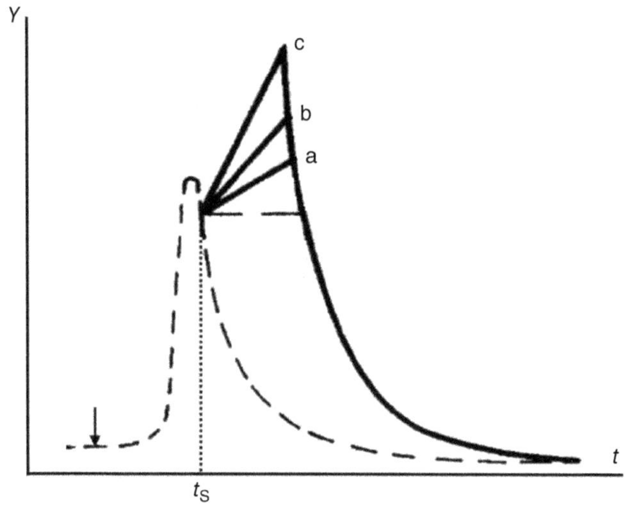

Figure 7.26 Characteristic measurement pattern obtained for stopped-flow technique; t_s – stopping moment of the sample and standard flow, a, b, c – changes in the injection peak profile caused by different analyte concentrations.

A good example of the benefits brought by the implementation of catalytic methods in FIA analysis is the introduced as early as 1982 [34]. In this approach, the sample zone is merged and mixed with a reagent in the conventional FIA system, and then it is desirable to stop the flow in the detector cell. By doing so the reaction can be observed as a change in the peak profile, which is presented in Figure 7.26. If no reaction occurs the signal is essentially a flat line parallel to the existing baseline. However, if the analyte is still produced due to the reaction, the observed signal is changed. From the measurement results obtained, the reaction rate with the analyte in the calibration solutions can be determined by the tangent method and the analyte can be determined by one of the calibration methods shown in Figure 7.22.

The method is characterized by very good sensitivity and selectivity. The key condition that determines the accuracy and precision of analytical results is the precise determination of the appropriate moment to stop the flow of standard and sample segments and the reproducibility of the moment at which this stop occurs. Due to its advantages, the stopped-flow technique is not only used for analytical purposes but finds general kinetic applications for studying fast chemical reactions [35].

Regardless of how kinetic analysis is implemented, analyte determinations are generally based on known, previously described calibration methods. However, one should not overlook the noted new calibration possibility revealed when reaction rates are determined for solutions prepared according to the E-SAM method in an indirect manner, i.e. based on measurements made for the reagent involved in the indicator reaction. This combination of the E-SAM method with the indirect method indicates that different calibration methods can be combined. This topic will be further developed in Chapter 8.

References

1 Skeggs, L.T. Jr. (1956). An automatic method for colorimetric analysis. *Clinical Chemistry* 2 (4): 241.
2 Trojanowicz, M. (ed.) (2008). *Advances in Flow Analysis*. Weinheim: Wiley-VCH.
3 Růžička, J. and Hansen, E.H. (1975). Flow injection analyses. 1. New concept of fast continuous-flow analysis. *Analytica Chimica Acta* 78 (1): 145–157.
4 Kościelniak, P. and Kozak, J. (2004). Review and classification of the univariate interpolative calibration procedures in flow analysis. *Critical Reviews in Analytical Chemistry* 34 (1): 25–37.
5 Kościelniak, P. and Kozak, J. (2006). Review of univariate standard addition calibration procedures in flow analysis. *Critical Reviews in Analytical Chemistry* 36 (1): 37–40.
6 Agudo, M., Rios, A., and Valcarcel, M. (1992). Automatic calibration and dilution in unsegmented flow systems. *Analytica Chimica Acta* 264 (2): 265–273.
7 Kościelniak, P., Herman, M., and Janiszewska, J. (1999). Flow calibration system with the use of fully rotary directive valve. *Laboratory Robotics and Automation* 11 (2): 111–119.
8 Tyson, J.F., Appleton, J.M.H., and Idris A.B. (1983). Flow injection calibration methods for atomic absorption spectrometry. *Analytica Chimica Acta* 145: 159–168.
9 Reis, R.F., Jacintho, A.O., Mortatti, J. et al. (1981). Zone-sampling processes in flow injection analysis. *Analytica Chimica Acta* 123: 221–228.
10 Tyson, J.F. and Bysouth S.R. (1988). Network flow injection manifolds for sample dilution and calibration in flame atomic absorption spectrometry. *Journal of Analytical Atomic Spectrometry* 3(1): 211–215.
11 Kościelniak, P. and Janiszewska, J. (1997). Design of the fully rotary valve for calibration purposes in flow analysis. *Laboratory Robotics and Automation* 9 (2): 47–54.
12 Wójtowicz, M., Kozak, J., Danielewska, D. et al. (2009). Flow injection titration basing on the merging-zone technique. *Talanta* 79 (4): 1006–1010.
13 Bettmer, J., Heilmann, J., Kutscher, D.J. et al. (2012). Direct μ-flow injection isotope dilution ICP-MS for the determination of heavy metals in oil samples. *Analytical and Bioanalytical Chemistry* 402 (1): 269–275.
14 Olsen, S., Růžička, J., and Hansen, E.H. (1982). Gradient techniques in flow injection analysis. Stopped flow measurement of the activity of lactate dehydrogenase with electronic dilution. *Analytica Chimica Acta* 136: 101–112.
15 Fang, Z., Harris, J., Růžička, J. et al. (1985). Simultaneous flame photometric determination of lithium, sodium, potassium, and calcium by flow injection analysis with gradient scanning standard addition. *Analytical Chemistry* 57 (7): 1457–1461.
16 Fan, S. and Fang, Z. (1990). Compensation of calibration graph curvature and interference in flow injection spectrophotometry using gradient ratio calibration. *Analytica Chimica Acta* 241 (1): 15–22.

17 Kościelniak, P., Sperling, M., and Welz, B. (1996). High efficient calibration procedure for flow injection flame atomic absorption spectrometry. *Chemia Analityczna (Warsaw)* 41 (4): 587–600.

18 Kościelniak, P. (1998). Calibration procedure for flow injection flame atomic absorption spectrometry with interferents as spectrochemical buffers. *Analytica Chimica Acta* 367 (1–3): 101–110.

19 Růžička, J., Hansen, E.H., and Mosbaek, H. (1977). Flow injection analysis. Part IX. A new approach to continuous flow titrations. *Analytica Chimica Acta* 92 (2): 235–249.

20 Kozak, J., Gutowski, J., Kozak, M. et al. (2010). New method for simultaneous determination of Fe(II) and Fe(III) in water using flow injection technique. *Analytica Chimica Acta* 668 (1): 8–12.

21 Růžička, J. and Hansen, E.H. (1988). *Flow Injection Analysis*, 2e. New York: Wiley.

22 Wieczorek, M., Kozak, J., Kościelniak, P. et al. (2012). Critical approach to flow injection gradient titration as a calibration method. *Talanta* 96: 34–38.

23 Guyard, M.A. (1876). On the formation of aniline black, by means of vanadium salts, and the theory of the formation of aniline black (in French). *Bulletin de la Société Chimique de France* 25: 58–64.

24 Sandell, E.B. and Kolthoff, I.M. (1934). Chronometric catalytic method for the determination of micro quantities of iodine. *Journal of the American Chemical Society* 56: 1426.

25 Krumholz, P. and Watzek, H. (1937). Über die katalytische Wirkung von Gold bei der Reduktion von Silbersalzen (in German). *Microchimica Acta* 2: 80–84.

26 Poluektov, N.S. (1941). Colorimetric determination of rhenium by a catalytic reaction (in Russian). *Journal of Applied Chemistry* 14: 695–702.

27 Yatsimirskii, K.B., Fedorova, Y.I., and Y.I. (1962). Catalimetric titration. *Proceedings of the USSR Academy of Sciences* 143: 143–143.

28 Yatsimirskii, K.B. (1967). *Kinetic Methods of Analysis (in Russian)*, 2e. Moscow: Goschimizdat.

29 Mark, H.B. Jr. and Rechnitz, G.A. (1968). *Kinetics in Analytical Chemistry*. New York: Interscience.

30 Kawashima, T. and Nakano, S. (1992). Flow-injection analysis of trace elements by use of catalytic reactions. *Analytica Chimica Acta* 261 (1, 2, 182): 167.

31 Műller, H. (1995). Catalytic methods of analysis: characterization, classification and methodology. *Pure and Applied Chemistry* 67 (4): 601–613.

32 Abramovic, B.F., Marinkovic, M.M., and Gaal, F.F. (1990). Titrimetric determination of gold in some pharmaceutical preparations. *Analyst* 115 (1): 79–83.

33 Armstrong, W.D. and Singer, L. (1954). Fluoride determination by electron transfer catalysis. *Analytical Chemistry* 26 (6): 1047–1049.

34 Růžička, J. and Hansen, E.H. (1979). Stopped flow and merging zones – new approach to enzymatic assay by flow injection analysis. *Analytica Chimica Acta* 106 (2): 207–224.

35 Gomez-Hens, A. and Perez-Bendito, D. (1991). The stopped-flow technique in analytical chemistry. *Analytica Chimica Acta* 242 (2): 147–177.

8
Complex Calibration Approaches

Chapters 5, 6, and 7 have presented the various calibration methods used in quantitative analysis. They differ from each other in many respects. Each has its own characteristic advantages and disadvantages that determine its applicability in different analytical circumstances. Despite the large number of these various approaches, however, their applicability in many unforeseen situations that may arise in analytical practice is limited.

This raises the question of whether calibration methods can be combined with each other and how to do this so that the combination is reasonable and beneficial? It is all about such combinations of methods that reveal more of the advantages of their component methods and less of their disadvantages and limitations. The creation of new calibration methods, even the most conceptually ingenious, but not showing in practice any new advantages over the methods already developed does not make sense in such a practical field of chemistry as chemical analysis. Several proposals for calibration methods assembled together in different ways are presented below. They show the various possibilities of creating new calibration approaches with their own specific properties and using these features to overcome various analytical problems. They are limited to those solutions that have found their way into the literature and have been experimentally proven. Thus, they certainly do not exhaust all the possibilities of creating complex calibration methods and still leave considerable room for analysts' own invention in this regard.

8.1 Extrapolative Methods

Among the calibration approaches discussed earlier, the method of particularly high analytical utility, and at the same time well-known and popular among analysts, is certainly the extrapolative standard addition method (E-SAM). As is well known, it offers the possibility of compensating for proportional interference effects, so commonly found in various analytical systems and posing an obstacle to obtaining analytical results of satisfactory accuracy. It has been shown that any attempt to change the E-SAM method to an interpolative or indicative version causes it to lose this analytical characteristic feature. Therefore, it seems that another way to improve the E-SAM method is to combine it with methods of an interpolative nature and thus create extrapolative versions of them.

Calibration in Analytical Science: Methods and Procedures, First Edition. Paweł Kościelniak.

This concept is quite attractive. It can be assumed that it will allow combining the advantages of E-SAM method with the benefits offered by other methods such as the possibility of using an extended range of analytical and measurement methods (indirect method), increasing the precision of determinations (internal standard method), or minimizing the complex interference effects (dilution method). Of course, it must be taken into account that the combination of methods will also reveal their drawbacks and limitations, but the extent to which this will occur and what the end result will be can best be seen experimentally.

As will be shown, the realization of the idea of transforming interpolation methods into their extrapolation versions is quite very simple and, in principle, requires only some changes at the preparative stage taking into account the characteristic way of preparing calibration solutions in the E-SAM method and in the methods combined with it. Of course, these changes consequently impose the necessity to make certain measurement and transformation modifications, but they do not generally go conceptually beyond the way in which these calibration steps are carried out in the two methods that are combined.

8.1.1 Extrapolative Indirect Method

When discussing calibration methods in kinetic analysis, attention was drawn to the possibility of performing calibrations by the E-SAM method in such a way that the reaction rate for the sample solutions and the sample with the addition of a standard was determined indirectly from measurements made for the reagent involved in the indicator reaction. This method was called the **extrapolative indirect method**. How can this calibration procedure be generalized by adapting it to steady-state analysis?

When starting the extrapolation calibration by the SAM in the basic version, reagent R in equal concentration and volume is added to the sample and standard sample solutions are prepared according to the E-SAM method. This procedure is shown schematically in Figure 8.1.

The amount of reagent added should be such that it is capable of reacting with the total amount of analyte contained in the sample and in the sample with the standard

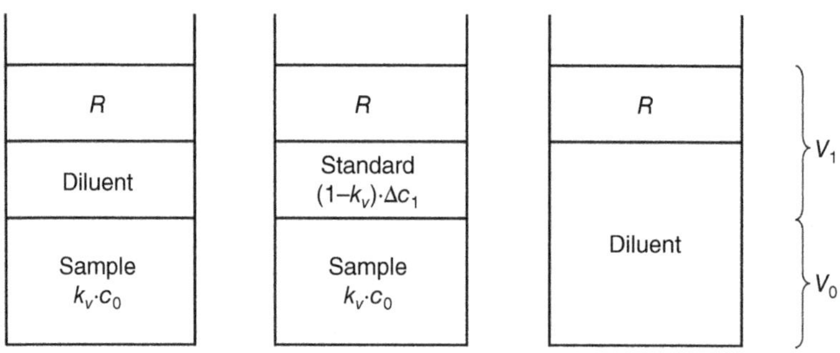

Figure 8.1 Scheme of the basic form of the extrapolative indirect method at the preparative stage of the calibration process.

added. If the sample and standard solutions initially contained the analyte at concentrations of c_0 and c_1, respectively, after dilution in the calibration solutions the concentrations take on the values $k_v \cdot c_0$ i $(1-k_v) \cdot \Delta c_1$, where k_v is the degree of dilution of the sample.

The reagent added to the calibration solutions should satisfy the same conditions as for the conventional indirect interpolation method. A chemical reaction involving the analyte and the reactant described by the general Eq. (5.33) then takes place in the prepared solutions. After the reaction is complete, measurements are made of the signal intensity Y_r for the reaction product AnR or that amount of the reactant R_r of the reactant that remains after the reaction in the sample (Y_{r0}) and in the samplc with the addition of the standard (Y_{r1}). The calibration graphs obtained in both cases are shown in Figure 8.2.

As can be seen, the method is extrapolative in nature due to the way the calibration solutions are prepared. When the signal is recorded for the AnR product (Figure 8.2a), the calibration graph is ascending and extrapolated as in the E-SAM method, i.e. until the signal reaches zero (after possibly correcting the blank signal beforehand). In the situation where the signal source is the R_r reagent (Figure 8.2b), the calibration graph is decreasing, making it impossible to extrapolate the analytical result in a similar way. So how to proceed in this situation?

Well, consider that in the E-SAM method the extrapolation is made to the signal corresponding to the solution containing no analyte. Thus, in analogy, it should be assumed that the decreasing calibration graph should be extrapolated to the Y_{rr} signal measured for the solution without the presence of the analyte, that is, containing the reactant itself [1]. In this case, there is, therefore, a need for a solution of this composition for the calibration procedure, which is included in

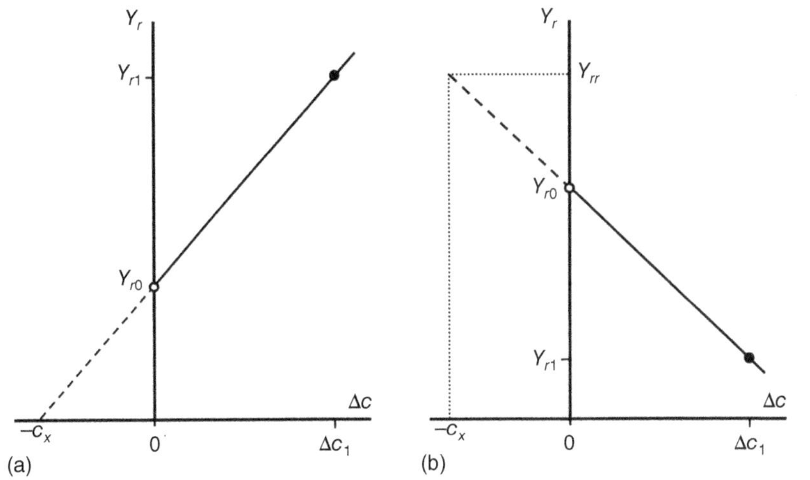

Figure 8.2 Extrapolative indirect method: calibration graphs constructed on the basis of signals, which are measured for the reaction product AnR (a) or for an unreacted amount R_r of reagent R (b) in the sample, Y_{r0}, and in the sample with standard, Y_{r1}, leading to the analytical result, c_x, in the extrapolative way.

Figure 8.1. This solution should, of course, contain the same amount of reactant as the other calibration solutions.

It can be shown that in both cases the analytical result, c_x, is determined by extrapolation using a formula analogous to formula (6.3) for calculating the analytical result in the E-SAM method:

$$c_x = \left| \frac{Y_0}{Y_1 - Y_0} \right| \cdot \frac{1 - k_v}{k_v} \cdot \Delta c_1 \tag{8.1}$$

where $k_v = V_0/(V_0 + V_1)$ is the degree of dilution of the sample in the calibration solutions.

The procedure of the extrapolative indirect method in the multipoint version is quite obvious. It requires the preparation of a series of sample solutions with the addition of standard with increasing concentration of the analyte, while the determined volumes of the sample (V_0), sample with the addition of standard (V_1) and regent (V_2) solutions should be retained in all calibration solutions, regardless of how they are prepared and added standard solutions. The concentration of the analyte in the sample is then determined from the formula:

$$c_x = \left| \frac{\hat{Y}_0}{\hat{Y}_n - \hat{Y}_0} \right| \cdot \frac{1 - k_v}{k_v} \cdot \Delta c_n \tag{8.2}$$

where $\hat{Y}_0$ and $\hat{Y}_n$ are the signal intensities, determined respectively for the sample and for the sample with the addition of the standard with the highest concentration of the analyte, Δc_n, resulting from the model function fitted (e.g. by the least squares method) to the measurement points.

As can be seen, the described calibration procedure combines the procedure of the E-SAM method with the procedure characteristic of the indirect method. It requires preparation of calibration solutions in a way characteristic of both methods, but it does not require making more solutions than is necessary for both component methods. It does not require a larger number of measurements and is relatively simple in terms of interpretation and calculation.

It can be proved that **the contribution of the E-SAM method to the extrapolative indirect method is manifested in that it becomes theoretically robust to proportional uncontrolled effects**. Moreover, the basis of the method is a chemical reaction that makes it possible to perform calibrations on the basis of signals generated by a component of the calibration solutions other than the analyte. Thus, it can be assumed that in practice the method is able to reveal both basic advantages of the component methods. **On the other hand, the indirect method brings in all the previously described problems associated with the additional participation of the reactant in the calibration process.**

Therefore, the question arises: what is the real value of the indirect extrapolation method and is it worth including it in the group of methods recommended for use in laboratory practice? It is difficult to answer this question unequivocally in the absence of sufficiently large experimental material of a research and comparative character. Certainly, the method is occasionally used by analysts, but it does not receive the main attention from the methodological side, especially when the resulting calibration graph is ascending (as in the case of kinetic analysis).

If the signal source is the reactant R_r remaining after the reaction (Figure 8.2b) and the calibration graph is decreasing, the practical application of this method may be hindered by the problem of the above described unusual interpretation of this graph at the stage of determining the analytical result. Therefore, this particular version of the extrapolation indirect method was of interest to the author and his team (as probably the first) from the experimental side [1].

The FAAS technique (using the FIA system) was performed with alkaline earth metals (Ca, Mg, Ba), which are known to be potential interferents both with respect to the analyte (phosphate ions) and with respect to each other when acting as a reactant. The results of the analyses of model samples in different reagent–interferent systems are presented in Table 8.1.

The method fulfills its basic purpose by acting as a calibration method: it allows the determination of analytes with satisfactory precision and accuracy in the absence of interferents in the sample. All three test metals can be used as reagents, however, leading to results of varying precision, depending mainly on the sensitivity of the reagent (the sensitivity of measurements performed for barium was much lower than for the other two reagents). Regardless of the type of reagent used and the possible presence of an interferent, the extrapolative transformation of the measurement results is associated, as expected, with a slight decrease in precision of the analytical results.

It was also found that the extrapolation method allows, in the system studied, to obtain results of greater or comparable accuracy to that achieved by using this method in the interpolation version. A clear improvement in the accuracy of phosphate determination is observed when strontium and calcium were used alternately as reactant and interferent. However, more accurate results cannot be counted on when one of these elements is present in the sample and barium is the reagent (although the relatively low repeatability of results observed in this case makes it difficult to draw clear conclusions).

Table 8.1 Results of the FAAS determination of phosphate obtained by the interpolative and extrapolative indirect calibration methods.

	Sample composition (mg l^{-1})		Analyte concentration found (mg l^{-1})	
Reagent (mg l^{-1})	**Analyte**	**Interferent**	**Interpolative mode**	**Extrapolative mode**
Sr, 10.0	PO_4^{3-}, 2.00	—	1.97 ± 0.05	1.87 ± 0.09
Sr, 10.0	PO_4^{3-}, 2.00	Ca, 10.0	1.10 ± 0.07	1.97 ± 0.20
Ca, 5.0	PO_4^{3-}, 2.00	—	1.94 ± 0.07	1.91 ± 0.12
Ca, 5.0	PO_4^{3-}, 2.00	Sr, 10.0	1.59 ± 0.06	1.96 ± 0.15
Ba, 200.0	PO_4^{3-}, 20.00	—	19.1 ± 2.5	19.4 ± 4.0
Ba, 200.0	PO_4^{3-}, 20.00	Sr, 10.0	16.2 ± 6.0	18.3 ± 7.3
Ba, 200.0	PO_4^{3-}, 20.00	Ca, 10.0	17.3 ± 5.1	25.5 ± 9.8

Source: Kościelniak and Wieczorek [1], table 3 (p. 430)/Taylor & Francis.

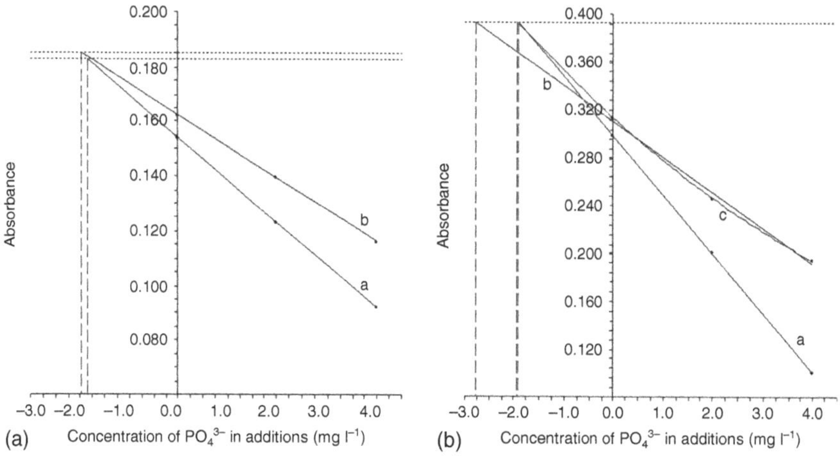

Figure 8.3 Extrapolative indirect method applied to the FAAS determination of phosphate with strontium (A) and calcium (4) as reagents as well as with no interferents (a) and with calcium (b) and strontium (c, *c′*) as interferents present in the samples. Source: Kościelniak and Wieczorek [1], fig 4 (p. 429)/with permission of Taylor & Francis.

As assumed, the condition for the effectiveness of the indirect method in the extrapolation version when the interference effect is present is its proportional character. As can be seen in Figure 8.3, an effect of this nature is observed in both analytical systems tested with calcium and strontium as reagent and interferent. Calcium produces a linear effect that can be compensated very well by extrapolation (Figure 8.3a). In the case of strontium, a pronounced nonlinear effect is observed (Figure 8.3b), which, however, when taken into account by means of a model function of appropriate form can, as can be seen, also be completely compensated after extrapolation of this function.

Based on the tests performed, it can be assumed that the proportional effects seen in Figure 8.3 are due to the interference of the interferents with the analyte, as they increase with increasing analyte concentration. Of course, there may be overlap of other effects, which is all the more likely since the reactant and interferent have similar chemical properties and both are interferents with respect to the analyte. The magnitude and nature of the final effect certainly also depend on the relative concentrations of all three chemical entities (analyte–reagent–interferent) making up the analytical system.

As it seems, however, the use of the extrapolative indirect method in analysis by such measurement techniques as FAAS is quite limited due to the risk of common interference effects and the difficulty in predicting the nature of these effects. This is evidenced not only by the results shown in Table 8.1 for the determination of phosphate using barium as reactant, but also by those shown in Figure 5.26 obtained with the interpolative indirect method. They prove that the use of the extrapolation version is not likely to be successful also when phosphates are determined with calcium and strontium in the presence of silicon and vanadium.

Table 8.2 Results of the spectrophotometric determination of chlorite in water samples obtained by the interpolative and extrapolative indirect calibration methods.

	Sample composition (mg l^{-1})		Analyte concentration found (mg l^{-1})	
Reagent (mg l^{-1})	**Analyte**	**Interferent**	**Interpolative mode**	**Extrapolative mode**
Fe^{2+}, 500	ClO_2^-, 0.200	—	0.209 ± 0.008	0.207 ± 0.008
Fe^{2+}, 500	ClO_2^-, 0.200	NH_4^+, 900	0.136 ± 0.009	0.195 ± 0.012
Fe^{2+}, 500	ClO_2^-, 0.400	NH_4^+, 900	0.331 ± 0.010	0.392 ± 0.012

Source: Kościelniak and Wieczorek [1], table 4 (p. 433)/Taylor & Francis.

In the analysis by molecular spectrometry methods, the interference effect that poses the greatest analytical problems is the spectral additive effect. This effect cannot, of course, be compensated for by the extrapolation calibration of the indirect method, nor can any other calibration method that has been discussed so far. Under established chemical conditions, proportional effects are rare, although, because of the chemical reaction underlying the indirect method, their manifestation is more likely in this method than, for example, in the external standard method.

An example of a strong proportional chemical effect is the effect of ammonium ions on the determination of chlorate ions, ClO_2^-, in water based on the reaction:

$$ClO_2^- + 4\,Fe^{2+} + 4\,H^+ = Cl^- + 4\,Fe^{3+} + 2\,H_2O \tag{8.3}$$

The signal is recorded under conditions characteristic of Fe^{2+} ions remaining after reaction with the analyte. The analytical results obtained are shown in Table 8.2.

In this case, the extrapolation method passed the test very well, producing much more accurate results with similar precision compared to the interpolation method. The calibration graphs leading to these results are shown in Figure 8.4.

Thus, it can be generally said that the extrapolative interpolation method works practically in accordance with the theoretical complexes and is undoubtedly an analytically useful calibration method in certain cases. In view, however, of the increased number of sources of interference effects of potentially different character, the possibility of its application in a particular case should be confirmed, however, by the results of previously performed experiments in the analytical system currently under investigation. Perhaps, in such systems where the analyte and reactant have similar chemical properties, a single, appropriately selected special reagent may be an effective means of eliminating interference effects, acting simultaneously on the various sources of these effects. Research in this area would certainly be of great interest.

8.1.2 Extrapolative Internal Standard Method

The internal standard method is much better documented in its extrapolation version than the indirect method. Suffice it to say that already the first application of the standard additive method in 1936 (for spectrographic determination of lead) was

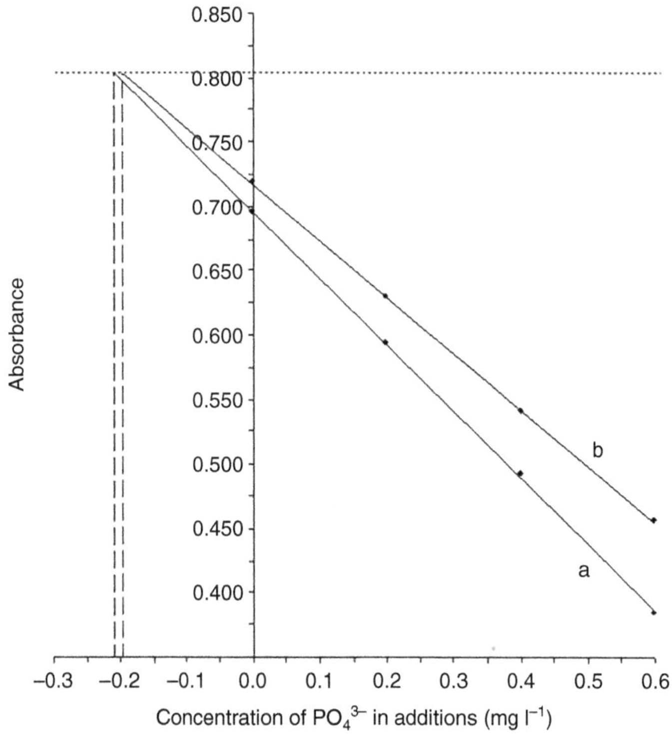

Figure 8.4 Extrapolative indirect method applied to the spectrophotometric determination of chlorite in the samples containing ammonium ions in concentration of 0 (a) and 900 (b) mg l^{-1}. Source: Kościelniak and Wieczorek [1], fig 6 (p. 432)/with permission of Taylor & Francis.

accompanied by the use of an internal standard (Mg) [2]. Thus, the extrapolation version of the internal standard method was developed on the basis of the interpolation version of this method, which, as is known, already at that time had a long tradition and was routinely used in spectrography. In later years, one can find reports of the use of this concept in analysis by chromatographic techniques, X-ray spectrometry, and ICP spectrometry, among others.

In its classical and basic version, the **extrapolation internal standard method** involves a similar way of preparing calibration solutions as the extrapolation indirect method: equal amounts of the internal standard IS are added to sample and sample solutions with analyte standard, prepared according to the E-SAM method, and all solutions are made up to the same volume. This is illustrated schematically in Figure 8.5.

Measurements for both solutions are performed as in the internal standard interpolation method: in parallel (or successively) under conditions characteristic of the analyte, Y_0, Y_1, and of the internal standard, Y_{w0} and Y_{w1}. The response of the measuring instrument to the analyte concentration is expressed as ratios of the intensities of the signals obtained for the analyte and the internal standard in the sample, $R_0 = Y_0/Y_{w0}$, and in the sample with the analyte standard, $R_1 = Y_0/Y_{w1}$, and from these values, the calibration graph shown in Figure 8.6 is constructed.

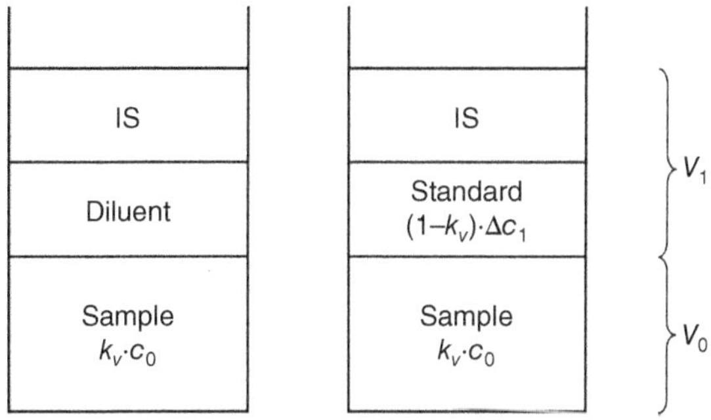

Figure 8.5 Scheme of the basic form of the extrapolative internal standard method at the preparative stage of the calibration process.

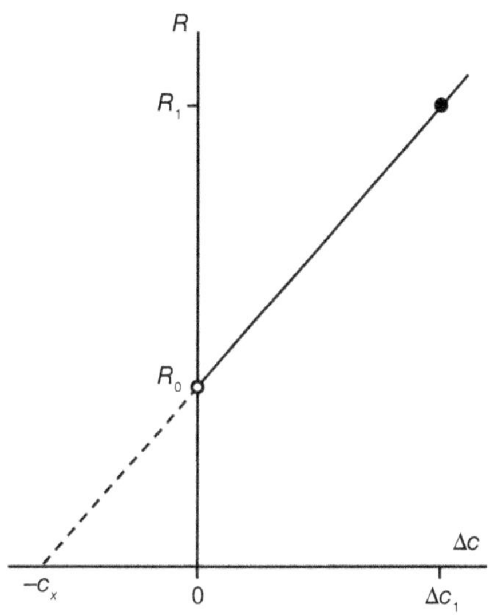

Figure 8.6 Calibration in accordance with extrapolative internal standard method: extrapolative transformation of the ratio of analyte to internal standard signals obtained for the sample, R_x, and for the sample with the analyte standard, R_1, to the analytical result, c_x.

The analytical result is calculated from:

$$c_x = \frac{R_0}{R_1 - R_0} \cdot \frac{1 - k_v}{k_v} \cdot \Delta c_1 \tag{8.4}$$

where k_v is the degree of dilution of the sample in the calibration solutions. In the multipoint extrapolation version of the internal standard method, formula (8.4) takes the form:

$$c_x = \frac{\hat{R}_0}{\hat{R}_n - \hat{R}_0} \cdot \frac{1 - k_v}{k_v} \cdot \Delta c_n \tag{8.5}$$

where $\hat{R}_0$ and $\hat{R}_n$ are the ratios of signal intensities, determined respectively for the sample and for the sample with the standard with the highest analyte concentration, Δc_n, resulting from the model function fitted (e.g. by the least squares method) to the measurement points.

As it is known, the internal standard method theoretically allows compensation of random errors, which should compensate for some limitations of the E-SAM method in this respect. On the other hand, the contribution of the E-SAM method provides an opportunity to strengthen the internal standard method from the point of view of its resistance to interference effects. Does indeed the extrapolated internal standard method show these positive features in analytical practice? Reports of authors who have tested this calibration method with different measurement methods do not allow a clear answer to this question.

Positive performance of the extrapolation internal standard method was recorded, for example, in the determination of aromatic compounds in gasoline samples by gas chromatography technique [3], as well as in the simultaneous determination of Cu, Zn, Rb, and Sr in the samples of spinach and orchard leaves by proton-induced X-ray emission (PIXE) spectroscopy [4]. However, the results obtained were not compared with those obtained by other calibration methods sufficiently to fully evaluate the analytical performance of the tested extrapolation method.

Much more information of this type is provided by studies on the determination of cadmium in urine samples performed by the ICP-OES technique [5]. An electrothermal technique (tungsten coil vaporizer) was used to vaporize and atomize the sample. It was assumed that in such a rather complex instrumental system it is advantageous to use an internal standard (bismuth). At the same time, it was expected that the use of the interpolation method would not be sufficient to improve the accuracy of the determinations due to the large potential systematic errors that can be induced by the complex matrix of biological samples. The results obtained with the interpolative internal standard method, shown in Table 8.3, confirmed these concerns. However, using the extrapolation version of the method allowed the determination of cadmium with much better accuracy and very good precision.

Slightly different conclusions arise from the application of the ICP-OES technique and the extrapolation internal standard method to the multicomponent analysis of geological materials [6]. This is shown by the results of the determination of rare earth metals using Sc as an internal standard, presented in Table 8.4. Although the calibration method tested allowed for significant compensation of interference

Table 8.3 Results of the ICP-OES determination of cadmium in the urine samples (IS = Bi) using interpolative and extrapolative internal standard methods; reference method: GFAAS with calibration by the external standard method.

	Concentration found ($\mu g\, l^{-1}$)		
Sample	**Reference method**	**Interpolative mode**	**Extrapolative mode**
A	1.34 ± 0.45	<0	1.10 ± 0.21
B	2.46 ± 0.34	1.6 ± 3.6	1.49 ± 0.09
C	4.93 ± 0.16	<0	4.85 ± 0.55
D	1.63 ± 0.14	1.3 ± 0.4	1.36 ± 0.03
E	2.31 ± 0.21	11.4 ± 1.8	3.17 ± 1.33
F	0.54 ± 0.26	1.6 ± 0.6	0.47 ± 0.04

Source: Adapted from Davis et al. [5].

Table 8.4 Results of the ICP-OES determination of rare earth elements (IS = Sc) in a rock-certified reference material obtained by different calibration methods (data arbitrarily selected).

		Content found ($\mu g\ g^{-1}$)		
Analyte	**Certified content ($\mu g\ g^{-1}$)**	**ESM**	**E-SAM**	**Extrapolative internal standard method**
Ce	467 ± 19	395 ± 15	473 ± 17	420 ± 17
Pr	47.1 ± 3.1	38.5 ± 1.6	46.2 ± 1.7	57.0 ± 1.4
Sm	23.8 ± 1.4	39.2 ± 1.5	24.6 ± 0.6	28.2 ± 1.0
Gd	20.3 ± 1.6	29.2 ± 0.8	21.6 ± 0.9	23.0 ± 0.9
Dy	18.9 ± 1.6	11.1 ± 0.5	18.7 ± 0.6	16.6 ± 0.5
Er	11.6 ± 1.1	19.5 ± 0.7	12.0 ± 0.4	14.2 ± 0.6
Yb	11.4 ± 1.0	17.6 ± 0.5	12.1 ± 0.3	14.0 ± 0.5
Y	95.5 ± 3.1	70.0 ± 2.9	97.6 ± 2.6	90.3 ± 3.4

Source: Yousefi et al. [6], table 3 (p. 151)/Taylor & Francis.

effects, it did so to a lesser extent than the E-SAM method without the internal standard. Furthermore, the use of the internal standard method did not result in improved precision of the analytical results. The authors explain this by the numerous potential sources of uncontrolled effects in the analysis by the ICP-OES technique, resulting, among others, from alteration in sample introduction efficiency due to different viscosity or surface tension of the samples and calibration standards, decreases in plasma power due to introduction of samples with high salt content, and the ionizable effect in axially viewing the plasma when a sample with easily ionizable elements are introduced to the plasma. They also point out that the interference effect can vary in nature for different analytes and their compensation with a single internal standard is unrealistic.

The extrapolative internal standard method was also verified in analyses by the ICP-MS technique on the example of determination of different selenium isotopes. The results of these studies, shown in Table 8.5, can be compared with those in Table 5.5 for the use of the interpolation version of the internal standard method. It is apparent that the combination of the E-SAM method using both internal standards, ^{72}Ge and ^{103}Rh, in conjunction with the E-SAM method generally does not contribute to a reduction in the precision of the analytical results. This is particularly evident for the determination of selenium in the presence of interferents, which can be attributed to the reduction in sensitivity of the measurements of all analytes by the interferents. Uniquely, an improvement in the precision of the determination results for all the isotopes tested was achieved by the extrapolation method using ^{103}Rh as an internal standard. The use of this method also resulted in an improvement in the precision of the ^{82}Se isotope determination, however, after a deeper analysis of the results, should be attributed to the contribution of the internal standards to the calibration procedure rather than to the E-SAM method.

Table 8.5 Results of the ICP-MS determination of selenium (20 mg l^{-1}) in model samples without and with sodium and potassium (320 and 280 mg l^{-1}) as interferents using the standard addition method (with no internal standard) and the extrapolative internal standard method with ^{72}Ge and ^{103}Rh as internal standards.

		No		^{72}Ge		^{103}Rh	
Analyte	**Interferents**	c_x **(mg l^{-1})**	**RSD**	c_x **(mg l^{-1})**	**RSD**	c_x **(mg l^{-1})**	**RSD**
^{77}Se	No	20.69	4.20	20.66	3.20	20.05	4.70
^{78}Se	No	20.24	4.10	20.15	4.65	18.92	6.20
^{82}Se	No	20.31	2.75	20.30	3.65	19.73	3.40
^{77}Se	Na + K	22.29	5.40	22.98	6.95	22.94	2.40
^{78}Se	Na + K	18.82	10.60	25.13	8.45	24.88	4.40
^{82}Se	Na + K	20.07	10.70	19.84	12.30	19.71	5.80

Source: Adapted from Wieczorek et al. [7].

Thus, based on the experimental examples presented, it can be said that the extrapolative internal standard method can be analytically useful, but only in exceptional, well-verified cases. Its usefulness depends, of course, on the nature of the interference effects and the effectiveness of the internal standards in the measurement method used for a given analysis. However, the method is governed by so many unpredictable, detailed factors, the overlap of which is contradictory, that the theoretical assumptions of the method, otherwise correct and sensible, may not be confirmed in analytical practice.

8.1.3 Extrapolative Dilution Method

The idea of combining the SAM method with the dilution method was first presented in 1989 [8], but the classical version of the extrapolative dilution method was developed in 1993 [9]. A few years later, it was used to study interference effects in combination with the interpolative dilution method [10], and then its effectiveness was more thoroughly tested in a version adapted for flow analysis [11].

The preparative step of the **extrapolation dilution method** procedure is as similar to its interpolation version as the two versions are similar for the indirect method and the internal standard. Two initial calibration solutions of sample volume V_0 and standard volume V_1 are prepared according to the E-SAM method, and then both solutions are gradually and equally diluted to successive values of dilution degrees k_i by adding diluent of successive volumes V_i, where:

$$k_i = \frac{V_0 + V_1 - V_i}{V_0 + V_1} \tag{8.6}$$

Proceeding in this way, the ratio of the sample volume to the total volume of sample and added standard, determined by the k_v step, remains the same. This situation is shown in Figure 8.7.

At each dilution step (i.e. for successive values of the dilution degree k_i), the signal intensities for the sample, Y_{xi}, and for the sample with standard addition, Y_{1i},

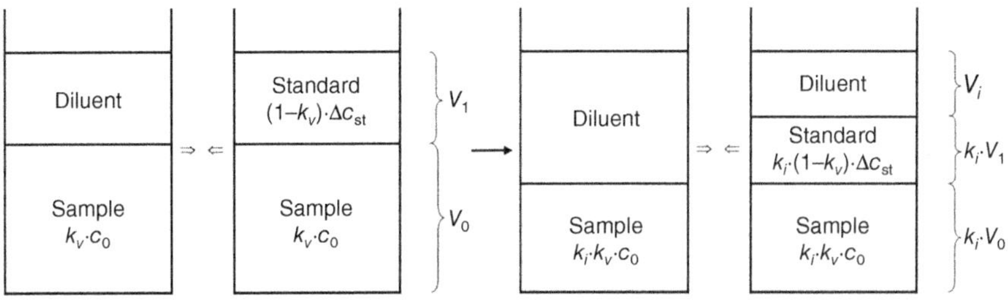

Figure 8.7 Scheme of the extrapolative dilution method at the preparative stage of the calibration process.

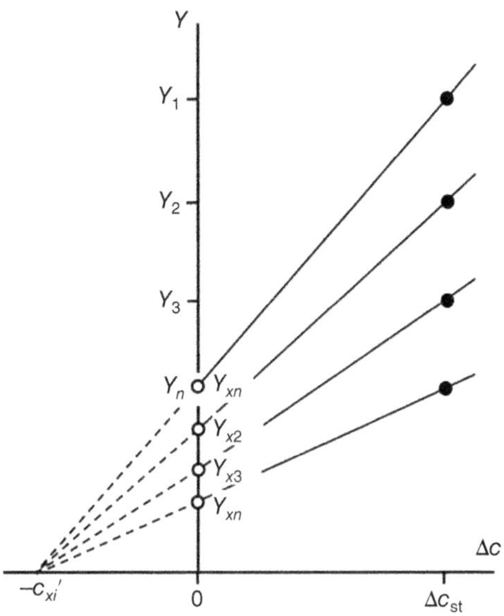

Figure 8.8 Calibration in accordance with extrapolative dilution method: a set of two-point calibration graphs allowing for determination of an analyte in a sample on the basis of a series of apparent analyte concentrations c_{xi}' ($i = 1, 2, \ldots, n$) obtained in the extrapolative way at different relative dilution values, k_{vi}.

are measured and a series of calibration graphs connecting pairs of measurement points obtained at the same dilution step is created. This procedure is illustrated in Figure 8.8. These graphs are then linearly extrapolated and the apparent concentrations are determined by extrapolation from the formula:

$$c_{xi}' = \frac{Y_{xi}}{Y_i - Y_{xi}} \cdot c_{st} \tag{8.7}$$

As with the interpolative dilution method, successive values of apparent concentrations are expected to approach the true value as dilution occurs, i.e.:

$$c_0 = \lim_{k_i \to 0} c_{xi}' \tag{8.8}$$

To determine the analytical result, c_x, which is a measure of the real concentration, c_0, the apparent concentration values are represented as a function of the dilution degree values, k_i (Figure 5.14, Chapter 5) and the function representing this relationship is extrapolated to $k_i = 0$.

What is the rationale and purpose of performing calibrations using such combined methods? The answer is quite obvious: it is supposed that there will be an increase in the ability of the dilution method to eliminate uncontrolled effects due to the ability of the E-SAM method to compensate for these effects. Of course, one must also take into account the additional problems associated with, and primarily caused by, the extrapolative way of determining analytical results by the E-SAM method, especially since the analytical result in the dilution method is also obtained by extrapolation.

To experimentally verify this supposition, the gradient technique is used in the gradient ratio calibration method (GRCM), the principle of which is illustrated in Figure 7.15. The only modification is that the two injection peaks to be compared with each other are recorded for the sample solution and the sample with the addition of the standard [11]. Further, the procedure remains unchanged: from the pairs of measurement points recorded on the falling parts of the two peaks after equal time after injection of both solutions, the values of apparent concentrations are determined from Eq. (8.7) and the results are presented as a function of dilution degree k. This procedure was consequently named the gradient ratio standard addition method (GRSAM).

Example experimental results obtained in this way are shown in Figures 8.9–8.11. It can be seen that in the GRSAM method, changes in the apparent concentration of an analyte in a sample can be tracked very accurately using multiple experimental points. However, large random fluctuations of these points occur earlier than in the interpolation version of the dilution method (compare with Figure 7.16, resulting in an increase in the extrapolation area that must be "overcome" to determine the final value of the analyte concentration in the sample (for $k = 0$).

The results shown in Figure 8.9 demonstrate that the extrapolative dilution method allows, like its interpolative version, to compensate for the nonlinear detection effect caused by the addition of a standard solution with too high analyte concentration (compare with Figure 5.15). Once the linear function is fitted to the experimental points, an analytical result with very good accuracy can be obtained. It is worth noting that if the calibration by the E-SAM method was only for the

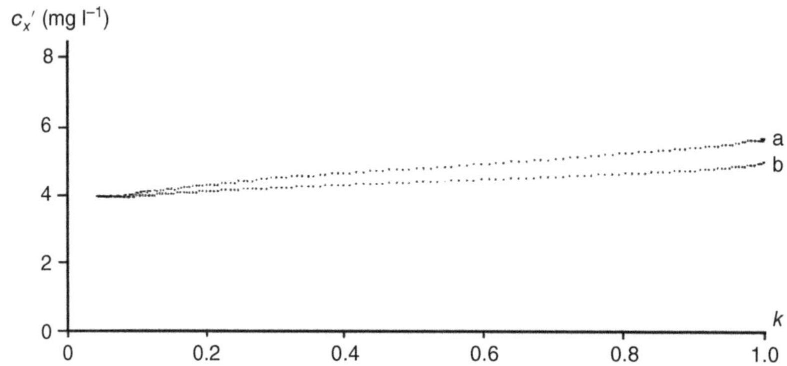

Figure 8.9 Calibration by the extrapolative dilution method: results obtained for FAAS determination of calcium in the synthetic samples containing 4 mg L^{-1} Ca with addition of the standard solutions containing 20 (a) and 25 (b) mg L^{-1} Ca. Source: Kościelniak and Kozak [11], fig 4 (p. 241)/with permission of Elsevier.

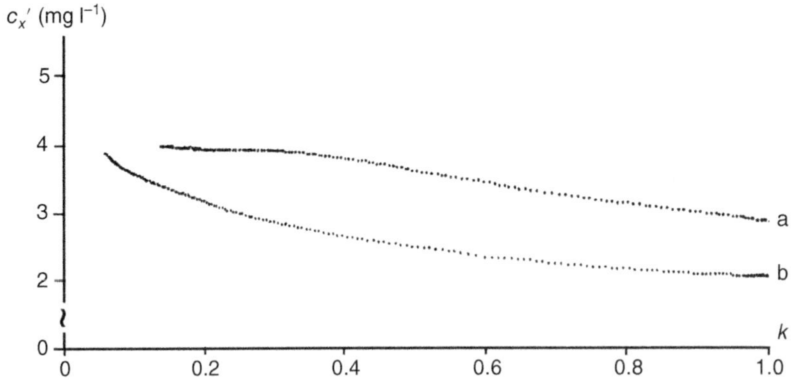

Figure 8.10 Calibration by the extrapolative (a) and interpolative (b) dilution method: results obtained for FAAS determination of calcium in the synthetic samples containing 4 mg l^{-1} Ca and 5 mg l^{-1} V with the use of standard solution containing 5 mg l^{-1} Ca. Source: Kościelniak and Kozak [11], fig 5 (p. 241)/with permission of Elsevier.

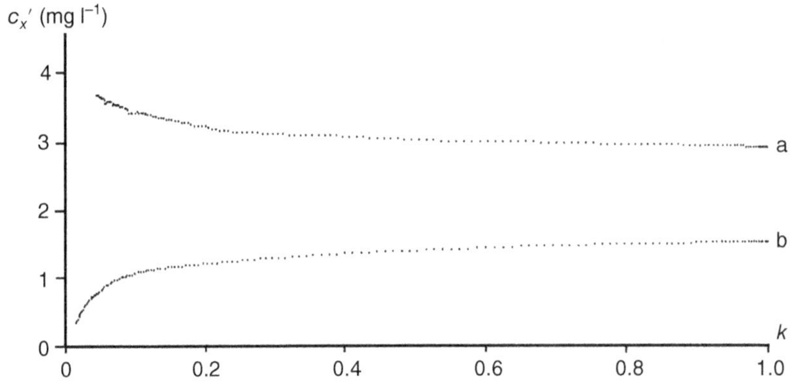

Figure 8.11 Calibration by the extrapolative (a) and interpolative (b) dilution method: results obtained for FAAS determination of calcium in the synthetic samples containing 4 mg l^{-1} Ca and 80 mg l^{-1} Al with the use of standard solution containing 15 mg l^{-1} Ca. Source: Kościelniak and Kozak [11], fig 6 (p. 242)/with permission of Elsevier.

undiluted sample ($k = 1$), calcium would have been determined with an error in the range of 10–20%.

The crucial question, however, is whether the extrapolation version of the method allows compensation for interference effects and, if so, whether it is more effective in this respect than the interpolation form. Some light is shed on this problem by the results shown in Figures 8.10 and 8.11, for the determination of calcium in the presence of strong interferents by the FAAS technique using both flow calibration methods, GRCM and GRSAM [11].

In both of these particularly difficult cases, the extrapolative dilution method proved to be much more effective than its interpolative version. Although the effect of vanadium on calcium (Figure 8.10) without the E-SAM method is compensated for as the sample is diluted, this change occurs gradually and nonlinearly throughout the dilution process, which obviously makes it much more difficult to obtain an accurate analytical result. Using the E-SAM method reduces this error in the undiluted sample by about 20%, and diluting the sample compensates for it almost completely. Moreover, the complete compensation of the interference effect occurs far before the end of the dilution, which allows the accurate determination of the analytical result after fitting a linear function to the experimental points.

In the case of the effect of aluminum on the calcium determination, so difficult to control by various calibration methods, the use of the extrapolative dilution method gives an even more spectacular result. This effect without the E-SAM method increases, as seen again in Figure 8.11, as the sample is diluted, leading to completely erroneous results. By combining the E-SAM method with the dilution process, the effect of aluminum, initially also very large (about 25%) is gradually compensated with a tendency to be compensated completely.

The results of the determination of calcium by the FAAS technique in synthetic and natural samples presented in Tables 8.6 and 8.7, confirm the very high efficiency of the extrapolative dilution method compared to other methods in compensating for strong interference effects. The random errors made by this method are, as expected, clearly larger than those characterizing the external standard method, but comparable to those made by the interpolative dilution method.

Improvements in the analytical usefulness of the extrapolation dilution method can be achieved by including the contribution of special reagents in the calibration procedure. For the flow-through version (GRSAM), the natural way to introduce a reagent into calibration solutions is, as already shown, to use it as a carrier solution. The effectiveness of this procedure is illustrated in Figure 8.12 with the example of the effect of titanium on the determination of calcium by the FAAS technique [11]. The compensation for this effect without the use of an additional reagent is revealed, as can be seen, only after about a 10-fold dilution of the sample, giving no chance for an accurate determination of the analyte concentration. A small amount of lanthanum introduced into the calibration solutions in the carrier solution dramatically changes the conditions of the determination.

Table 8.6 Results obtained for FAAS determination of calcium ($c_0 = 10\ \text{mg l}^{-1}$) in synthetic samples containing strong interferents.

Interferent (μg ml^{-1})	Concentration found (μg g^{-1}) ESM	Dilution method	Extrapolative dilution method
4 Si	7.28 ± 0.07	10.06 ± 0.17	9.87 ± 0.23
50 P	5.97 ± 0.05	9.91 ± 0.22	9.93 ± 0.23
5 V	4.01 ± 0.07	10.32 ± 0.09	10.12 ± 0.19
8 Al	7.23 ± 0.11	7.11 ± 0.22	9.45 ± 0.32
80 Al	4.18 ± 0.09	2.11 ± 0.38	9.56 ± 0.29

Source: Adapted from Kościelniak and Kozak [11].

Table 8.7 Results obtained for FAAS determination of calcium ($c_0 = 10\,mg\,l^{-1}$) in the iron ore samples (CRMs).

	Main sample components (%)					Analyte amount found (%)		
Sample code	Ca	Al	Fe	Si	Mg	ESM	Dilution method	Extrapolative dilution method
2.53	0.286	0.37	66.1	1.90	0.17	0.172 ± 0.011	0.209 ± 0.028	0.281 ± 0.030
2.54	2.423	2.97	41.5	6.07	2.21	1.809 ± 0.024	1.532 ± 0.083	2.513 ± 0.096
2.55	0.529	0.51	64.8	2.45	0.89	0.378 ± 0.016	0.456 ± 0.056	0.545 ± 0.038
2.57	0.051	0.71	83.9	1.15	0.02	0.017 ± 0.002	0.014 ± 0.009	0.045 ± 0.011

Source: Kościelniak and Kozak [11], table 2 (p. 244)/with permission of Elsevier.

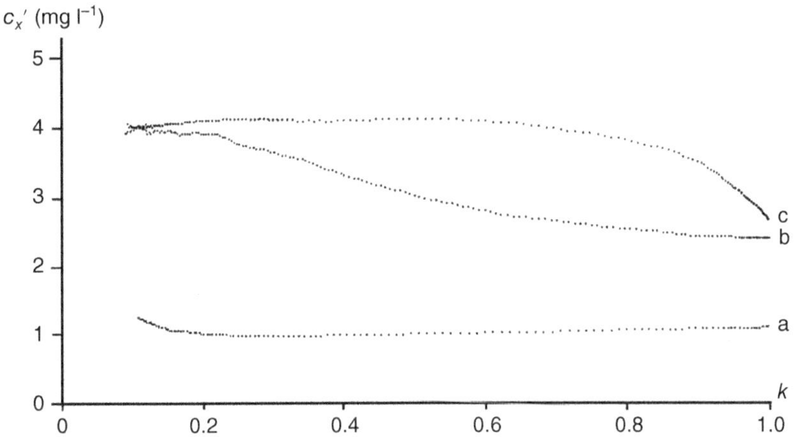

Figure 8.12 Calibration by the extrapolative dilution method: results obtained for FAAS determination of calcium in the synthetic sample contained $4\,mg\,l^{-1}$ with $20\,mg\,l^{-1}$ Ti and diluted with strontium solution of concentrations 0 (a), 1 (b), and 20 (c) $mg\,l^{-1}$: concentration of Ca in the standard solution: $5\,mg\,l^{-1}$. Source: Kościelniak and Kozak [11], fig 7 (p. 242)/with permission of Elsevier.

In Table 8.8 the results show the effectiveness of the compensation of strong interference effects shown by the extrapolative dilution method against the performance of the external standard method and the interpolative dilution method under the same conditions. It is noteworthy that the accuracy of the results can be improved with a very small amount of special reagent even when analyte determination errors reach −90%. As explained earlier, this is because after the sample is injected, the reagent gradually penetrates the sample segment, and an increasing amount of reagent is accompanied by a decreasing amount of analyte and thus also a decreasing amount of interferent (see Figure 7.14). The presence of gradients of reagent and interferent in opposite directions in the sample segment means that the interference effect is compensated for by the minimum amount of reagent that is necessary under the conditions to make this compensation.

Table 8.8 Results obtained for FAAS determination of calcium ($c_0 = 10\,\text{mg}\,\text{l}^{-1}$) in synthetic samples in the presence of strong interferents using different releasing reagents.

		Content found (μg g^{-1})		
Interferent (μg ml^{-1})	**Releasing element (μg ml^{-1})**	**ESM**	**Dilution method**	**Extrapolative dilution method**
20 Ti	10 Sr	5.62 ± 0.11	7.78 ± 0.19	10.22 ± 0.15
20 Ti	10 La	5.81 ± 0.07	8.26 ± 0.22	9.76 ± 0.18
20 Ti + 10 V	10 Sr	0.92 ± 0.06	3.32 ± 0.39	9.75 ± 0.21
20 Ti + 10 V	10 La	1.10 ± 0.07	4.02 ± 0.25	10.43 ± 0.12
20 Ti + 80 Al	10 Sr	4.45 ± 0.12	6.95 ± 0.13	9.87 ± 0.23
20 Ti + 80 Al	10 La	4.88 ± 0.09	7.06 ± 0.19	10.13 ± 0.19

Source: Adapted from Kościelniak and Kozak [11].

Another way of introducing the special reagent into the analytical system was proposed in the method originally named "standard addition and successive dilution method" (SASD) [8]. On the basis of a very careful theoretical and empirical analysis of the interference effects occurring in the FAAS method, it was concluded that, regardless of the mechanism of interaction of the interferent with the analyte, the most effective way to assist the dilution method in interference compensation is to keep the concentration of special reagents constant during the dilution process. For this purpose, the selected reagent was initially added to the calibration solutions at a well-defined concentration and then diluted with a solution of that reagent at the same concentration.

The procedure used does indeed work very well, as confirmed by the example of several analyses performed. Both compounds showing release activity and interferents acting as buffers were tested as special reagents. In each case, the SASD method produced more accurate results than other calibration methods. Figure 8.13 and Table 8.9 show selected examples of the results of the analyses performed.

By keeping the concentrations of the special reagents constant, their effect is already partially manifested in undiluted samples, and as the calibration solutions are diluted the apparent concentrations of the analyte decrease linearly, which facilitates obtaining accurate analytical results. It should be noted, however, that the concentrations of the special reagents used are relatively very large – about 1000 times (!) greater than the concentrations determined in the previously described extrapolation procedure of the dilution method. Such a large amount of reagent, particularly if it is an interferent acting as a buffer, can nevertheless produce its "own" interference effect, as can be seen in Figure 8.13.

Although the experimental results presented are few and obtained by a single measurement method, it can be said that the extrapolative dilution method is an example of such a composite calibration approach that shows greater advantages

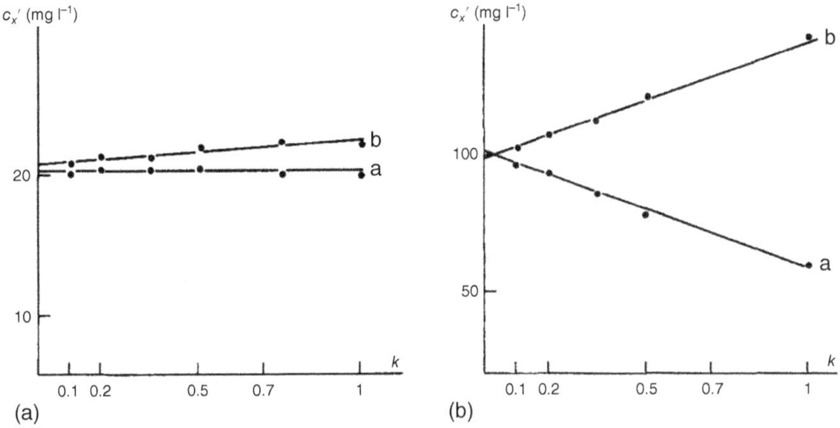

Figure 8.13 Results obtained for FAAS determination of Sr and Mo using SASD method: (A) 20 mg l^{-1} Sr in the presence of 20 mg l^{-1} V without special reagent (a), with 5 mg ml^{-1} Al (b) (c), and (B) 100 mg l^{-1} Mo in the presence of 3 mg ml^{-1} Ca without special reagent (a), with 5 mg ml^{-1} V. Source: Pszonicki and Skwara [8], fig 9 & 11 (p. 1271–1272)/with permission of Elsevier.

Table 8.9 Results of the FAAS determination of MgO in the iron ore samples (CRMs) using external standard method in relation to all samples and using the E-SAM method in relation to the selected sample while external standard method to the remaining samples (from own materials).

			Contents found, c_x, (%)				
				Sample selected for calibration by E-SAM			
No.	**Sample**	**Certified contents (%)**	**ESM**	**I**	**II**	**III**	**IV**
I	Pellets ore	0.28	0.29	0.27	0.27	0.26	0.29
II	Roasted siderite	3.67	4.29	3.66	3.75	3.61	3.94
III	Kiruna ore	1.44	1.68	1.38	1.41	1.41	1.46
IV	Venezuelan ore	0.035	0.039	0.032	0.033	0.032	0.039

than each of the two-component methods without multiplying their limitations and disadvantages. Its essential advantage is the possibility of eliminating uncontrolled effects (especially interference effects) of different nature to a degree undoubtedly greater than that demonstrated by the interpolation version of the dilution method. In other words, **the probability of obtaining an accurate analytical result by the dilution method is greater when applied in the extrapolation version than when applied in the interpolation version**. Although the participation of the E-SAM method increases the area of extrapolation of the dependence of the apparent analyte concentration on relative dilution, this difficulty is of little importance in view of the possibility of approximating this dependence by a linear or nonlinear function with a relatively small and well-defined degree of nonlinearity. Nevertheless, it should be remembered that the initial concentration of analyte in the sample should be sufficiently large that the arrangement of experimental points clearly indicates the type of function to be fitted to those points.

8.2 Mixed Methods

The different resistance of calibration interpolation and extrapolation methods to interference effects can be exploited in yet another way than that shown in Chapters 5 and 6. It consists in determining a specific analyte by the E-SAM method, that is, by extrapolating the created calibration graph to a signal corresponding to zero analyte concentration, and then using this graph in its full range (i.e. extended by the extrapolation range) to determine the same analyte by the ESM method in another sample (or in a series of other samples). An illustration of the procedure of such a method, which can be called a **mixed method** (E-SAM/ESM), is shown in Figure 8.14.

By performing analyses in this manner, it is assumed that the model function determined for the selected sample is a correct representation of the true function for the other samples. This is because the position of the calibration graph is due to compensation for matrix effects, i.e. the combined effect of all components present in the sample on the analytical signal, and this graph can be used to effectively compensate for effects of the same or similar magnitude present in other samples. In other words, the calibration graph determined for a selected sample by extrapolation can be used in both its experimental and extrapolation parts to determine the analyte in the other samples.

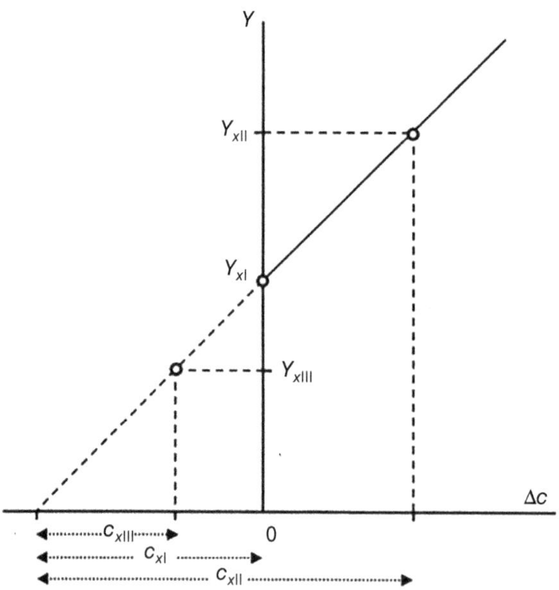

Figure 8.14 Application of the mixed method to analysis of samples of similar composition: calibration graph obtained for sample I in accordance with the E-SAM method serves for estimation of the analyte concentration in sample I (c_{xI}) in the extrapolative way and in samples II (c_{xII}) and III (c_{xIII}) in the interpolative way.

This calibration approach is justified, for example, when one analyzes a series of samples of similar composition in terms of the potential or known interferents present. The described method, simple and fast, can therefore be very useful when analyzing a series of samples of the same type (e.g. river water, human blood, etc.) and with defined interferents that can be compensated by the E-SAM method. Although the calibration approach itself is not new, it is very difficult to find experimental examples of its application.

Table 8.9 shows the results of the determination of MgO in the iron ores obtained by this route by the author's research team. It is evident that the accuracy of the analytical results obtained in this way can be significantly better than the results obtained by ESM. However, it should be remembered that, on the one hand, the sample selected for calibration by this E-SAM method should have an average analyte concentration in relation to the concentrations in the other samples, and on the other hand, the highest concentration of analyte added to the sample should be at least equal to the analyte concentration in the sample. It is of course also important that the model function fitted to all experimental points covers a linear range of the true function.

A limitation of this method is the dependence of analytical results obtained by interpolation on a single result obtained by extrapolation. It should be noted (see Figure 8.2) that all random and systematic errors made during the analysis of a selected sample affect the analyte determination errors in the other samples. This is especially true for systematic errors characteristic of the E-SAM method, e.g. resulting from the different forms of the real function in the experimental and extrapolation areas. E-SAM calibration should therefore be carried out with particular care, observing all the rules that apply to it and using its multipoint version.

Clearly, there is no obstacle to mixing other interpolation and extrapolation calibration methods in the manner described above. If an analyte needs to be determined by an indirect method or by an internal standard method, then the analysis of the selected sample shall be performed by an extrapolation version of one of these methods and the analyte in other samples shall be determined by an interpolation version of the same calibration method. Determinations performed in this way are illustrated in Figure 8.15.

The mixed method has been ingeniously adapted to speciation analysis for the determination of an analyte present in a sample at two oxidation levels [12]. The idea is that a calibration graph formulated from the determination of one form of an analyte (I) by E-SAM is used to determine another form (II) of the same analyte by interpolation after reduction or oxidation of form II to I. This concept is illustrated in Figure 8.16. From a formal point of view, this approach is an example of yet another combination of calibration methods than those discussed so far, namely the combination of the E-SAM method with an indirect interpolation method (since the form II of the analyte as a product of the redox reaction is indirectly indicative of the concentration of form I).

The approach was tested and verified on the example of spectrophotometric determination of the analytes in pairs Fe^{2+}–Fe^{3+} and Cr^{3+}–Cr^{6+}. To rationalize the methods implementation, the analyses were performed in the flow mode with dedicated

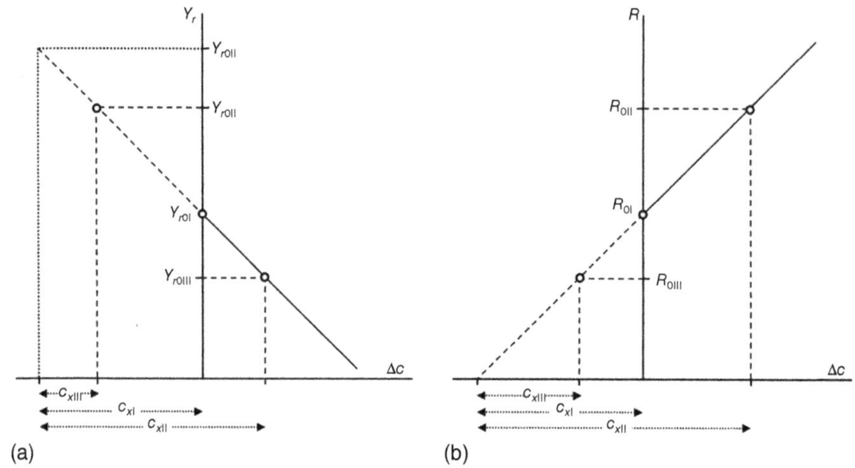

Figure 8.15 Application of the mixed method to analysis of samples (I, II, III) of similar composition in accordance with the extrapolative (sample I) and interpolative (samples II and III) versions of the indirect method (a) and internal standard method (b).

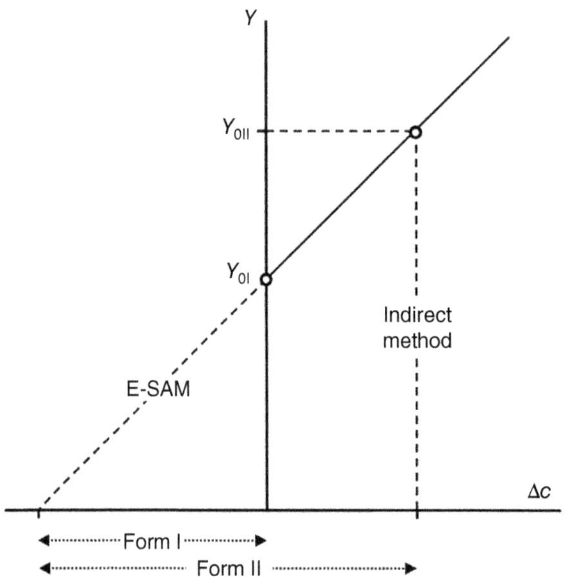

Figure 8.16 Application of the mixed method to speciation analysis of samples containing an analyte present in two forms at two oxidation levels: calibration graph obtained for form I by the E-SAM method serves for estimation of the form II in the interpolative way by the indirect method (after reduction or oxidation to form I). Source: Adapted from Paluch et al. [12].

systems has been developed and used. The results presented in Table 8.10 for the determination of Fe^{2+} and Fe^{3+} in water samples using ascorbic acid as the reductant and 1,10-phenanthroline as chromogenic reagent indicate the full utility of the method. The very good accuracy and precision of the results were confirmed by the certified values and the results obtained with the ICP method as a reference method.

A major advantage of the method is the ability to compensate for potential interference acting on the analyte in form I regardless of whether or not form II is exposed

Table 8.10 Results of determination of **Fe(II) and Fe(III)** Fe^{2+} and Fe^{3+} in certified reference materials, ES-H-2 and EU-H-3, of ground and wastewater, respectively and in three samples (1, 2, 3) of artesian water; CI – confidence interval ($n = 3$, $\alpha = 0.05$).

Water sample	Mixed method			Certified value	ICP-OES
	Fe^{2+} (mg l^{-1})	Fe^{3+} (mg l^{-1})	Fe (sum) (mg l^{-1})	Fe (total) (mg l^{-1})	
ES-H-2	0.433 ± 0.034	0.754 ± 0.040	1.187 ± 0.052	1.170 ± 0.050	—
EU-H-3	0.424 ± 0.003	0.077 ± 0.007	0.501 ± 0.005	0.491 ± 0.015	—
1	1.105 ± 0.035	0.138 ± 0.015	1.224 ± 0.038	—	1.196 ± 0.011
2	3.545 ± 0.027	0.480 ± 0.050	4.025 ± 0.057	—	4.033 ± 0.367
3	2.946 ± 0.078	0.512 ± 0.079	3.459 ± 0.111	—	3.457 ± 0.013

Source: Paluch et al. [12], Table 4 (p. 280)/Elsevier.

to its "own" interference. The basic condition of obtaining the accurate analytical results is the accurate determination of form I by E-SAM and conversion of form II into the form I with efficiency close to 100%. Moreover, the reagent(s) used in the chemical treatment of form II (added to the sample in excess) should neither react with other sample component(s) to give analytical signal measured in the conditions of analysis nor to give a signal itself in these conditions.

The latter problem was encountered in the determination of Cr^{3+} and Cr^{6+} with 1,5-diphenylcarbazide as chromogenic reagent. When Ce^{4+} was added to the sample as oxidizing reagent, its excess caused the interference effect during Cr^{3+} determination. The more Cr^{3+} was present in the sample, the less Ce^{4+} was involved in the reaction with 1,5-diphenylcarbazide. Hence, the interference effect caused by Ce^{4+} decreases with the increase of Cr^{3+} concentration in the sample. Considerable systematic errors due to this effect are evident from the data collected in Table 8.11.

Table 8.11 Results of simultaneous determination of in different synthetic samples (1–7) using indirect method mixed with E-SAM method; c_x and c_x' has been obtained for Cr^{3+} before and after correction (for details see text).

Water sample	Concentration of Cr^{3+} (mg l^{-1})			Concentration of Cr^{6+} (mg l^{-1})	
	Expected, c_0	Obtained, c_x	Obtained, c_x'	Expected, c_0	Obtained, c_x
1	0.125	0.172	0.130	0.125	0.128
2	0.125	0.184	0.122	0.250	0.243
3	0.250	0.322	0.267	0.250	0.241
4	0.250	0.305	0.254	0.375	0.360
5	0.375	0.422	0.373	0.125	0.122
6	0.375	0.433	0.390	0.250	0.243
7	0.500	0.529	0.490	0.500	0.511

Source: Paluch et al. [12], Table 3 (p. 279)/Elsevier.

Excess Ce^{4+} can be removed with sodium azide, but the authors have proposed an efficient procedure for correcting these errors without the need for additional reagents. The corrected results are, as seen in Table 8.11, in very good agreement with the expected results.

The method presented for combining calibration methods is so simple, fast, and versatile, yet with such tangible benefits, that it seems it can and should find widespread laboratory use.

8.3 Combined Methods

The complex calibration methods presented so far have been based on relatively simple and known procedures for preparing calibration solutions. In the case of extrapolation methods, solutions are prepared according to a specific interpolation method (indirect, internal standard, dilutions) based on the sample and the sample with the addition of the standard, thus giving them an extrapolation character. In mixed methods, the procedure is even simpler – calibration solutions are prepared according to the chosen extrapolation method, and the model function thus formulated is used to determine the analyte in other samples or in other forms by extrapolation.

Are these all possibilities for combining calibration methods with each other, and if not, how can this be done in other ways? Some examples of other, more complex calibration procedures leading to various specific analytical advantages are given below.

8.3.1 Integrated Calibration Methods

Underlying the idea of integrated calibration methods (ICMs) is an observation derived from the previously presented experiments: basic empirical calibration methods can be implemented in interpolation and extrapolation versions, and both versions of the same method differ in their robustness to analytical uncontrollable effects. Therefore, it can be assumed that an appropriate combination (integration) of two methods of different nature can lead to at least two analytical results with accuracy depending on the occurrence of uncontrolled effects. Thus, **the integration of methods is intended to improve the accuracy of the analytical result through a preliminary diagnosis of the magnitude of systematic errors committed in a given analysis**.

The concept of the **integrated calibration method** was first more widely presented in its simplest form in 2001 [13], and a more complex version of it, called the "complementary dilution method (CDM)," was described and developed several years later [14].

8.3.1.1 Simple Integrated Method

In the **simplest version of the integrated method**, four calibration solutions are prepared with the composition shown in Figure 8.17. The first is a solution used to

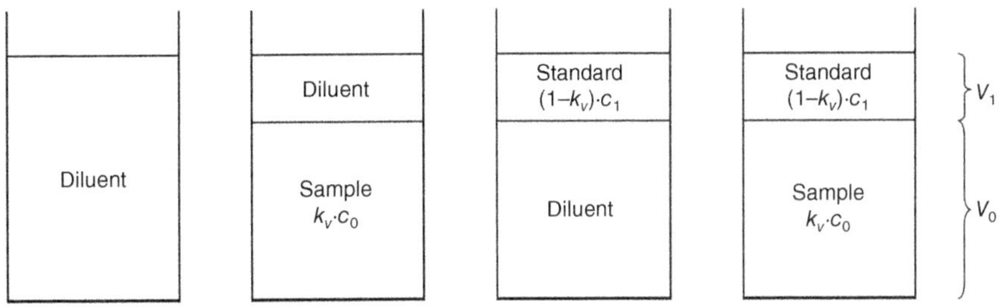

Figure 8.17 Scheme of the simple integrated calibration method (ICM) in the basic form at the preparative stage.

dilute the sample and standard in two other calibration solutions, and the next three contain a sample solution of volume V_0 and a standard solution of volume V_1, which are filled up with each other or with a diluent solution to volume $V_0 + V_1$. All these solutions are subjected to measurements under identical experimental conditions, yielding signals of intensities Y_0, Y_x, Y_1, and Y_2, respectively.

The signal for the solution containing the diluent alone is set to zero ($Y_0 = 0$). If a blank effect is present in the sample and standard solutions, solutions containing the substances causing the effect in the appropriate concentration (taking into account the dilution of the sample and standard in the calibration solutions) must be prepared in advance, measured for them and the signals Y_x, Y_1, and Y_2 adjusted accordingly. This note applies to all integrated method procedures described in this chapter.

Based on the obtained measurement results, two calibration graphs are constructed as shown in Figure 8.18. One is specific to the ESM method and the other to the E-SAM method. Both graphs allow the concentration of analyte in the sample, c_0, to be estimated using two analytical results, c_{1x}' and c_{2x}', determined by interpolation and extrapolation, respectively. Because of the potentially different accuracy of the determined concentrations, these results should be treated as apparent concentrations.

If the analyte in the sample is not subject to any effects that cause a systematic change in the analytical signal and the real function is linear over the analyte concentration range considered, then the relationship between these signals and the analyte concentration in the sample, c_0, and in the standard solution, c_{st}, can be expressed by the equation:

$$Y_x = B_1 \cdot k_v \cdot c_x \tag{8.9}$$

$$Y_1 = B_1 \cdot (1 - k_v) \cdot c_{st} \tag{8.10}$$

$$Y_2 = B_1 \cdot k_v \cdot c_x + B_1 \cdot (1 - k_v) \cdot c_{st} \tag{8.11}$$

where $k_v = V_0/(V_1 + V_0)$, and values c_{1x}' and c_{2x}' are calculated from:

$$c_1' = \frac{Y_x}{Y_1} \cdot \frac{(1 - k_v)}{k_v} \cdot c_{st} \tag{8.12}$$

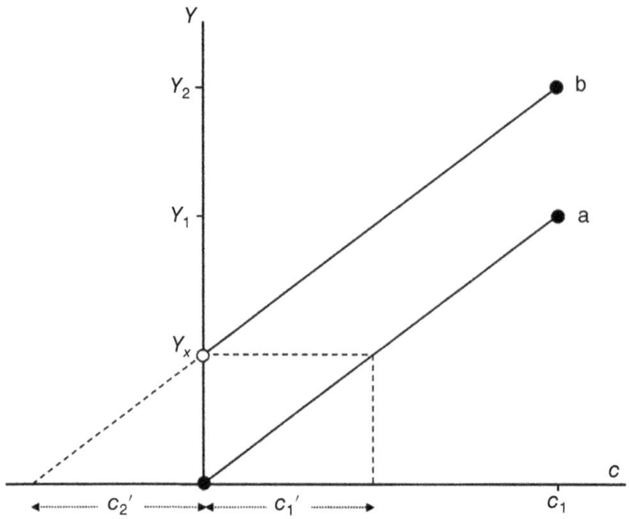

Figure 8.18 Principle of the integrated calibration method in the basic form: calibration curves constructed in accordance with the external standard method (a) and the standard addition method (b) lead to two values of apparent concentrations, c_x' and c_2', estimating independently the analytical result, c_x, in interpolative and extrapolative ways, respectively.

$$c_2' = \frac{Y_x}{Y_2 - Y_x} \cdot \frac{(1 - k_v)}{k_v} \cdot c_{st} \tag{8.13}$$

Thus, the advantage of the described approach is **the possibility to obtain in a single calibration procedure two independent values of the analytical result, whose values testify to the accuracy of the determination performed**. If uncontrolled effects are not present in the analytical system, then the two results, c_{1x}' and c_{2x}', should be equal (including random errors), but if they are present, then they should be significantly different from each other. **This increased analytical information distinguishes the integrated method from all other calibration methods developed to date**.

An example of the interpretation of analytical results obtained by this method is shown in Table 8.12. As can be seen, in the absence of an interferent in the sample, the two results obtained by interpolation and extrapolation are close to each other, while the presence of an interferent causes a clear difference between them, which suggests that appropriate actions should be taken to eliminate the source of error.

However, the presented description of the obtained results is obvious when the qualitative and quantitative composition of the sample is known. But how to interpret these results assuming that they concern analysis of samples of unknown composition? The proposed sequence of proceedings is then as follows:

- the pair of results obtained for each sample is compared with each other statistically on the basis of random errors;
- if the differences between the results are statistically insignificant, both results are assumed to be exact (close to the real result) and the concentration of the analyte in the sample is expressed as the arithmetic mean of these results;

Table 8.12 Results of the determination of magnesium in the presence of no interferent and of aluminum, obtained by the interpolative (a) and extrapolative (b) modes of the ICM method.

Magnesium concentration ($\mu g\, ml^{-1}$)	Aluminum concentration ($\mu g\, ml^{-1}$)	Magnesium concentration found ($\mu g\, ml^{-1}$)			
		Without special reagent		With 1 $\mu g\, l^{-1}$ Sr	
		a	b	a	b
2	0	2.08 ± 0.04	2.04 ± 0.07	—	—
4	0	3.98 ± 0.05	4.01 ± 0.09	—	—
2	10	1.66 ± 0.05	1.95 ± 0.06	2.04 ± 0.04	1.96 ± 0.08
4	10	2.72 ± 0.08	4.03 ± 0.09	4.06 ± 0.02	4.04 ± 0.09

Source: Adapted from Kościelniak and Wieczorek [1].

- if the two results are statistically different from each other and there are indications that this may be due to a proportional interference effect, it is assumed that the concentration of the analyte in the sample can be expressed by the result obtained by extrapolation;
- if it is not absolutely certain that the significant difference between the results is due to interference, the analysis of the sample shall be repeated, e.g. with the aid of a special reagent; the two results being close to each other (and to the result obtained previously by extrapolation) confirms that the interference effect has been eliminated and the concentration of the analyte in the sample is then expressed by the arithmetic mean of the two results;
- in any other case, the analysis shall be repeated under different experimental conditions, modified in such a way as to improve the accuracy of the results to the best of the analyst's knowledge and experience.

Considering the calibration method described above from a formal point of view, it can be called a **conventional integrated method** because it combines two well-known and widely used methods of interpolation (ESM) and extrapolation (E-SAM). In a similar way, the interpolation and extrapolation versions of the indirect method and the internal standard method can be combined in single calibration procedures to form the **integrated indirect method and the integrated internal standard method**, respectively.

The preparation of solutions for calibration by the two latter methods is shown in Figure 8.19.

After measurements are made according to the indirect method (under conditions characteristic of the reactant) and the internal standard method (under conditions characteristic of the analyte and the internal standard), the calibration graphs shown in Figure 8.20 are obtained. The analytical results are calculated in both methods from the formulas:

$$c_x{}' = \frac{Y_{rx}}{Y_{r1}} \cdot \frac{(1 - k_v)}{k_v} \cdot c_{st} \tag{8.14}$$

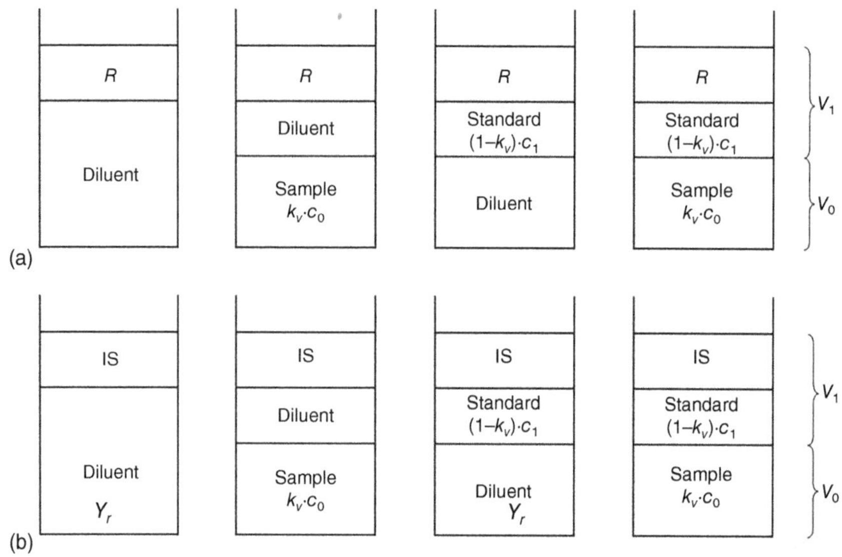

Figure 8.19 Scheme of the basic form of the integrated indirect method (a) and the integrated internal standard method (b) at the preparative stage.

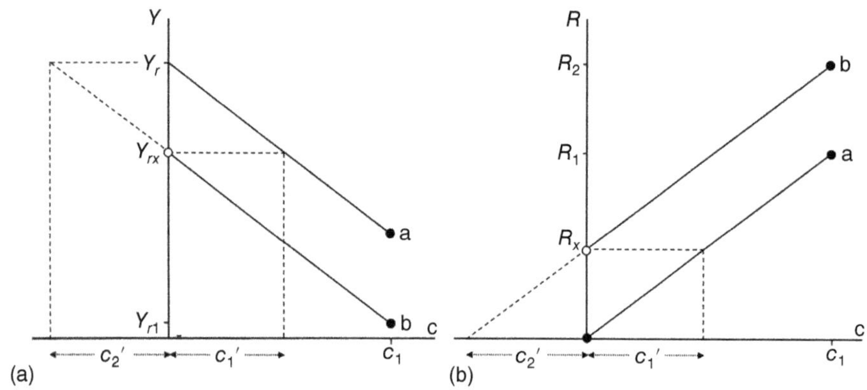

Figure 8.20 Transformation of the measurement results to the apparent concentrations, c_1' and c_2', in the indirect integration method (A) and in the internal standard integration method (B) (for details see text).

$$c_2' = \left| \frac{Y_{rx}}{Y_{r2} - Y_{rx}} \right| \cdot \frac{(1 - k_v)}{k_v} \cdot c_{st} \tag{8.15}$$

$$c_1' = \frac{R_x}{R_1} \cdot \frac{(1 - k_v)}{k_v} \cdot c_{st} \tag{8.16}$$

$$c_2' = \frac{R_x}{R_2 - R_x} \cdot \frac{(1 - k_v)}{k_v} \cdot c_{st} \tag{8.17}$$

Calibration by both methods brings in practice the benefits characteristic of the indirect method and the internal standard method, respectively. Integrating the interpolation and extrapolation versions of each method additionally allows for a preliminary diagnosis of the presence of uncontrolled effects in the analytical system under study. Regardless of the integrated method used (conventional, indirect, internal standard), this diagnosis can help in the selection of an appropriate calibration method that will provide sufficiently good accuracy for the subsequent analytical procedure: the interpolation version of a given method (in the absence of uncontrolled effects) or its extrapolation version (in the presence of proportional interference effects).

8.3.1.2 Complementary Dilution Method

Some previously presented calibration methods (and in particular the dilution method) demonstrate that the process of diluting the sample can result in a change in uncontrolled effects and consequently a change in the systematic analytical error. On this basis, it was decided to increase the efficiency of the simple integrated method by preparing the calibration solutions in such a way as to further vary the degrees of dilution of the sample and standard in them. Since the sample and standard are diluted with complementary volumes of dilution solutions, this form of integrated method has been called the **complementary dilution method (CDM)** [14].

In the CDM method, a diluent solution and six calibration solutions of the composition shown in Figure 8.21 are prepared. The first three solutions are the same in terms of the reciprocal volume of sample and standard, i.e. the sample is in volume V_0 and the standard is in volume V_1. The second series of solutions are prepared in an analogous manner except that the sample takes up the volume V_1 in the calibration solutions and the standard takes up the volume V_0. Thus, the sample, standard, and sample with added standard appear in the calibration solutions at two degrees of dilution: $P = k_v$ and $Q = (1 - k_v)$.

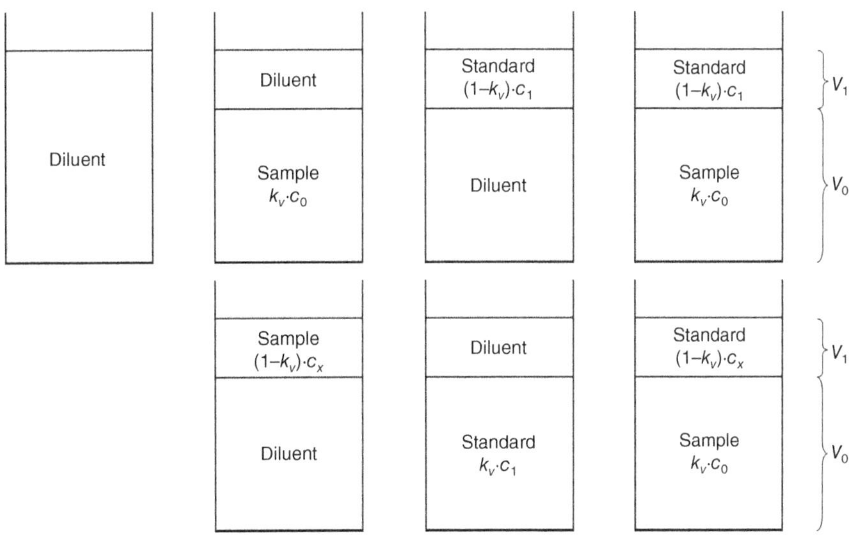

Figure 8.21 Scheme of the conventional integrated method in the CDM form at the preparative stage.

After subjecting the calibration solutions to measurements under identical experimental conditions, six nonzero intensity signals are obtained for the sample, Y_{xP}, Y_{xQ}, the standard, Y_{1P}, Y_{2Q}, and the sample with the addition of the standard, Y_{2P}, Y_{2Q} diluted to P or Q. From the results of these measurements, four linear calibration graphs (aP, aQ, bP, bQ) are constructed that provide an estimate of the analyte concentration in the sample using four independent apparent concentration values c_{1P}', c_{1Q}', c_{2P}', and c_{2Q}'. This is shown in Figure 8.22.

When the real function is linear and there are no uncontrolled effects, the relationship between the measured signal intensities and the analyte concentration in the sample, c_x, and in the standard solution, c_{st}, can be expressed by the equations:

$$Y_{xP} = B_1 \cdot Q \cdot c_x \tag{8.18}$$

$$Y_{1P} = B_1 \cdot P \cdot c_{st} \tag{8.19}$$

$$Y_{2P} = B_1 \cdot Q \cdot c_x + B_1 \cdot P \cdot c_{st} \tag{8.20}$$

$$Y_{xQ} = B_1 \cdot P \cdot c_x \tag{8.21}$$

$$Y_{1Q} = B_1 \cdot Q \cdot c_{st} \tag{8.22}$$

$$Y_{2Q} = B_1 \cdot P \cdot c_x + B_1 \cdot Q \cdot c_{st} \tag{8.23}$$

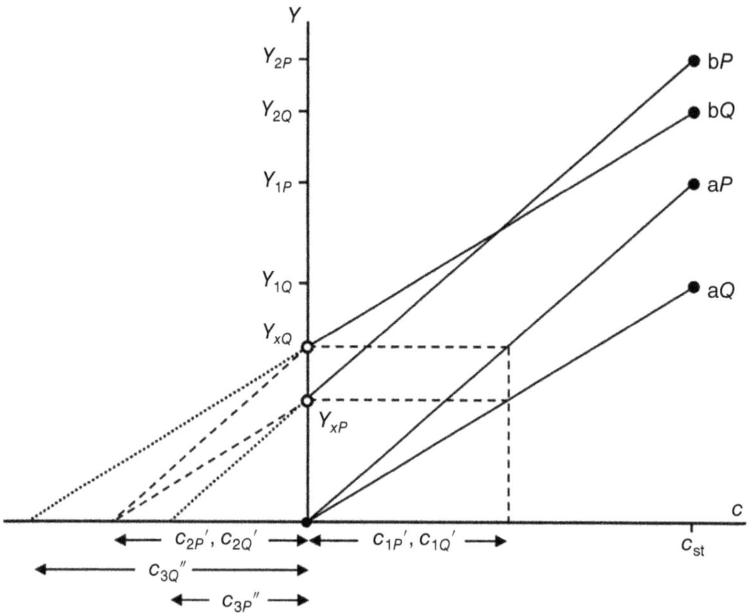

Figure 8.22 Principle of the integrated calibration method in the CDM form: calibration curves constructed in accordance with the external standard method (aP, aQ) and the standard addition method (bP, bQ) lead to six values of apparent concentrations, c_{1P}', c_{1Q}', c_{2P}', c_{2Q}', c_{3P}'', and c_{3Q}'', estimating the analytical result, c_x, in interpolative (c_{1P}', c_{1Q}'), quasi-extrapolative (c_{2P}', c_{2Q}') and extrapolative (c_{3P}'', c_{3Q}'') ways.

It follows from these equations that the apparent concentrations, c_{1P}', c_{1Q}', c_{2P}', and c_{2Q}', can be calculated from:

$$c_{1P}' = \frac{Y_{xQ}}{Y_{1P}} \cdot c_{st} \tag{8.24}$$

$$c_{1Q}' = \frac{Y_{xP}}{Y_{1Q}} \cdot c_{st} \tag{8.25}$$

$$c_{2P}' = \frac{Y_{xQ}}{Y_{2P} - Y_{xP}} \cdot c_{st} \tag{8.26}$$

$$c_{2Q}' = \frac{Y_{xP}}{Y_{2Q} - Y_{xQ}} \cdot c_{st} \tag{8.27}$$

The complementarity of the CDM method manifests itself not only at the preparative stage but also at the transformation stage, since each of the four concentrations, c_{1P}', c_{1Q}', c_{2P}', and c_{2Q}', is calculated from the signals obtained under both dilution conditions, P and Q. This is graphically expressed by the fact that, as can be seen in Figure 8.21, the signal obtained under condition P relates interpolatively to the graph obtained under condition Q and vice versa. Similarly, the graph obtained under P conditions extrapolates according to the slope of the graph obtained under Q conditions and vice versa. Due to this complementarity, the apparent concentrations c_{1P}', c_{1Q}', c_{2P}', and c_{2Q}' can be calculated without knowing the exact values of the dilution degrees of P and Q under the condition that these values are complementary to each other ($P + Q = 1$).

It should be noted, however, that due to the unconventional extrapolative calibration approach leading to the c_{2P}' and c_{2Q}' values, the standard additive method implemented in this way may lose its ability to compensate for proportional effects. Therefore, the results of c_{2P}' and c_{2Q}' should be regarded as "quasi-extrapolative" (see Figure 8.22). To better control the situation, two other results, c_{3P}' and c_{3Q}', obtained by "purely" extrapolative means, i.e. after extrapolation according to the slope of the calibration graphs in the experimental area (see Figure 8.22), can be used as an aid and are calculated with the formulas:

$$c_{3P}'' = \frac{Y_{xP}}{Y_{2P} - Y_{xP}} \cdot c_{st} \tag{8.28}$$

$$c_{3Q}'' = \frac{Y_{xQ}}{Y_{2Q} - Y_{xQ}} \cdot c_{st} \tag{8.29}$$

It should be emphasized that the concentrations of c_{3P}' and c_{3Q}' are not independent quantities for estimating the analytical result. It may be noted that whether or not uncontrolled effects are present in a given analytical system, there is a simple relationship between quasi-extrapolation and extrapolation results:

$$c_{2P}' \cdot c_{2Q}' = c_{3P}' \cdot c_{3P}' \tag{8.30}$$

This relationship, therefore, allows the experimentally obtained values of apparent concentrations to be used to control the correctness of the preparation of calibration solutions and, in particular, to control the strict behavior of the P and Q parameters during this process.

In accordance with the presented CDM procedure of conventional integrated method, the interpolative and extrapolative versions of both the indirect method and the internal standard method can, of course, be integrated creating integrated indirect method and integrated internal standard method.

An important practical limitation of the CDM method is its rather tedious process of preparing calibration solutions and the resulting relatively long time required to perform a single analysis. This problem can be solved by appropriately designed flow systems. The construction of an FIA system dedicated to the CDM method is shown in Figure 8.23 [14].

The system is operated in accordance to the merging zone technique. The standard, st, and sample, S, solutions are introduced together with the carrier solution, C, into the system with flow rates, p, q, and r, respectively. The injection valve, V, is equipped with four loops, α, β, γ, and δ, of equal capacity. If the valve is initially adjusted to the position presented in Figure 8.23 (position I), the carrier only reaches the detector and the zero signal (Y_0) is detected. After turning the valve rotor 45° counterclockwise to position II the standard is injected from loop δ to the carrier in tube L_1 with flow rate p, but the sample is injected from loop β to the carrier in tube L_2 with flow rate q. Due to different lengths of tubes L_1 and L_2 both zones are able to be partially overlapped in tube L_3 in such a manner presented in Figure 8.23, that they are mixed with the carrier and with each other. As a consequence, four different signals can be measured now: for the carrier (Y_0), for the standard diluted with the carrier (Y_{1P}), and with the sample (Y_{2P}), as well as for the sample diluted with the carrier (Y_{xQ}).

When the valve is rotated back to position I shown in Figure 8.22, the standard is injected from loop γ to the carrier in tube L_2 with flow rate q, but the sample is

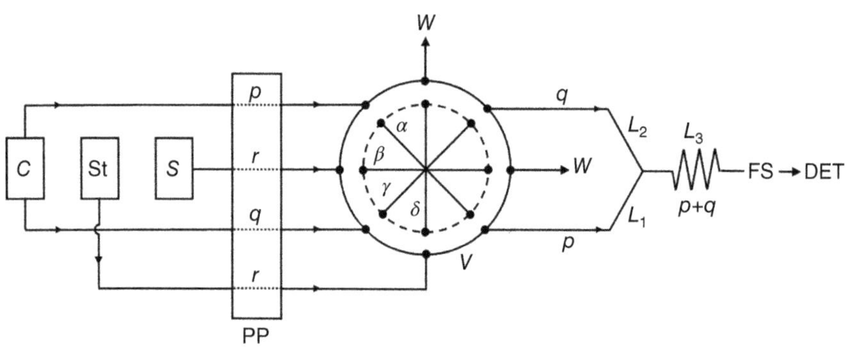

Figure 8.23 The flow injection calibration module dedicated to the integrated method in the CDM form: st – standard, S – sample, C – carrier, W – waste, PP – peristaltic pump, V – two-positional injection valve, α, β, γ, δ – injection loops; L_1, L_2, L_3 – tubes transporting calibration solutions before (L_1, L_2) and after (L_3) their mutual and complementary dilution; p, q, r – flow rates, FS – flow system, DET – detection system. Source: Adapted from Kościelniak et al. [14].

injected from loop α to the carrier in tube L_1 with flow rate p. Both zones are overlapped again in loop L_3, but now they flow with different rates than previously and, consequently, they are diluted with the carrier and with each other to different degree. Thus, four signals are measured again: for the carrier (Y_0), for the sample diluted with the carrier (Y_{xP}), and with the standard (Y_{2Q}), as well as for the standard diluted with the carrier (Y_{1Q}).

The parameters of the flow system parameters, i.e. the loop volumes, lengths L_1 and L_2, and flow rates p and q, should be established at such values, that the standard and sample zones are of volumes great enough to measure the signals in the form of steady-state signals, as shown in Figure 8.23.

The basic features of the FIA system so constructed are as follows:

- the total flow rate of the zones reaching the detector is still constant $(p+q)$,
- in one of the valve's position the standard and sample solutions are diluted with factors $P = p/(p+q)$, and $Q = q/(p+q)$, respectively,
- in another valve's position the standard and sample solutions are diluted with factors Q and P, respectively,
- dilution factors are complementary, i.e. $P + Q = 1$.

As a consequence, the FIA system fulfills the conditions required by the CDM method, i.e. signals measured, Y_{1P}–Y_{1Q}, can serve for construction of the calibration graphs shown in Figure 8.6 and for calculation the apparent concentrations in accordance with Eqs. (8.24)–(8.29).

A number of examples of flow systems for CDM calibration can be found in the literature, which is based on different modes of flow analysis, including sequential injection (SIA) [15], continuous (CFA) [16], multicommuted (MCFA) [17], multi pump (MPFA) [18], and stepwise injection (SWIA) [19] analysis. All of them allow efficient, automatic preparation of calibration solutions according to the scheme shown in Figure 8.24. Some characteristic features of the instrumental solutions are summarized in Table 8.13. These data can be used to compare the systems and facilitate the selection of one of them to be applied alternatively to calibration by CDM depending on the current conditions and needs.

In particular, the CDM method was used to analyze a synthetic sample several times using standards of equal and different analyte concentrations [14]. In each analysis, statistically, equal values of the apparent concentrations, c_{1P}, c_{2P}, c_{1Q}, and c_{2Q}, were obtained as expected (shown in Table 8.14), allowing the analytical results to be calculated from their mean value. Each result was obtained with very good precision (RSD = 2%) and accuracy (RE = ±11%). The results obtained from each analysis varied randomly. All this demonstrates the high analytical value and usefulness of the CDM method.

From these results, it was noted that the mean value of the apparent concentrations obtained in a single determination is always less accurate (farther from the expected value) than at least one of these results. However, a thesis that is easy to prove theoretically has been verified, namely, that if the results obtained in an n-fold analysis of a given sample are subject to statistically equal random errors, then as the

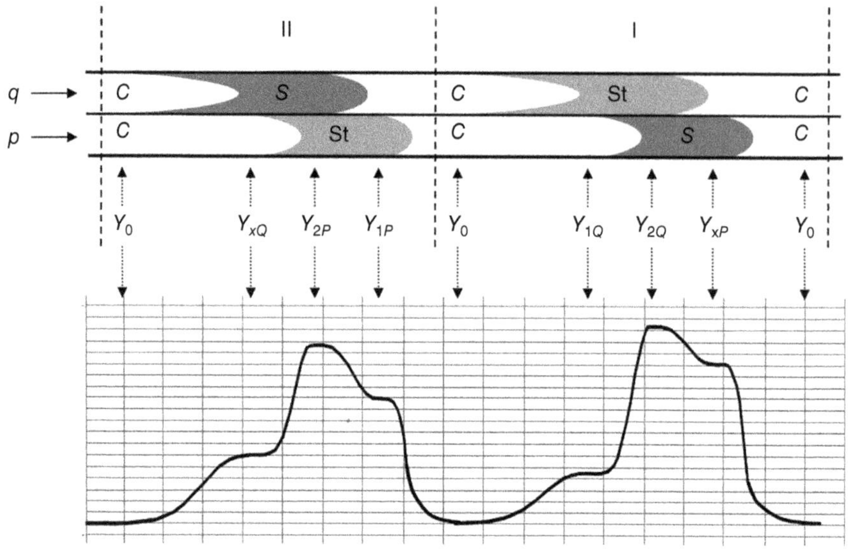

Figure 8.24 The CDM method realized with the use of the FIA system: the standard (st) and sample (*S*) zones flow to the detector with different rates (*p* and *q*) and they are diluted with each other and with the carrier solution (*C*) to produce seven different analytical signals, $Y_0 = 0$, Y_{xQ}, Y_{2P}, Y_{1P}, Y_{1Q}, Y_{2Q}, and $Y_{xP,}$ in two positions, I and II, of the injection valve (compare Figure 8.22). Source: Adapted from Kościelniak et al. [14].

Table 8.13 The indicative comparison of the various flow systems developed for the ICM method (+++, ++, + denote very good, good, and acceptable feature, respectively).

Feature	FIA	SIA	CFA	MCFA	MPFA	SWIA
Form of registered signal	Two three-part peaks	Six single peaks	Six steady-state signals	Six single peaks	Six single peaks	Six steady-state signals
Precision	+++	+	+++	+	++	++
Time of individual analysis	+++	+	++	++	+++	++
Sample consumption, ml	++	+++	+	++	+++	+

number of analyses performed increases, the probability that the mean accuracy of a single result of each analysis is greater than the mean accuracy of the mean value of the results of those analyses becomes less and less. An illustration of this thesis for $n = 4$ and random error RSD = 3% is Figure 8.25.

The results shown in Table 8.14 confirm the theory: it can be seen that after seven analyses, the analyte was able to be determined by the integrated method with

Table 8.14 Results of the spectrophotometric determination of iron (2.50 mg ml^{-1}) in the synthetic samples with the use of the CDM method; the FIA system was used.

C_{st} (mg ml^{-1})	c_0 (μg ml^{-1})	Apparent concentrations (μg ml^{-1}) c_{1P}'	c_{1Q}'	c_{2P}'	c_{2Q}'	c_x (μg ml^{-1})	$\overline{\|RE\|}$ (%)
5.00	2.50	2.51 (0.81)	2.53 (1.23)	2.51 (0.44)	2.51 (1.22)	2.51	0.47
5.00	2.50	2.49 (0.37)	2.51 (0.48)	2.49 (0.57)	2.54 (0.85)	2.51	0.41
5.00	2.50	2.51 (0.40)	2.51 (0.47)	2.49 (0.40)	2.47 (0.92)	2.49	0.26
5.00	2.50	2.49 (0.42)	2.47 (0.49)	2.49 (0.24)	2.49 (0.49)	2.49	0.60
5.00	2.50	2.53 (0.44)	2.51 (0.56)	2.53 (0.44)	2.54 (1.46)	2.53	1.17
4.00	2.50	2.50 (1.55)	2.49 (0.71)	2.50 (1.47)	2.49 (1.65)	2.50	0.13
3.00	2.50	2.52 (0.51)	2.48 (0.60)	2.52 (0.51)	2.50 (0.91)	2.50	0.19
$\overline{\|RE\|}$ (%)		0.54	0.52	0.52	0.87	0.44	

Source: Kościelniak et al. [14], table 1 (p. 1245)/Taylor & Francis.

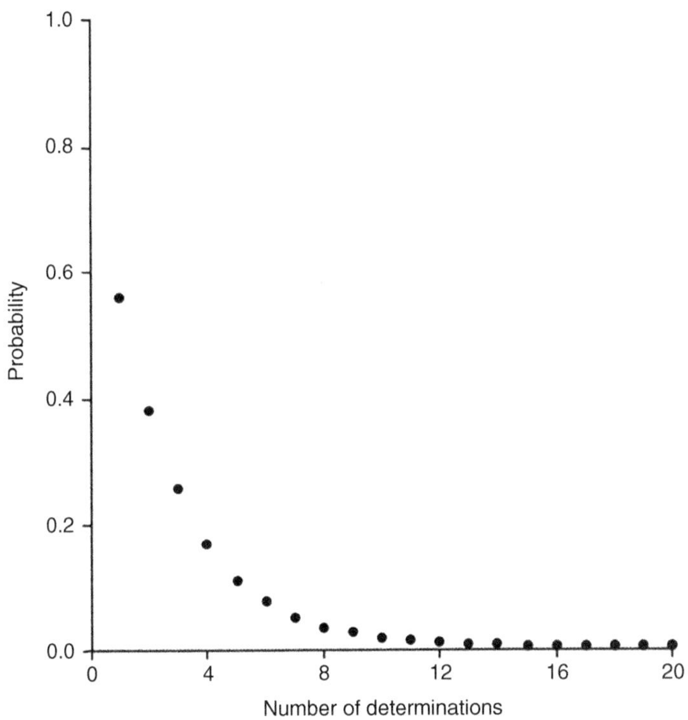

Figure 8.25 Probability (P) of that after successive determinations the cumulative accuracy of one from among four results obtained in the CDM method with the same random error (RSD = 3%) is greater than the cumulative accuracy of the mean value of all four results.

Table 8.15 Results of the FAAS determination of Ca under conditions causing the nonlinear effect; the analytical results, c_x, are calculated from the bold values; the FIA system was used, $P = 2Q$.

		Apparent concentrations (μg ml^{-1})						
c_0 (μg ml^{-1})	C_{st} (μg ml^{-1})	c_{1P}'	c_{1Q}'	c_{2P}'	c_{2Q}'	c_{3P}'	c_{3Q}'	c_x (mg ml^{-1})
2.00	15	**2.35**	**2.32**	2.61	2.47	2.50	2.58	2.33
4.00	15	**4.40**	**4.39**	5.18	4.92	5.20	4.90	4.40
6.00	15	**6.26**	**6.26**	8.07	7.14	8.07	7.13	6.26

Source: Adapted from Kościelniak et al. [14].

higher accuracy ($\overline{|RE|} = 0.44\%$), than by the component methods: the ESM method ($\overline{|RE|} = 0.54\%$ and 0.52%) and the E-SAM method ($\overline{|RE|} = 0.52\%$ and 0.87%). This leads to the important conclusion that **the integration of interpolation and extrapolation calibration methods according to the CDM procedure can increase the accuracy of the results compared to the results obtained by the same methods without integration.**

What information can the CDM method provide when uncontrolled effects are present in the analytical system under study? The answer to this question is facilitated by the experimental data presented below.

Table 8.15 shows the results obtained over a wide nonlinear range of analyte concentrations in the standard solutions. It can be seen that the nonlinearity of the real function due to the detection effect is revealed by a small difference between the interpolation results, c_{1P}', c_{1Q}', much smaller than between the extrapolation results, c_{3P}', c_{3Q}', and the quasi-extrapolation results, c_{2P}', c_{2Q}'. This is due to the known fact that the ESM method is more robust than the E-SAM method to the effect of nonlinearity. The average value of the interpolation results, $c_x = (c_{1P}' + c_{1Q}')/2$, should therefore be treated as closer to the expected result, although it can be expected that they are not determined with sufficiently good accuracy ($c_x \neq c_0$).

When interference effects are present in the sample, the effect of different dilutions of sample and standard (P and Q) in the calibration solutions is generally revealed in the form of a relatively large scatter of quasi-extrapolation values due to the characteristic manner of determining these values described earlier. Simulation calculations based on model (4.10) showed that this scatter is always larger than the scatter of extrapolation values obtained under the same conditions. Moreover, the quasi-extrapolation values can differ even when the extrapolation-determined concentrations are equal (i.e. when the dilution of sample and standard does not affect the magnitude of the interference effect). Such a situation is illustrated in Figure 8.26. Furthermore, as can be seen, the interrelationship between quasi-extrapolation and extrapolation values does not depend on the magnitude of the interference.

Experimental studies have also shown that in the case of the proportional interference effect, the mean value of the concentrations determined by the

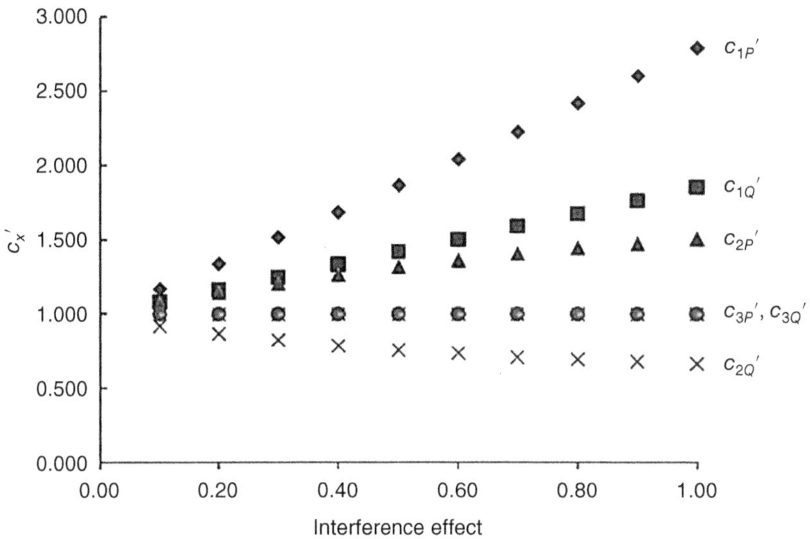

Figure 8.26 Apparent concentration values calculated from Eqs. (8.24)–(8.29) (4.10) when a proportional interference effect of different magnitudes occurs and acts on the analytical signal in accordance with Eq. (4.10).

quasi-extrapolation route, regardless of the difference of these concentrations, is generally not significantly different from the mean value of the extrapolation concentrations. In addition, the results obtained by the interpolation route are characterized by the fact, that the differences between them are much larger than in the case of the nonlinear effect and are obviously significantly different from the other results. This is shown by the experimental results in Tables 8.16 and 8.17 obtained for samples with known (expected) and unknown analyte concentration in the sample, respectively. Thus, in such cases, the analytical result is calculated as the mean value of the apparent concentrations obtained by quasi-extrapolation and extrapolation, $c_x = (c_{2P}' + c_{2Q}' + c_{3P}' + c_{3Q}')/4$. In a situation where all the apparent concentrations are statistically equal in value (see sample 5 in Table 8.16), the analytical result is determined as the mean value of all these apparent concentrations. Analytical results calculated in this way have the potential, as seen in Table 8.16, to be an accurate estimate of the expected value.

Based on a number of experimental results and simulation calculations, guidelines have been formulated to assist in effect recognition and selection of an appropriate route for determining the analytical result in the CDM method. They are summarized in Table 8.18.

It is clear that in analytical practice, a variety of more complicated situations may become apparent (as predicted in Table 8.18) that are difficult to interpret properly from the small amount of data obtained by the CDM method. Nevertheless, although the CDM method provides a limited diagnosis of the state of the analytical system in terms of uncontrolled effects, this diagnosis is broader and more certain than that obtained by the basic integrated method. As has been shown, in many

Table 8.16 Results of the GF-AFS determination of Se in the presence of Cu in the synthetic samples; the analytical results, c_x, are calculated from the bold values, the MCFA system was used, $P = 2Q$.

			Apparent concentrations (mg ml^{-1})						
Sample	c_0 (mg ml^{-1})	c_m (mg ml^{-1})	c_{1P}'	c_{1Q}'	c_{2P}'	c_{2Q}'	c_{3P}'	c_{3Q}'	c_x (mg ml^{-1})
1	50.0	20000	27.28	31.44	**40.61**	**62.13**	**48.43**	**52.10**	50.89
2	50.0	2000	39.55	37.73	**46.00**	**56.25**	**48.87**	**52.90**	51.00
3	12.5	5000	6.64	7.31	**9.58**	**15.95**	**11.84**	**12.90**	*12.57*
4	12.5	500	11.55	11.91	**11.28**	**13.34**	**11.71**	**12.85**	*12.30*
5	32.0	?	**32.80**	**32.41**	**31.00**	**32.19**	**31.37**	**31.80**	*31.92*
6	14.7	?	10.72	12.24	**13.56**	**17.96**	**15.73**	**15.48**	*15.68*

Source: Adapted from Wieczorek et al. [17].

Table 8.17 Results of the HPLC determination of glutamate in the cerebrospinal fluid samples after microdialysis; the analytical results, c_x, are calculated from the bold values, the manual technique was used, $P = 2Q$.

	Apparent concentrations (nM)						
Sample code	c_{1P}'	c_{1Q}'	c_{2P}'	c_{2Q}'	c_{3P}'	c_{3Q}'	c_x (nM)
I1	0.55	0.54	**0.66**	**0.68**	**0.69**	**0.65**	0.67
I2	2.83	2.77	**3.58**	**3.79**	**3.77**	**3.67**	*3.70*
I3	1.41	1.54	**1.47**	**1.75**	**1.61**	**1.60**	*1.61*
I4	1.25	1.56	**1.86**	**2.27**	**2.05**	**2.01**	*2.05*

Source: Adapted from Świt et al. [20].

cases the comparison of six values of apparent concentrations determined in different mathematical ways and under different conditions of the sample and standard dilution is entirely sufficient to provide an accurate estimate of the concentration of the true analyte in the sample. If in other cases this is not the case, in any case, the information obtained is very useful for taking further steps toward an accurate analytical result.

8.3.2 Generalized Calibration Strategy

The results shown in Figure 8.26 show that as the interference effect diminishes, all values of the six apparent concentrations tend to gradually approach a single, common analytical result. It can therefore be assumed that the combination of an integrated method with a dilution-compatible procedure, i.e. interference compensation by progressive dilution of sample and standard, can lead to an indication of a single, accurate concentration of analyte in the sample by all values of the

Table 8.18 Rules of diagnosis of the uncontrolled effects calculation of the analytical result in the CDM method.

Indication	Diagnosis	Calculation of the analytical result
$c_{1P}' = c_{1Q}' = c_{2P}' = c_{2Q}' = c_{3P}' = c_{3Q}'$	No uncontrolled effects	$(c_{1P}' + c_{1Q}' + c_{2P}' + c_{2Q}' + c_{3P}' + c_{3Q}')/2$
c_{1P}' and $c_{1Q}' \neq c_{2P}', c_{2Q}', c_{3P}'$ and c_{3Q}'	Uncontrolled effects	–
$c_{1P}' = c_{1Q}' \neq c_{2P}' \neq c_{2Q}' \neq c_{3P}' \neq c_{3Q}'$	Nonlinear effect	$(c_{1P}' = c_{1Q}')/2$
$c_{1P}' \neq c_{1Q}'$ and $(c_{2P}' + c_{2Q}')/2 = c_{3P}' = c_{3Q}'$	Linear interference effect	$(c_{2P}' + c_{2Q}' + c_{3P}' + c_{3Q}')/2$
$c_{1P}' = c_{1Q}'$ and $(c_{2P}' + c_{2Q}')/2 = c_{3P}' = c_{3Q}'$	Nonlinear or complex interference effect	–
Another situation	Unknown effect	–

apparent concentrations. This assumption is the basis of the so-called **generalized calibration strategy (GCS)** [21]. The generalized nature of the GCS method is expressed both in its ability to integrate all of the basic and most commonly used calibration methods and in its full use of the basic processes that are available for analytical calibration without the need for additional chemical reagents: the addition of standard to sample and the dilution of sample and standard.

To rationalize GCS, it is proposed that its additional element should be the realization of the preparative step by means of a dedicated flow system. The system should operate so that the manually prepared individual sample and standard solutions, when introduced into the system, are combined with each other and with the carrier solution according to the integrated method procedure, and the created calibration solutions (e.g. of the composition shown in Figure 8.24) are then processed according to the required analytical procedure and measured.

As shown experimentally, some of the systems that are used for CDM calibration, namely FIA [14], MPFA [18], and SWIA [19] systems, can be used to implement the GCS method. Particularly convenient is the FIA system, shown in Figure 8.27, which not only allows the efficient preparation of calibration solutions according to the CDM method but, when suitably modified, enables their efficient, automatic, and controlled dilution according to the manipulative circulation technique (see Figure 7.3).

Now, if injection valve V is rotated 45° counterclockwise and back, in the subsequent operation cycles the sample (S) and standard (st) solutions are injected to the carrier solution (C) and transported towards detector, and at the same time solution C is injected to solutions ST and S and transported to their reservoirs. As a result, the sample and standard solutions are progressively diluted.

As in each calibration cycle, n, the valve is situated in positions I and II when sample and standard solutions flow alternately with flow rates p and q (see Figure 8.10), they are diluted with dilution factors of different values, k_{2n-1} and k_{2n}. The current dilution factor value can be accurately calculated when knowing initial volume of both sample and standard solutions in their reservoirs, volumes of the carrier zones injected, and previous dilution factor value. Consequently, the apparent concentrations have to be calculated adequately considering values k_{2n-1} and k_{2n}.

The measurement results and apparent concentrations obtained in the first cycle of the GCS procedure in the indirect method version are shown in Figure 8.28. It can be seen that the measurement points are shifted due to the sample and standard solutions flowing at different rates.

The closed FIA system streamlines the procedure of the GCS method by allowing multistep dilution of calibration solutions to a point where fluctuations of the measurement signals still allow reliable determination of apparent concentrations. This makes it possible to proceed in the transformation step according to the procedure of the dilution method, i.e. to present the apparent concentrations calculated at each dilution step as a function of the value of the dilution factor, k, and to determine the analytical result from the values of these concentrations in a sample of infinite high dilution. An example of this method of presenting results obtained by the GCS method is shown in Figure 8.29.

Figure clearly shows that in both cases (of the nonlinear and interference effects), the apparent concentration values are successively closer and closer to each other and to the true concentration value. The interposition of all experimental points allows the two effects to be easily distinguished, and a deeper analysis of the position of the points according to the rules in Table 8.18 allows the type and nature of the interference effect to be recognized. Most importantly, however, the results obtained after diluting the sample and standard several times allow the analytical result to be determined with increased accuracy. Such a result can only be determined by comparing the apparent concentration values obtained at the end of the dilution process without the need to fit model functions to experimental data and extrapolate them to a zero dilution step value.

On the basis of many experiments, it has been found that a 10-fold dilution of solutions is in most cases sufficient to determine the analyte concentration with

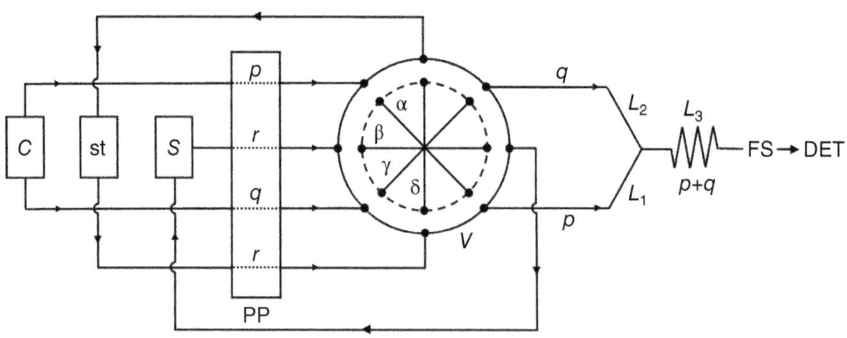

Figure 8.27 The FIA module dedicated to the GCS method; meaning of symbols as in Figure 8.22.

very good accuracy. Sometimes a significant improvement in the accuracy of analytical results can occur even after two dilutions of sample and standard, as exemplified by the results shown in Figure 8.30. It is, of course, also possible that the results obtained by extrapolation are an accurate measure of the analyte concentration already in the undiluted sample (see Table 8.19, sample I); the results obtained after successive dilutions of sample and standard then confirm this fact.

The problem that still remains is the decreasing precision of the apparent concentrations determined in a sample of increasing dilution. The initial concentration of the analyte in the sample should therefore be high enough so that, after such a dilution of the sample as to permit its determination with acceptable precision, the apparent concentrations will permit the indication of its exact value. An example of failure to maintain this condition is provided by the results shown in Table 8.19 for sample II: after an acceptable dilution of this sample from the point of view of precision four times, further dilution of the sample would, it would seem, be desirable for the sake of the accuracy of the determination.

The general characteristics of the GCS method can be expressed as follows:

- due to the integration mode applied, an analytical result is estimated by six values of different origin, since the uncontrolled effects can be detected and the systematic error can be avoided,
- due to the SAM included, linear interference effect can be compensated,

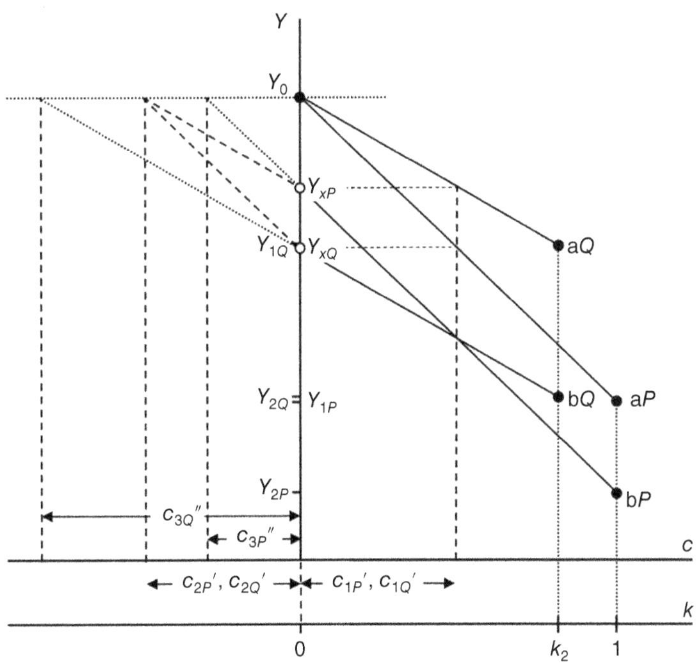

Figure 8.28 Measurement results obtained in the first cycle of the indirect GCS method; Y_0 – signal measured for the reagent, meaning of the remaining symbols as in Figure 8.21.

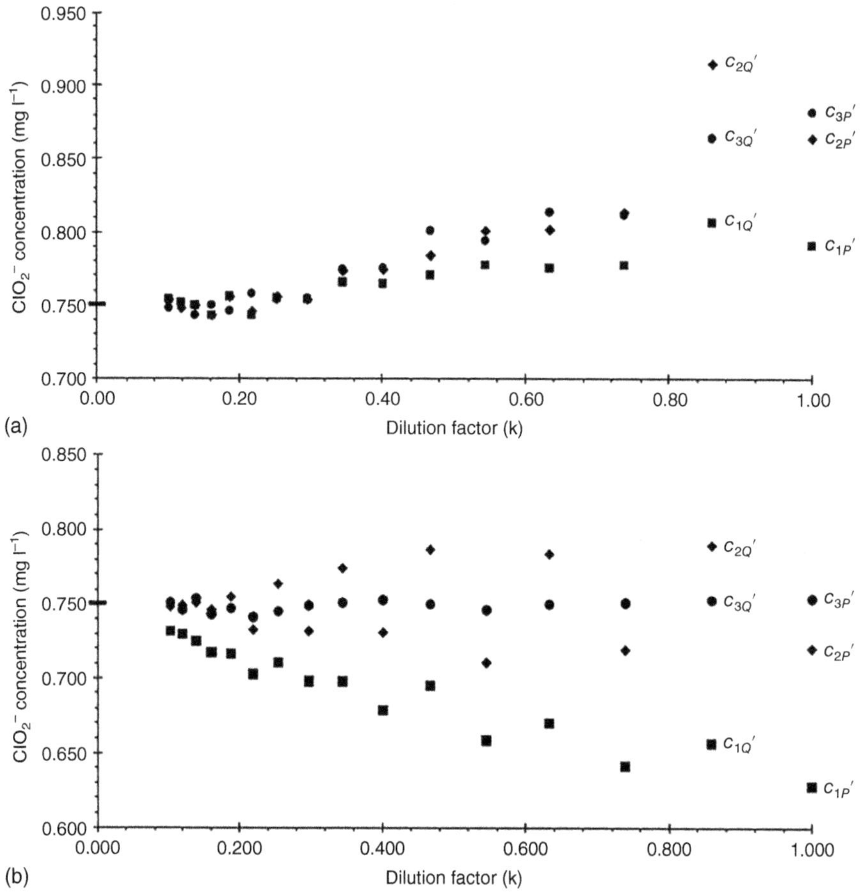

Figure 8.29 Application of the GCS method to the spectrophotometric indirect determination of chlorites ($c_0 = 0.75$ mg l^{-1}) in water samples: (a) – under conditions causing the nonlinear effect (without ammonium), (b) – in the presence of ammonium (900 mg l^{-1}); the obtained results, c_x, are equal to 0.753 and 0.748 mg l^{-1}, respectively. Source: Adapted from Świt et al. [20].

- due to the dilution method included, the nonlinear and complex interference effect can be compensated and the nonlinear effect can be eliminated,
- due to the single solution used, errors in preparation of a set of standards are avoided,
- due to any additional chemicals used, the green analytical chemistry rule is fulfilled,
- due to the flow injection system used, the calibration procedure is efficient and accelerated as well as the analytical results have a chance to be better in terms of precision,
- due to the calibration module being independent of the chemical module, different analytical and measurement methods can be adapted.

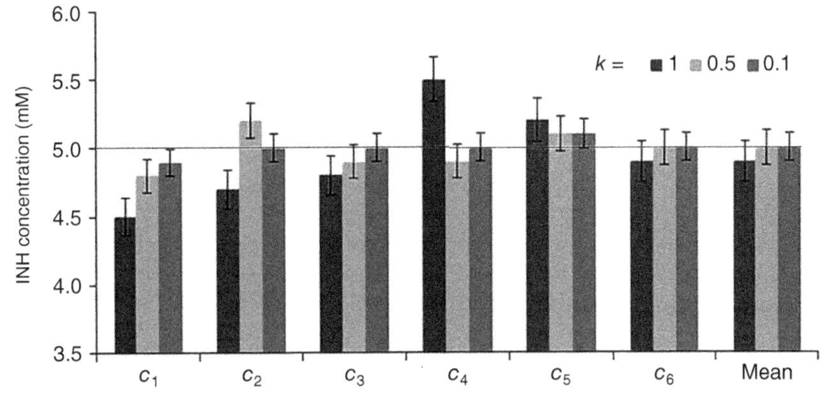

Figure 8.30 Results of the spectrophotometric determination of isoniazid spiked in concentration 5 mM to the urine sample, c_1–c_6 – apparent concentrations, the SWIA system was used. Source: Medinskaia et al. [19]/Elsevier.

Table 8.19 Results of the FAAS determination of Ca in natural samples of: slim milk powder (I, certified material), and "oscypek" cheese (II); the final results are calculated from the bold values, the MPFA system was used.

			Apparent contents (mg g^{-1})					
Sample	c_0 **(mg g^{-1})**	**Dilution degree, k**	c_{1P}'	c_{1Q}'	$(c_{2P}' + c_{2Q}')/2$	c_{3P}'	c_{3Q}'	c_x **(mg g^{-1})**
I	13.49	1.00	8.23	10.12	**14.05**	**12.85**	**14.26**	13.72
	13.49	0.75	8.82	11.24	**13.82**	**12.06**	**14.90**	13.59
	13.49	0.50	9.79	11.90	**13.77**	**12.97**	**14.01**	13.58
	13.49	0.25	10.30	12.19	**13.74**	**13.05**	**14.02**	13.66
II	?	1.00	6.37	7.25	**7.27**	**6.93**	**7.43**	7.21
	?	0.75	6.82	7.11	**7.68**	**7.51**	**7.79**	7.66
	?	0.50	7.13	8.24	**8.38**	**7.80**	**8.83**	8.34
	?	0.25	8.64	10.10	**10.64**	**9.58**	**11.67**	10.63

Source: Adapted from Wieczorek et al. [18].

The method is of course not free from certain limitations, the most significant of which are:

- it is adapted to the analysis of liquid samples with relatively high analyte concentrations,
- similarly to other calibration methods discussed so far, it is not capable of eliminating additive effects,
- despite the use of a flow system, the calibration procedure is relatively long (analysis of a single sample takes from several to several dozen minutes).

Despite its limitations, the GCS method is the most versatile calibration method in terms of its ability to handle uncontrolled effects problems. It can be applied to analyses carried out by a wide variety of analytical and measurement methods. It also seems that the situations in which it can fail are exceptional, although, of course, one should always remember that the type and nature of uncontrolled effects are unpredictable, just as the mechanisms of phenomena occurring in nature are unpredictable.

8.3.2.1 Versatile Flow Injection Calibration Module

The question very important from a practical point of view is whether the procedures of all the above-described calibration methods can be carried out automatically by means of a single, versatile analytical and measuring instrument. Well, it turns out that such a possibility is offered by the FIA calibration module, whose operation and use were described in Chapter 7. Adapting it to even more versatile use requires only a small instrumental modification [22]. It consists in mounting two additional valves that, depending on the current need, direct the sample and standard solutions either outside the system or to the vessels from which these solutions are taken. The **versatile flow injection calibration module (VFICM)** constructed in this way is shown in Figure 8.31.

As known, injection valve V can be rotated in turn from position shown in Figure 8.31 (position I) to position changed of 45° counterclockwise (position II) and back allowing in the subsequent operation cycles successive injection of solutions ST and S to solution C as well as of solution C to solutions ST and S. Depending on the positions of valves V_1 and V_2 the manifold can work in different configurations, i.e.:

- opened (OP) when the solutions aspirated from both reservoirs, ST and S, are directed to the waste (see Figure 8.23),
- half-closed (HC), when solution ST is directed back to its reservoir but solution S is directed to the waste,

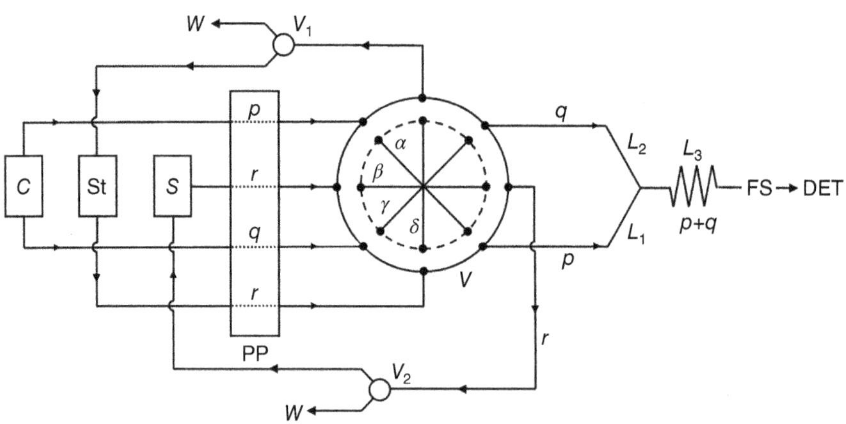

Figure 8.31 Versatile calibration flow injection manifold: C, st, S, C – standard, sample, and carrier solutions, respectively; W – waste; PP – pump; V – two-positional injection valve; V_1, V_2 – two-positional directive valves; $\alpha, \beta, \gamma, \delta$ – injection loops; L_1, L_2, L_3 – transfer tubes; p, q, r – flow rates, FS – flow system, DET – detector.

- closed (CL), when both solutions, ST and *S*, are directed back to their reservoirs (see Figure 8.27).

In configuration OP, the sample and standard solutions are not able to be gradually diluted with the carrier solution (as in the CDM method) while in configuration CL both calibration solutions are diluted (as in the GCS method). In configuration HC, the sample is not diluted and the standard is able to be progressively diluted with the carrier. This provides the opportunity to analyze a single sample using a series of standard solutions with progressively smaller analyte concentrations.

Irrespective of the configuration, different instrumental parameters of the module can be controllably changed, including flow rate values p and q, length of tubes L_1, L_2, and L_3, and the volumes of injection loops α, β, γ, and δ. Depending on the length of the tubes L_1 and L_2 and on the volumes of loops α, β, γ, and δ the injected zones of solutions ST and *S* can be directed to tube L_3 separately, or they can be merged with each other partly, as presented in Figure 8.32.

After reaching tube L_3 both zones are diluted either with the carrier (when they are in position A) or with each other (in positions B and C). If $p = q$, they are diluted equally with dilution factor $P = Q = 0.5$ in both positions, I and II, of the injection valve. If $p \neq q$, the sample and standard zones are diluted with different dilution factors, $P = p/(p+q)$ and $Q = q/(p+q)$, whereby if in one valve position the sample zone is diluted with factor P and the standard zone with factor Q, in the other valve position they are diluted inversely.

When the sample and standard segments reach the detector at a certain position of the injection valve, two separate measurement signals (if the segments are in position A, see Figure 8.2) or superimposed signals (if the segments are in position B or C) are obtained. Depending on the volumes of loops α, β, γ, and δ the signals take the form of peaks or fixed signals. If the segments are in position A, it is acceptable for the signals to be in the form of peaks and the measure of the intensity of the signals is then the maximum peak height (or area under the peaks). If the segments are in

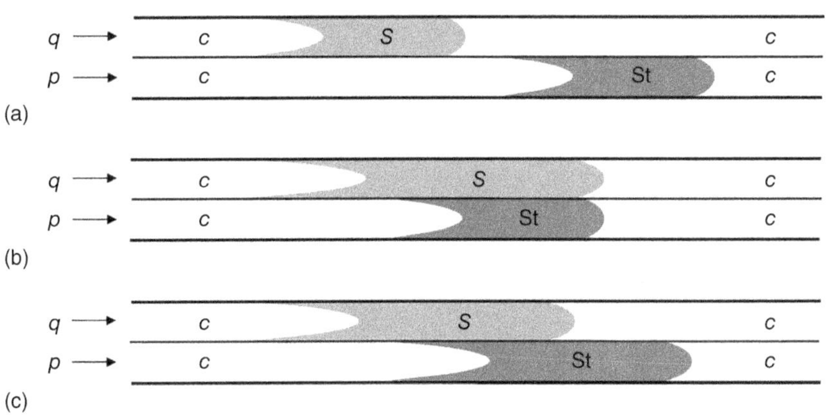

Figure 8.32 The standard (ST) and sample (*S*) zones injected into the carrier solution (*C*) in separate (a) and partly overlapped (b, c) mutual positions depending on instrumental parameters (e.g. the flow rates values, *p* and *q*) established in the calibration module.

position B or C, then loops of sufficient capacity must be installed so that signals for the sample, standard, and sample with standard can be recorded as signals composed of fragments of fixed height (see Figure 8.8) and their intensities are measured from these heights.

It should be noted that if the system operates in the OP configuration and the segments are transported at equal speeds ($p = q$), then regardless of their mutual position, the measurement information obtained for the sample and standard at both positions of the injection valve is duplicated. In the case of the HC configuration, this applies only to the sample. In any other case, the information is different, because the change in position of the injection valve is then associated with a different dilution of the two segments.

By setting the appropriate configuration and instrument parameter values, it is possible to adapt the module to realize different calibration methods. General guidelines to allow calibration with single, extrapolation, and complex calibration methods are presented in Tables 8.20–8.22, respectively. In each case, preliminary optimization studies should be performed to determine the most appropriate instrumental parameter values. First of all, the idea is that the measured signals should be as intense as possible and appropriately positioned with respect to each other. On the other hand, it should of course be ensured that the procedure of the respective calibration method is carried out as quickly as possible.

For single calibration methods (see Table 1.9), the sample and standard segments should be transported to the detector generally separately (position A in Figure 8.2) at equal speeds. The exception is the SAM, which requires the segments to overlap in the B way. The dilution method is implemented in the closed configuration of the VFICM module. The other methods can be implemented in the OP or HC configuration, allowing them to be used in the basic and multipoint versions, respectively. If the sample, standard, and carrier are present in the respective vessels, then the calibration follows the external standard method, the SAM, or the dilution method. If the reagent or internal standard is added in equal concentrations to all vessels,

Table 8.20 General conditions enabling calibration by the single methods with the use of VFICM.

Calibration method	Version	St reservoir	*S* reservoir	*C* reservoir	Module configuration	Zone position	Flow rates
External standard method	Basic	St	*S*	*C*	OP	A	$p = q$
External standard method	Multipoint	St	*S*	*C*	HC	A	
Indirect method	Basic	St + *R*	*S* + *R*	*R*	*OP*	A	
Indirect method	Multipoint	St + *R*	*S* + *R*	*R*	*HC*	A	
Internal standard method	Basic	St + IS	S + IS	IS	*OP*	A	
Internal standard method	Multipoint	St + IS	S + IS	IS	*HC*	A	
Standard addition method	Basic	St	*S*	*C*	*OP*	B	
Standard addition method	Multipoint	St	*S*	*C*	*HC*		
Dilution method	–	St	*S*	*C*	*CL*	A	

Table 8.21 General conditions enabling calibration by the extrapolation methods with the use of VFICM.

Calibration method	Version	St reservoir	*S* reservoir	*C* reservoir	Module configuration	Zone position	Flow rates
Extrapolative indirect method	Basic	St + *R*	S + *R*	*R*	OP	B	$p = q$
Extrapolative indirect method	Multipoint	St + *R*	S + *R*	*R*	HC	B	
Extrapolative internal standard method	Basic	St + IS	*S* + IS	IS	*OP*	B	
Extrapolative internal standard method	Multipoint	St + IS	*S* + IS	IS	*HC*	B	
Dilution method	–	St	*S*	*C*	*CL*	B	

Table 8.22 General conditions enabling calibration by the combined methods with the use of VFICM.

Calibration method	Mode	St reservoir	*S* reservoir	*C* reservoir	Module configuration	Zone position	Flow rates
Basic integration method	Conventional	St	S	*C*	OP	*C*	$p = q$
Basic integration method	Direct	St + *R*	*S* + *R*	*R*	OP	*C*	
Basic integration method	Internal standard	St + IS	S + IS	IS	OP	*C*	
CDM	Conventional	St	S	*C*	OP	*C*	$p \neq q$
CDM	Direct	St + *R*	S + R	*R*	OP	*C*	
CDM	Internal standard	St + IS	*S* + IS	IS	OP	*C*	
GCS	Conventional	St	*S*	*C*	CL	*C*	
GCS	Direct	St + *R*	*S* + R	*R*	CL	*C*	
GCS	Internal standard	St + IS	*S*+ IS	IS	CL	*C*	

then the calibration procedures follow the indirect method or the internal standard method, respectively.

Extrapolation methods (Table 8.10) are implemented similarly to single calibration methods. The extrapolation character is given by the partial overlap of the sample and standard segments in the B manner, which is characteristic of the SAM.

The adaptation of the calibration module to implement the complex CDM and GCS methods has been discussed previously in this chapters. By setting the flow velocity of the sample and standard segments to be equal ($p = q$), the different measurement information obtained in the CDM method at both positions of the injection valve is reduced to equal information delivered twice under the same conditions. Consequently, calibration then occurs according to the basic integrated method.

The presented ways of adapting the VFICM module for realization of calibration methods do not exhaust all its possibilities in this field. Other combinations of different configurations of the module and its instrumental parameters can lead to other, interesting and perhaps completely new calibration procedures, characterized by their own unique analytical features. Without wishing to devote too much space and time to this subject in this chapter, the author encourages the readers to undertake their own conceptual and experimental attempts.

8.3.3 Standard Dilution Analysis

Calibration methods can, of course, be combined even differently than the one presented above. A very good example of invention in this regard is the called **standard dilution analysis (SDA)** [23]. In this case, the processes of adding standard to the sample and dilution are used to integrate the E-SAM method with the internal standard method.

The preparative procedure of the SDA method is shown in Figure 8.33. The basic calibration solution is a sample solution of volume V_0 with the analyte standard and internal standard mixed and added to the sample in a well-defined volume V_1. The sample is then added to this solution diluted with diluent in the same ratio as it is diluted with the addition of standards in the first solution. By this procedure, the concentration of the analyte in the sample, $k_v{\cdot}c_0$, remains the same in both calibration solutions, while the concentration of the standards decreases in degree K as a result of their dilution with the sample solution.

The solutions thus prepared are subjected to signal intensity measurements under conditions characteristic of the analyte, Y, and of the internal standard, Y_w. If the signals are not subject to uncontrolled effects, their dependence on the concentration of the analyte in the sample, c_0, and in the analyte standard, Δc_1, and on the concentration of the internal standard, c_w, by the following formulas:

$$Y_0 = B_1 \cdot [k_v \cdot c_0 + (1 - k_v) \cdot \Delta c_1] \tag{8.31}$$

$$Y_1 = B_1 \cdot \left[k_v \cdot c_0 + (1 - k_v) \cdot \frac{1}{K+1} \cdot \Delta c_1\right] \tag{8.32}$$

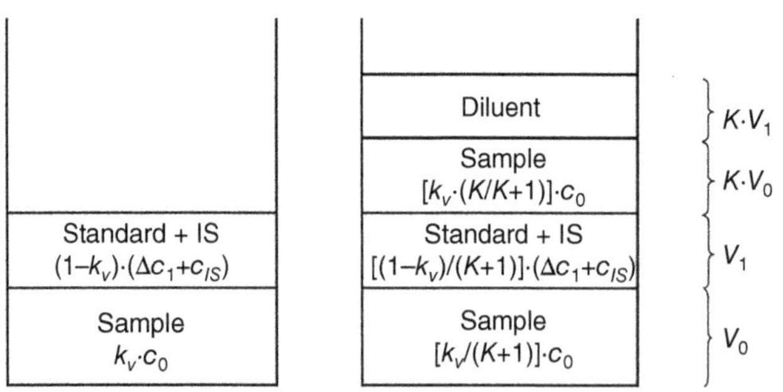

Figure 8.33 Scheme of the basic form of the SDA method at the preparative stage of the calibration process (for details see text).

$$Y_{w0} = B_2 \cdot (1 - k_v) \cdot c_w \tag{8.33}$$

$$Y_{w1} = B_2 \cdot (1 - k_v) \cdot \frac{1}{K+1} \cdot c_w \tag{8.34}$$

where B_1 and B_2 are the measurement sensitivity values of the analyte and the internal standard. According to the internal standard method rule, a measure of the response of a measuring instrument to the amount of analyte in calibration solutions is the ratio of the signals: $R_0 = Y_0/Y_{w0}$ and $R_1 = Y_1/Y_{w1}$. From Eqs. (8.31)–(8.34) it follows that the analytical result is obtained from the formula:

$$c_0 = \frac{R_1 - R_0}{(K+1) \cdot R_0 - R_1} \cdot \frac{1 - k_v}{k_v} \cdot \Delta c_1 \tag{8.35}$$

Graphical interpretation of the SDA method in its basic version (for $k_v = 0.5$) is presented in Figure 8.34.

The process of diluting the sample calibration solution with the standards using the sample solution can of course be continued. As the dilution proceeds, the analyte concentration in the sample will remain the same all the time, and the analyte concentrations in the standard and the internal standard concentrations will become smaller and smaller. It can be shown that the change in the ratio of analyte and internal standard signals then occurs according to the following formula:

$$R_{nK+1} = \frac{B_1 \cdot \Delta c_1}{B_2 \cdot c_w} + \frac{B_1 \cdot c_0}{B_2 \cdot c_w} \cdot \frac{k_v}{1 - k_v} \cdot (nK + 1) \tag{8.36}$$

where R_{nK+1} is the signal ratio at $nK + 1$ dilution degree, and n is the number of calibration solutions prepared (i.e. dilutions made of the first calibration solution). Formula (8.36) can be represented in the simpler form of a linear function:

$$R_{nK+1} = A_r + B_r \cdot (nK + 1) \tag{8.37}$$

Function (8.37) is presented in Figure 8.35.

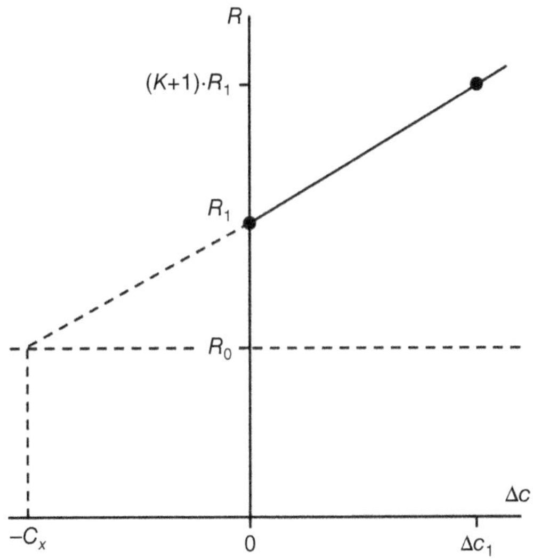

Figure 8.34 Calibration in accordance with the SDA method: extrapolative transformation of the ratio of analyte to internal standard signals, R_1 and R_2, obtained for two calibration solutions shown in Figure 8.33 to the analytical result, c_x.

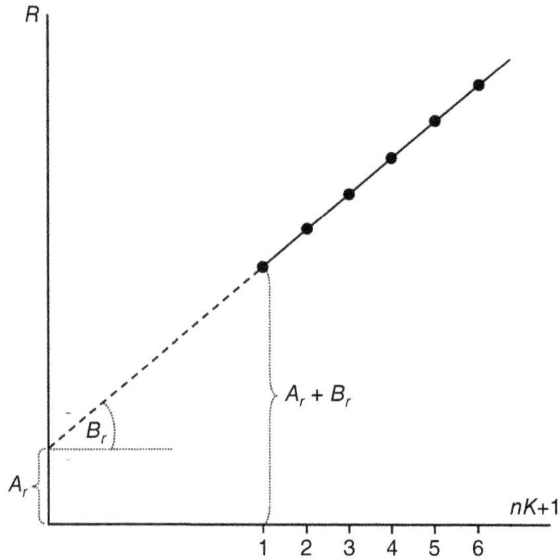

Figure 8.35 Calibration in accordance with the multipoint version of the SDA method: extrapolative transformation of the ratio of analyte to internal standard signals, R_1 and R_2, obtained for two calibration solutions shown in Figure 8.6 to the analytical result, c_x.

The concentration of the analyte in the sample is then calculated from:

$$c_0 = \frac{B_r}{A_r} \cdot \frac{1 - k_v}{k_v} \Delta c_1 \tag{8.38}$$

As can be seen, in the SDA method the analytical result is determined by extrapolation. The method differs from the "classical" extrapolation internal standard method essentially in that the concentration of the internal standard in successive calibration solutions is not fixed, but changes to the same extent and in the same direction as the concentration of the analyte in the analyte standard changes. The different ways of preparing the calibration solutions by gradually reducing the concentrations of the analyte standard and the internal standard result in a different interpretation of the measurement data, while maintaining the extrapolative nature of the method.

The original version of the SDA method follows the change in signal ratio R with the change in the inverse of the concentration of the internal standard, c_w [23]. As can be seen from Eq. (8.37), this does not change the form of the calibration plot and the way the analytical result is determined, but it requires the calculation of the concentration c_w at each dilution step of the sample with added standards. Example graphs, obtained experimentally for an analyte present in several samples at different concentrations, are shown in Figure 8.36.

The number of calibration solutions prepared, n, and therefore also the number of measurement points, affects the size of the experimental part of the calibration graph without changing the size of its extrapolation part. Thus, to achieve the optimal ratio of these parts (from 1 to 4), many calibration solutions must be made, as shown in Figure 8.36. However, the same goal can be achieved with less effort and time by diluting the first calibration solution with a standard solution of relatively larger volume (i.e. larger K value). Although a smaller number of points on a particular part of the calibration graph is then obtained, this is of little importance from the point of view of the precision of the analytical result.

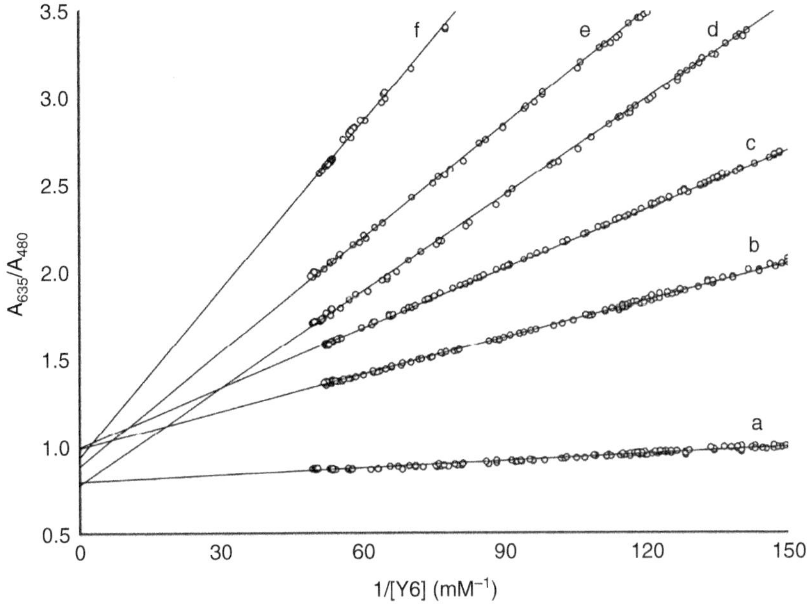

Figure 8.36 SDA graphs (including multiple replicates) obtained for a FD&C Blue No. 1 dye determined by UV/VIS spectrophotometry in the presence of a FD&C Yellow No. 6 (Y6) dye as the internal standard in six different mouthwash samples (a–f). Source: Jones et al. [23], fig 3 (p.D)/with permission of American Chemical Society.

The precision of the analytical result depends mainly on the slope of the graph and thus, as shown in Eq. (8.37), on the ratio of the concentration of the analyte in the sample, c_0, to the concentration of the internal standard, c_w at specific values of measurement sensitivity, B_1 and B_2, of both substances. The low concentration of c_w also favors the precise determination of the intercept, A_r. It must be remembered, however, that the role of the internal standard is to compensate for fluctuations in the signals measured for the analyte. Thus, for this compensation to be effective, the concentration should be comparable to the combined concentration of the analyte in the sample, c_0, and in the standard. The slope of the calibration graph can also be determined to some extent by setting an appropriate k_v value, the volume of sample to the volume of standards solution in the first calibration solution.

It is very important that the SDA method retains that condition of the E-SAM method that determines the ability to compensate for proportional interference effects: the concentration of sample components in successive calibration solutions is kept constant. Of course, it should not be expected to be resistant to interference effects of a different nature (complex and additive), since its component methods are not resistant to such effects either. At the same time, it is difficult to predict the extent to which this calibration method is effective as an internal standard method, that is, what is its ability to compensate for random variations in the signals measured for the analyte. It can be theoretically supposed that in such cases when in a given measurement method the signals fluctuate proportionally to the absolute values of these signals, the SDA method is more effective than the classical internal standard method.

Numerous experimental examples of the SDA method provide surprisingly positive support for this supposition. Representative examples are the results presented in Table 8.23. They show that the method allows the determination of analytes by ICP-OES generally with significantly lower random and systematic error than the E-SAM and internal standard methods. In comparison with other results obtained with ICP-OES technique with the use of internal standard, the significant improvement of precision of analytical results achieved with the use of single internal standard looks particularly spectacular. Similar effectiveness of the SDA method has been demonstrated for the analysis of samples with complex and diverse matrices by other spectrometric methods, including UV/VIS spectrophotometry, flame atomic emission spectrometry (FAES), FAAS, ICP-OES, ICP-MS, and Raman spectroscopy.

A certain limitation of the SDA method is the relatively long time to perform a single analysis, which is particularly troublesome when a sample wants to be analyzed with a large number of calibration solutions. However, the calibration procedure can be mechanized and automated using a gradient-based FIA system [24]. In this case, the flow system is designed so that the two calibration solutions shown in Figure 8.33 are injected from a relatively large capacity loop into the blank solution in the form of two segments in contact with each other. As they flow, these segments increasingly interpenetrate each other, creating a concentration gradient between the analyte standard and the internal standard. As a result, measurement results taken under conditions characteristic of the analyte and the internal standard are recorded in the form of complex peaks as shown in Figure 8.36. A calibration graph is constructed

Table 8.23 Average results obtained for the ICP-OES simultaneous determination of eight metals (with Y as internal standard) added to different matrices (water, nitric acid, mouthwash, cola) in concentration of 10 mg l^{-1} equal for each metal determined; standard addition method (E-SAM), internal standard method (ISA), and standard dilution analysis (SDA) were used for calibration.

	Concentration found (mg l^{-1})			Relative error (%)			Relative standard deviation (%)		
Analyte	**E-SAM**	**ISA**	**SDA**	**E-SAM**	**ISA**	**SDA**	**E-SAM**	**ISA**	**SDA**
Al	9.5	12.6	9.5	4.8	26.1	5.0	2.9	17.4	3.5
Cd	11.0	12.2	10.5	16.0	22.2	7.3	18.8	10.9	7.7
Co	10.2	12.3	10.0	11.5	23.3	4.2	13.9	8.0	5.1
Cr	10.3	11.6	10.2	10.4	16.4	4.3	12.7	5.9	4.5
Cu	9.5	12.3	9.7	7.8	23.4	3.8	7.6	0.7	3.3
Fe	10.5	11.8	10.3	10.8	17.6	4.5	13.5	12.7	5.0
Ni	10.4	11.7	10.2	12.4	17.5	4.9	14.9	9.4	5.8
Pb	10.1	11.6	9.9	12.0	15.8	4.0	13.9	9.3	4.9
Average	10.2	12.0	10.1	10.7	20.3	4.7	13.3	9.3	5.8

Source: Adapted from Jones et al. [23].

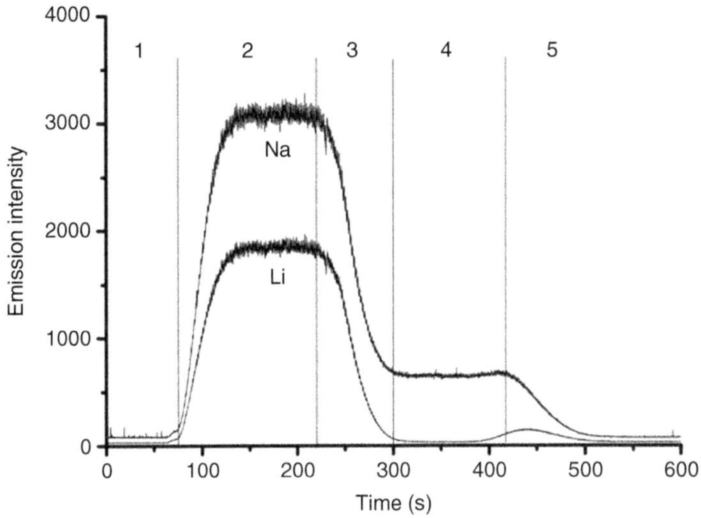

Figure 8.37 Transient signals measured for Na (analyte) and Li (internal standard) by the FIA-SDA system, consisted of five regions corresponding to: (1) the blank, (2) first calibration solution (see Figure 8.6), (3) SDA process, (4) second calibration solution, and (5) blank. Source: Fortunato et al. [24], fig 2 (p. 664)/with permission of Elsevier.

from the ratio of signal pairs forming the falling portion of the peaks (region 3 in Figure 8.37), and the analytical result is calculated from Eq. (8.38).

Example analytical results obtained in the above-described manner are summarized in Table 8.24 against results obtained with other calibration methods: ESM, E-SAM, and internal standard method (ISM) [24]. The application of the SDA method to the analysis of biological materials contributes, as can be seen, to the improvement of the quality of analytical results not so much because of their precision, but rather because of their accuracy. It is apparent that, the participation of the internal standard in the calibration process of the SDA method will allow the control of interference effects sometimes to a greater extent than can be achieved by the E-SAM method.

The technique presented here makes it easy and fast to perform SDA analyses, with the ability to generate dozens of measurement points that form a calibration plot. There is some difficulty in accurately determining from the injection peak that the signal which corresponds to the initial maximum concentration of the internal standard and is the basis for calculating successively smaller values of this concentration. This problem was solved by introducing a second internal standard into the dilution solution [25], for which signals of decreasing intensity are measured separately. The two internal standards are plotted against each other, and the maximum signal of the first internal standard is defined as the y-intercept of this graph.

The SDA method is a very good example of an ingenious and simple, yet analytically very effective combination of calibration methods. Thus, it constitutes a valuable addition to the numerous range of empirical complex calibration methods of quantitative analysis. Similar to some other previously presented methods, it has a

Table 8.24 Results obtained for the FAES determination of Na with Li as the internal standard in CRMs of biological materials using different calibration methods.

Sample	Certified values ($\%m\,m^{-1}$)	Found values ($\%m\,m^{-1}$)			
		ESM	E-SAM	ISM	SDA
Biodiesel	2.5[a)]	5.7 ± 0.3[a)]	2.7 ± 0.3[a)]	1.9 ± 0.1[a)]	2.5 ± 0.2[a)]
Bovine liver	0.242 ± 0.006	0.253 ± 0.010	0.249 ± 0.019	0.244 ± 0.011	0.241 ± 0.009
Mussel tissue	3.5 ± 0.1	3.6 ± 0.3	3.4 ± 0.1	3.5 ± 0.2	3.5 ± 0.1
Not fat milk	0.497 ± 0.010	0.576 ± 0.040	0.532 ± 0.060	0.528 ± 0.020	0.498 ± 0.010
Whole milk	0.356 ± 0.040	0.430 ± 0.020	0.363 ± 0.014	0.402 ± 0.020	0.365 ± 0.040

a) Values expressed as $mg\,kg^{-1}$.
Source: Fortunato et al. [24], table 1 (p. 666)/Elsevier.

real chance to find application in an automated form in routine analysis laboratories. However, this will only become a reality when flow analysis is a widely recognized and used tool in these laboratories.

References

1 Kościelniak, P. and Wieczorek, M. (2010). Extrapolative version of the indirect calibration method. *Analytical Letters* 43 (3): 424–435.

2 Foster, J.S., Langstroth, G.O., and McRae, D.R. (1936). Quantitative spectrographic analysis of biological material. I – A method for the determination of lead in cerebrospinal fluid. *Proceedings of the Royal Society of London Series A, Mathematical and Physical Sciences* 119 (814): 380–380.

3 Matisova, E., Krupcik, J., Cellar, P. et al. (1984). Quantitative analysis of hydrocarbons in gasolines by capillary gas–liquid chromatography: I. Isothermal analysis. *Journal of Chromatography* 303 (1): 151–157.

4 Yagi, M., Izawa, G., Omori, T. et al. (1987). A new internal standard method coupled with the standard addition method for proton-induced X-ray emission analysis and its application determination of Cu, Zn, Rb and Sr in biological materials. *Journal of Radioanalytical and Nuclear Chemistry* 116 (1): 213–222.

5 Davis, A.C., Alligood, B.W., Calloway, C.P. Jr. et al. (2005). Direct determination of cadmium in urine by electrothermal vaporizer-inductively coupled plasma analysis using a tungsten coil vaporizer. *Applied Spectroscopy* 59 (10): 1300–1303.

6 Yousefi, S.R., Zolfonoun, E., and Ahmadi, S.J. (2014). On-line standard addition technique for the determination of elements in the samples with high dissolved solids by ICP-OES: application to determination of rare earth elements in geological samples. *Analytical Letters* 47 (1): 146–154.

7 Wieczorek, M., Tobiasz, A., Dudek-Adamska, D. et al. (2017). Analytical strategy for the determination of selenium in biological materials by inductively coupled plasma – mass spectrometry with a dynamic reaction cell. *Analytical Letters* 50 (14): 2279–2291.

8 Pszonicki, L. and Skwara, W. (1989). The standard addition and successive dilution method for evaluation and verification of results in atomic-absorption analysis. *Talanta* 36 (12): 1265–1276.

9 Kościelniak, P. and Uberna, E. (1993). Dilution process as the basis of the calibration procedure in flame atomic spectrometry. *Universitatis Iagiellonicae Acta Chimica* 1092 (36): 39–51.

10 Kościelniak, P., Kozak, M., and Karocki, A. (1996). Detection and examination of interferences basing on the dilution process. *Chemia Analityczna (Warsaw)* 41 (3): 363–376.

11 Kościelniak, P. and Kozak, J. (2002). Calibration by the gradient ratio-standard addition method in flow injection flame atomic absorption spectrometry. *Analytica Chimica Acta* 460 (2): 235–245.

12 Paluch, J., Kozak, J., Wieczorek, M. et al. (2017). Novel approach to two-component speciation analysis. Spectrophotometric flow-based determinations of Fe(II)/Fe(III) and Cr(III)/Cr(VI). *Talanta* 171: 275–282.

13 Kościelniak, P. (2001). Univariate calibration techniques in flow injection analysis. *Analytica Chimica Acta* 438 (1, 2): 323–333.

14 Kościelniak, P., Kozak, J., Herman, M. et al. (2004). Complementary dilution method – a new version of calibration by the integrated strategy. *Analytical Letters* 37 (6): 1233–1253.

15 Kozak, J., Wójtowicz, M., Wróbel, A. et al. (2008). Novel approach to calibration by the complementary dilution method with the use of a monosegmented sequential injection system. *Talanta* 77 (2): 587–592.

16 Kościelniak, P., Wieczorek, M., Kozak, J. et al. (2011). The integrated calibration method – comparison of various flow approaches. *Analytical Letters* 44 (1–3): 398–410.

17 Wieczorek, M., Kościelniak, P., Świt, P. et al. (2014). Novel multicommuted flow manifold dedicated to the integrated calibration method. *Analytical Methods* 6 (23): 9276–9282.

18 Wieczorek, M., Kościelniak, P., Świt, P. et al. (2015). Solenoid micropump-based flow system for generalized calibration strategy. *Talanta* 133: 21–26.

19 Medinskaia, K., Garmonov, S., Kozak, J. et al. (2015). Stepwise injection determination of isoniazid in human urine samples coupled with generalized calibration method. *Microchemical Journal* 123: 111–117.

20 Świt, P., Kamińska, K., and Herian, M. (2020). Bioanalytical methodology for determination of glutamate and aspartate for use in pharmacological sciences with application of integrated calibration method. *Talanta* 206: 120237.

21 Kościelniak, P., Wieczorek, M., Kozak, J. et al. (2011). Generalized calibration strategy in analytical chemistry. *Analytical Letters* 44 (1–3): 411–430.

22 Kościelniak, P., Wieczorek, M., Kozak, J. et al. (2007). Versatile flow injection manifold for analytical calibration. *Analytica Chimica Acta* 600 (1, 2): 6–13.

23 Jones, W.B., Donati, G.L., Calloway, C.P. Jr. et al. (2015). Standard dilution analysis. *Analytical Chemistry* 87 (4): 2321–2327.

24 Fortunato, F.M., Bechlin, M.A., Gomes Neto, J.A. et al. (2016). Standard dilution analysis in flow system: sodium determination by flame atomic emission spectrometry. *Microchemical Journal* 124: 662–667.

25 Sloop, J.T., Bonilla, H.J.B., Harville, T. et al. (2019). Automated matrix-matching calibration using standard dilution analysis with two internal standards and a simple three-port mixing chamber. *Talanta* 205: 120160.

9

Calibration Approaches for Detection and Examination of Interference Effects

9.1 Introduction

Chapters 3–8 clearly demonstrated how important and difficult the problem of chemical analysis, and in particular quantitative analysis, is the interference effect. It is well known that this effect can be partially or even completely minimized by taking appropriate steps before subjecting the sample and standards for measurement. It has also been proven that, in addition, some calibration methods have the ability to compensate for these effects. Nevertheless, an important element of the correct analytical procedure still remains the preventive examination of the occurrence of these effects in each newly developed analytical procedure, as well as after any modifications to the routine already used procedure. This is necessary to find out if the effect is present at all, and if so, what its magnitude and nature are. This information allows for the appropriate way of its elimination and compensation.

Typically, interferences are investigated by "trial and error." It consists of the fact that a substance is added to the standard solution of the analyte in a certain concentration, which, according to the analyst's assumptions based on the initial information about the sample (as well as on his knowledge and intuition), may interfere with this analyte. No change in the signal of the analytical signal measured for the analyte itself under the influence of the added substance proves the lack of interference, and a significant change of the signal – about its interference effect. However, such a procedure is tedious and ineffective. It is practically impossible to check the influence of all potential sample components on the analytical signal, if only because the composition of the sample is never fully known. In addition, the magnitude of the effect exhibited by a particular component may vary depending on the concentrations of the analyte and interferent in the sample.

A separate analytical problem is the presence of an additive effect of an additive nature. A particular threat to the accuracy of the analytical result is the additive effect coming from native, often unknown components of the sample. If it is not detected and eliminated before measuring the sample and standards, none of the numerous calibration methods discussed previously can contribute to its compensation.

This chapter presents some simple ways to detect and study nonadditive effects, more effective and universal than the above-mentioned method. A separate section

Calibration in Analytical Science: Methods and Procedures, First Edition. Paweł Kościelniak.

is devoted to the detection and compensation of the additive effect. Although in both cases the procedures used are based on conventional calibration approaches, they are not widely known and used in practice.

9.2 Simple Procedures for Detection and Examination of Interference Effects

The simplest way **to detect the matrix effect**, i.e. the influence of all components present in the sample under analysis on the analytical signal of the component underdetermination, is **to follow the intensity of the analytical signal as the sample is progressively diluted**. As demonstrated by calibration methods based on the process of diluting the sample and standard (e.g. the dilution method or the standard addition and indicative dilution method), it is very likely that in the absence of interference effect the dependence of signal intensity on the degree of sample dilution will be strictly linear, while it will be nonlinear when interference (particularly chemical interference) is present. This can be demonstrated by the model formula (4.10): if there is no additive effect in the analytical system under study and the real function is described (at least approximately) by formula (4.10), it is easy to see that both situations are expressed respectively by the following formulas:

$$F(c) = B_1 \cdot c \tag{9.1}$$

$$F(c) = B_1 \cdot c + B_2 \cdot \frac{R_m \cdot c^2}{1 + a \cdot c + bR_m \cdot c} \tag{9.2}$$

where R_m is the ratio of the interferent concentrations to the analyte (c_m/c) that remains constant during the dilution process. Note that the function (8.2) is nonlinear even when the interference effect is linear ($a = 0$).

It should be noted, however, that this type of effect study is not completely sure because in certain situations the relationship studied may be linear despite the presence of interferents in the sample. A linear relationship may also manifest itself over a wide initial dilution range, changing to a nonlinear relationship towards the end of the process. The sample should therefore be diluted to the maximum extent acceptable from the point of view of measurement fluctuations.

The exact form of the analytical signal changes during the sample dilution process depends mainly on the type of measurement method used and the associated mechanism of influence of specific interferents on the analyte. The validity of the interference study in the manner described above and the evaluation of the results obtained require therefore a lot of experience of the analyst in working with the given measurement method.

A simple matrix effect test can be performed, for example,, based on the principle of integration of external standard method (ESM) and extrapolative standard addition method (E-SAM) presented earlier. The procedure is very simple and is recommended by IUPAC as the **test for general matrix**

effect [1]. It consists in measuring under identical experimental conditions the intensities of analytical signals in three solutions: in the sample, in the standard solution, and in the sample with the addition of the standard. The concentration of the analyte in the standard can be unknown, but then a prerequisite is that the concentration of the analyte in the sample and in the standard is kept equal before and after they are combined. Solutions are therefore prepared as shown in Figure 8.17, which is characteristic of the integrated method. The obtained measurement results are presented on a single coordinate system as shown in Figure 9.1 (if the blank effect is present in the sample and standard solutions, it must be determined beforehand and the intensities of the obtained signals corrected accordingly).

The next step is to analyze the mutual position of the calibration graphs a and b shown in Figure 9.1. Depending on whether or not the two lines are parallel to each other, it can be assumed that there is or is not a proportional (linearly multiplicative) interference effect in the sample, respectively. The magnitude of this effect can be determined from the difference between the slopes of the straight lines a and b with respect to the concentration axis.

Deeper conclusions can be based on additional premises or possibly on additionally obtained experimental data. The conclusion that there is no interference may be erroneous insofar as line b does not account for the presence of a possible nonlinear effect. On the other hand, the detection of an interference effect does not determine the type and nature of the effect, as it cannot be excluded that it is nonlinear or complex.

It is worth emphasizing that the interpretation of the measurement data does not require the determination of the analyte concentration in the sample, but only the study of the mutual position of the measurement points and not. An analogous way of detecting the matrix effect can be applied when the analytical procedure used requires calibration by an indirect method or an internal standard method:

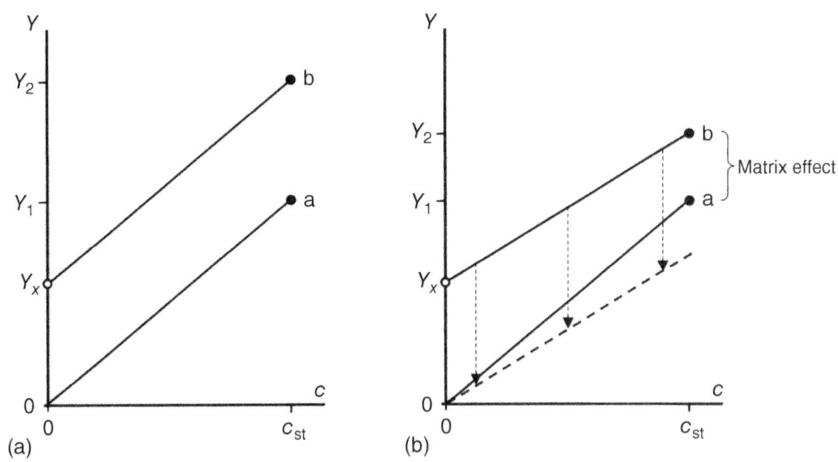

Figure 9.1 Two mutual locations of the calibration graphs prepared in accordance with the ESM *external standard method* (a) and E-SAM (b) methods indicating absence (a) and presence (b) of the matrix effect in the sample analyzed; Y_x, Y_1, and Y_2 – signals measured for sample, standard solution, and sample with standard solution, respectively.

the standard solution is then prepared according to the interpolation version of the method in question, and the sample and sample solutions with the standard according to the extrapolation version of the same method.

The procedure presented here provides information of great importance to the analyst. If he is convinced that the sample is free of matrix effect (and this is, after all, very often the case), then, avoiding the tedious examination of the influence of interferents by the "one by one" method, he can be almost certain that he can use calibration methods of an interpolative nature when proceeding to the actual analysis of that sample. Consistently, if the analysis involves a series of samples of similar composition, he can use the calibration graph once made to determine the analyte in all those samples. If, on the other hand, he detects a matrix effect, he knows that additional efforts must be made to overcome this obstacle (e.g. perform an E-SAM or mixed method calibration). It can find this out easily and relatively quickly even when the sample under investigation has a completely unknown composition.

An even more detailed situation prevailing in the analytical system under study in terms of interference effects is shown by the so-called **interference maps**. These are diagrams drawn for a specific magnitude of interference effect corresponding to different concentrations of analyte and interferent. An example of such a map is shown in Figure 9.2. This type of plot is particularly useful when the effect of a particular interferent on the analyte in a series of samples with different expected concentrations of both components needs to be investigated in detail.

Creation of interference maps by manual preparation of solutions with different analyte and interferent compositions is obviously very difficult and tedious in practice. It is much easier to collect the necessary measurement data using the FIA system in the same gradient manner as the previously described calibration method called gradient ratio calibration method (GRCM). The procedure, in this case, is as follows:

- Two solutions containing analyte in concentration c_a without interferent and with interferent of concentration c_m are prepared,
- both solutions are injected one after the other into the carrier solution from the same injection loop under identical experimental conditions,
- the peaks obtained are superimposed taking into account their widths at the base (see Figure 9.3),
- after a time t_i from the injection of each solution, the values of the signals obtained for the solutions of the analyte, $Y_{st,i}$, and the analyte with the interferent, $Y_{s,i}$, are read, taking as a measure of the interference effect, IE_i, the ratio, $(Y_{st,i} - Y_{s,i})/Y_{st,i}$.
- the determined interference effect value, IE_i, is assigned the apparent concentrations of analyte, $c_{a,i}'$, and interferent, $c_{m,i}'$, calculated from the formulas:

$$c_{a,i}' = \frac{Y_{st,i}}{Y_{max}} \cdot c_a \tag{9.3}$$

$$c_{m,i}' = \frac{Y_{st,i}}{Y_{max}} \cdot c_m \tag{9.4}$$

where Y_{max} is the signal corresponding to the maximum peak height obtained for the analyte alone.

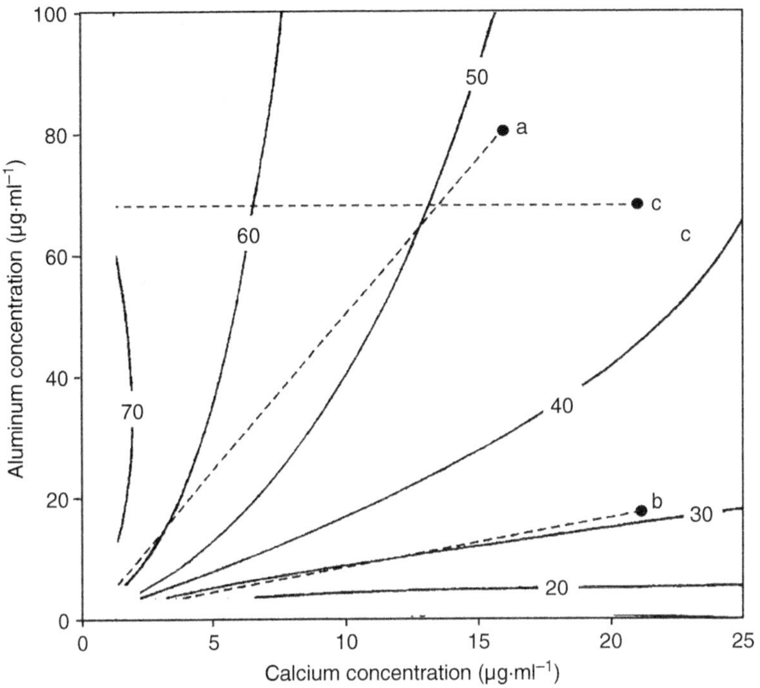

Figure 9.2 Interference map showing the effect of aluminum on the signal measured by FAAS for calcium; the dashed lines correspond to dilution process (a and b) and to calibration line created in the E-SAM method (c).

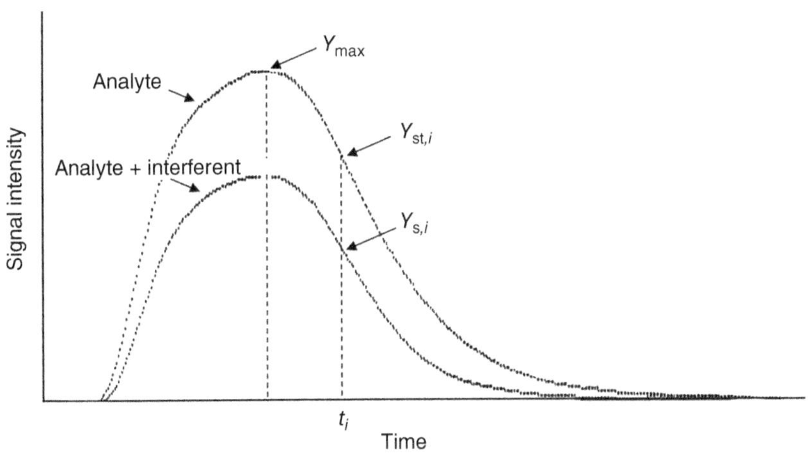

Figure 9.3 Pairs of the signals obtained for the sample, $Y_{s,i}$, and standard, $Y_{st,i}$, after well-defined delay time t_i are used for construction for construction of the interference map (for details see text).

The interference effect is determined in this way for a solution of analyte of concentration c_a and a series of solutions containing analyte at the same concentration and interferent at different concentrations. If the process of overlaying the peaks, as well as collecting and selecting the measurement data, is carried out with a suitable computer program, the creation of the interference map takes reasonable time.

How can such graphs be used? With the help of an interference map, one can easily follow the changes of the interference effect occurring with the dilution of a sample, i.e. the simultaneous decrease of analyte and interferent concentration in the sample. As can be seen in Figure 9.2, when diluting a sample with a relatively high concentration of aluminum (line a), one should expect a phenomenon that has been noted before, consisting of a gradual increase of the interferent effect on calcium. During the dilution of samples with lower aluminum concentration (line b), the interference effect remains constant (therefore, it cannot then be uniquely detected in the simplest way, i.e. by following the dependence of the analytical signal as the dilution process proceeds).

The interference maps also allow one to predict, to some extent, what changes in the interference effect can be expected in the system under study during calibration by the E-SAM method and, therefore, what the possibility of obtaining accurate analytical results by this method is. Line c in Figure 9.2, showing the change in calcium concentration in the presence of a constant concentration of aluminum, indicates a change in the interference effect, which allows us to assume that the calibration graph constructed according to the E-SAM method is nonlinear in this region of analyte and interferent concentrations. This phenomenon has been previously reported (e.g. see Figure 6.8a). The results developed in the form of the interference map therefore clearly confirm the complex, although expected from previous experiments, mechanism of aluminum influence on the calcium determination by FAAS.

In another way of detecting and testing interference, the dilution method can be used and in particular the phenomenon of **buffering the interference effect** caused by interferents present in the sample with another interferent. It has previously been shown, using the FAAS assay as an example, that this process occurs with particularly high efficiency if, in the FIA system, the sample and standard are injected into a buffering interferent contained in the carrier solution (see Table 5.3). It is observed that if the buffering interferent interacts with the analyte with greater force than the native interferents, the graph of the dependence of the apparent analyte concentration on the degree of dilution changes its course in such a way that it clearly indicates the presence of an interference effect. This is shown in Figure 9.4.

The observed phenomenon should be explained by the competing action of the interferents. As the sample is diluted, the buffering component with increasing concentration in the sample takes over the interfering action from the native interferents when a well-defined ratio of concentrations of both components is reached. If the buffer is much "stronger" than the interferents present in the sample, the change in interferent action may even manifest itself as a change in the direction in which interference compensation occurs (i.e. as in Figure 9.4). Very importantly, interference can be detected in this relatively simple way in a sample of completely unknown composition.

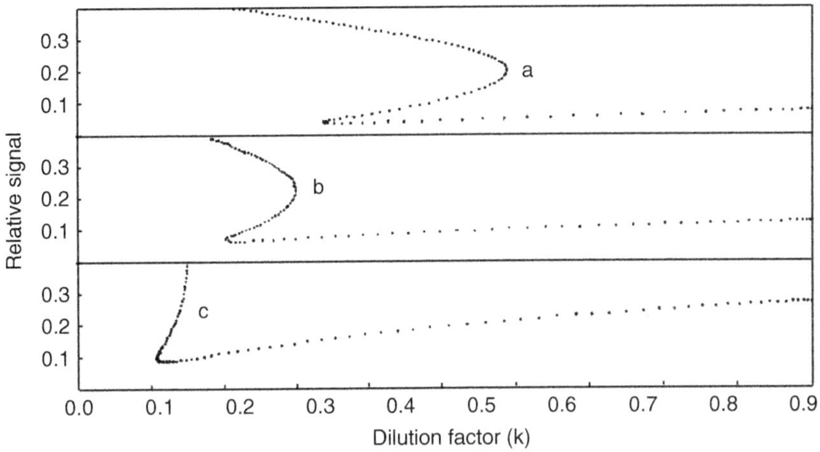

Figure 9.4 Compensation of interference caused by aluminum present in the sample in concentrations: 40 (a), 80 (b), and 160 (c) mg l^{-1} on calcium (10 mg l^{-1}) with the use of vanadium (10 mg l^{-1}).

It is worth noting that although the procedure described is based on reactions occurring in the flame of the FAAS spectrometer, it has many similarities to conventional gradient titration using reactions in sample solution. The buffer interferent acting as titrant does not reveal its presence as the concentration in the sample increases until it reaches a concentration relative to the native interferent concentration that is sufficient to replace it in the reaction with the analyte. This state corresponds, so to speak, to the end point of the titration. In Figure 9.4 it is apparent that the position of this point relative to the dilution step depends on the concentration of the interferent in the standard solution. This relationship can provide a basis for performing additional calibration and determining the concentration of the interferent in the sample being analyzed. Studies of this type have not yet been published, but are certainly worthy of interest and follow-up.

9.3 Detection and Compensation of Additive Interference Effect

The additive interference effect creates a very large analytical problem especially in quantitative analysis. If it is not recognized and eliminated before the sample is measured, it cannot be compensated for by any of the calibration procedures previously used. For this reason, it becomes often the source of serious systematic errors in the analytical result.

An additive effect occurs when sample components naturally present in the sample or introduced into the sample prior to measurement produce a signal under analyte-specific conditions. The interferents causing the additive effect also include chemical compounds not directly present in the sample, but present in the plasma of the ICP-MS spectrometer and forming adducts with the sample components with m/z values equal to those of the determined isotopes. Thus, by performing a signal intensity measurement under analyte-specific conditions, the instrument response corresponding to the combined concentration of analyte and interferents in the sample is obtained. If in the transformation step of any calibration method this signal is taken as a measure of the concentration of the analyte itself, the analytical result is subject to systematic error.

In general, three sources of the origin of additive interference can be distinguished, differing in the degree of difficulty of their detection and the possibility of compensating effects:

- substances of a known type and amount added to the sample at the stage of its preparation for measurement,
- substances of a known type (less often of a known amount) naturally present in the sample,
- substances of unknown type and quantity added (as a rule at random) to the sample or naturally present in the sample.

The easiest to control is the blank effect caused by known reagents added to the sample prior to measurement. Their presence in the sample is usually necessary to give the whole sample or its components (analyte, interferents causing additionally nonadditive interference effect) the physicochemical properties most suitable for the quality of the measurements to be made. In some cases, as shown earlier, it is sufficient to add the same substances to the standards so that the additive effect is compensated automatically during calibration. In other cases, the signals from the added reagents themselves must be measured under analyte-specific conditions and subtracted from the signals measured for the sample or standard with the added reagents.

It is much more difficult to eliminate or compensate for additive effects originating from substances naturally present in the sample or added to the sample during the analytical procedure in an uncontrolled manner. If these substances are unknown, a common way to manage this situation is to separate the analyte from the sample matrix using an appropriate technique (e.g. extraction, precipitation, etc.). Quite apart from the laboriousness of this procedure, it is imperfect because it involves the loss of a more or less unknown amount of analyte due, if only to the naturally limited ability of the techniques used to fully quantitatively separate substances.

Less difficulty is associated with the elimination of additive effects when the substances present in the sample are known. It is then possible to separate the interferents from the sample matrix without changing (at least theoretically) the amount of analyte in the sample. Alternatively, additional substances can be used to isolate (e.g. to mask) the analyte from the interferent (or vice versa) by purely chemical means.

A calibration method that does not require chemical intervention in the analytical system under study and that is useful in this situation is the multicomponent calibration method and the determination – in addition to the analyte – of the interferent causing the additive effect. Usually, however, the primary goal of such procedure is not to eliminate the additive effect, but to perform simultaneous determination of several equally treated sample components with less attention paid to the interference properties of each of them. It is therefore difficult to call multicomponent calibration a method of compensation of additive effects *sensu stricto*.

There is, however, another calibration approach whose main objective is the determination of a single analyte with the possibility of "freeing" it from the additive effect of another sample component. This method is described below and consists, in general, of applying a given calibration method at least twice under different, appropriately selected conditions to the same analyte. This method, which can generally be called **the multicalibration method** (MCM), is not widely known and is used relatively rarely, although it can give very good results both when the additive effect is due to a known sample component and when it is due to one or even many unknown substances native or added to the sample during the analytical procedure.

9.3.1 Interpolative Procedure

The **MCM method** can be performed using interpolative and extrapolative calibration methods. The **interpolation version of the MCM method** will be described using the external standard method (ESM) as an example.

Let us assume that the ESM method is used to perform two analyses of a sample of a given composition and analyte concentration, c_0, under two different experimental conditions (instrumental or chemical) defined by the parameters P_1 and P_2. Let us further assume that the only uncontrolled effects that can change the analytical signal measured for the analyte in the two samples are additive and proportional interference effects. The real function, $F(c)$, is therefore described by the formula:

$$F(c) = A(c_n) + B_1 \cdot c + B_2 H(c_m) \cdot c \tag{9.5}$$

where A is a measure of the additive effect induced by known or unknown components of the sample with a total concentration c_n, which are native components and/or components added to the sample in an uncontrolled manner, B_1 is a measure of the measurement sensitivity of the analyte, and B_2 is a measure of the multiplicative effect induced by interferents with a total concentration c_m, described by a hyperbolic function (according to Eq. (4.10)). The model function, $G(c)$, which does not include interference effects, then has the form:

$$G(c) = B_1 \cdot c \tag{9.6}$$

The intensities of the analytical signals obtained under the two measurement conditions P_1 and P_2 for the analyte in the sample, $Y_x^{(1)}$ and $Y_x^{(2)}$, and in the standard solution, $Y_1^{(1)}$ and $Y_1^{(2)}$, respectively, obtained for the analyte in the sample and in the standard solution, c_{st}, are given by the following formulas taking into account potential differences between the values of additive effects, $A^{(1)}$, $A^{(2)}$, measurement sensitivities, $B_1^{(1)}$ and $B_1^{(2)}$, and multiplicative effects, $B_2^{(1)}$ and $B_2^{(2)}$, occurring under the two conditions:

$$Y_x^{(1)} = A^{(1)} + B_1^{(1)} \cdot c_x' + B_2^{(1)} H \cdot c_x' \tag{9.7}$$

$$Y_x^{(2)} = A^{(2)} + B_1^{(2)} \cdot c_x'' + B_2^{(2)} H \cdot c_x'' \tag{9.8}$$

$$Y_1^{(1)} = B_1^{(1)} \cdot c_{\text{st}} \tag{9.9}$$

$$Y_1^{(2)} = B_1^{(2)} \cdot c_{\text{st}} \tag{9.10}$$

The values of apparent concentrations, c_x' and c_x'', which are the measures of the analyte concentration in the sample, are therefore calculated from the formulas:

$$c_x' = \frac{Y_x^{(1)} - A^{(1)}}{Y_1^{(1)} + B_2^{(1)}H \cdot c_{\text{st}}} \cdot c_{\text{st}} \tag{9.11}$$

$$c_x'' = \frac{Y_x^{(2)} - A^{(2)}}{Y_1^{(2)} + B_2^{(2)}H \cdot c_{\text{st}}} \cdot c_{\text{st}} \tag{9.12}$$

The difference between the values of the apparent concentrations, c_x' and c_x'', is obviously due (ignoring random errors) to the additive ($A^{(1)}, A^{(2)}$), and/or multiplicative ($B_2^{(1)}$ and $B_2^{(2)}$) effects of the interferents on the analyte and is to be expected regardless of whether the effects are different from each other or equal under the two experimental conditions. This situation is illustrated in Figure 9.5.

However, if the multiplicative effect does not exist, $B_2^{(1)} = B_2^{(2)} = 0$, and only the components causing the additive effect are present in the sample, and this effect has the same value in both measuring conditions $\text{AE} = A^{(1)} = A^{(2)}$, then the apparent concentrations are given by the formulas:

$$c_x' = \frac{Y_x^{(1)} - \text{AE}}{Y_{\text{st}}^{(1)}} \cdot c_{\text{st}} \tag{9.13}$$

$$c_x'' = \frac{Y_x^{(2)} - \text{AE}}{Y_{\text{st}}^{(2)}} \cdot c_{\text{st}} \tag{9.14}$$

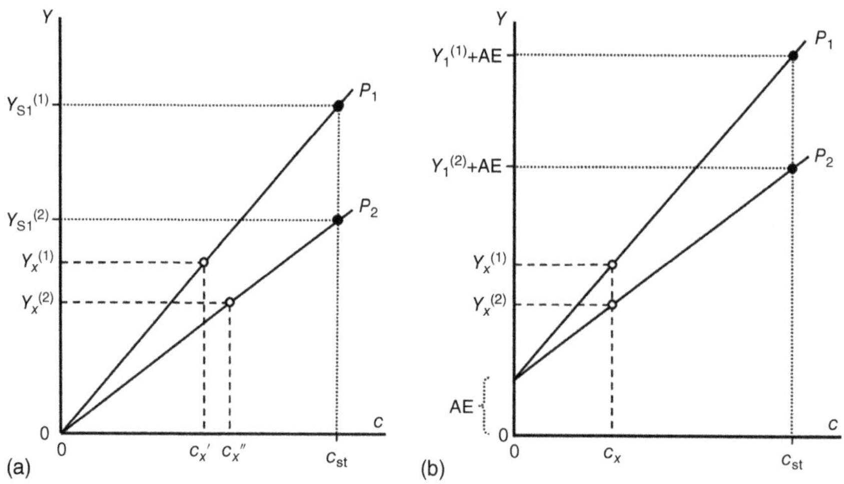

Figure 9.5 Interpolative MDM method: concentration of an analyte in a sample is estimated by two different values, c_x' and c_x'', and by single value, c_x, before (a) and after (b) correction of additive effect, AE, respectively (for details see text).

From the system of Eqs. (9.13) and (9.14), the analytical result, c_x, free of additive effect, and the value of the additive effect itself, AE, are calculated from the formulas:

$$c_x = c_x' = c_x'' = \frac{Y_x^{(1)} - Y_x^{(2)}}{Y_{st}^{(1)} - Y_{st}^{(2)}} \cdot c_{st} \tag{9.15}$$

$$\mathrm{AE} = \frac{Y_x^{(2)} \cdot Y_{st}^{(1)} - Y_x^{(1)} \cdot Y_{st}^{(2)}}{Y_{st}^{(1)} - Y_{st}^{(2)}} \tag{9.16}$$

The method of determining the additive effect and its compensation to obtain an accurate analytical result is interpreted graphically in Figure 9.5b. Figuratively speaking, it consists in shifting both calibration graphs along the signal intensity axis by such a value of AE that the concentrations c_x' and c_x'' reach the same value of c_x.

It should be noted that both formulas, (9.15) and (9.16), are valid when the signals obtained for the analyte in the sample, $Y_x^{(1)}$, $Y_x^{(2)}$, and in the standard, $Y_{st}^{(1)}$, $Y_{st}^{(2)}$, are different from each other, i.e. when changing the parameters P_1 and P_2 provides a change in the measurement sensitivity of the analyte. Thus, it can be definitively concluded that **if the sample assayed contains components that cause only an additive interference effect, this effect can be determined by performing the ECM calibration at least twice under different experimental conditions that ensure a change in the analyte measurement sensitivity, provided that the change does not cause a change in the value of this effect**. It is not difficult to prove that the same can be achieved by using at least twice another calibration method of an interpolative nature (in particular, the indirect method and the internal standard method).

Although the principle of the MCM method is simple, its practical implementation may cause considerable problems. The above conditions could most easily be met by inducing a multiplicative effect of varying value in the standard solutions; the method does not accept the occurrence of an effect of this nature. On the other hand, changing instrumental parameters does not assure that the required conditions are achieved when the analyte and interferent trigger the signal under identical measurement conditions. If even a change in conditions causes a change in the analyte signal, it is difficult to predict whether the signal measured for the interferent then remains the same.

It is certainly easier to find suitable conditions for the calibration compensation of the additive effect when the type of analyte responsible for this effect in the analyzed sample is known. In a spectrometric analysis, two separate spectra can then be prepared for the analyte and the interferent and, based on these, two appropriate wavelengths ($P_1 = \lambda_1$, $P_2 = \lambda_2$) can be selected at which measurements should be made. This procedure is shown in Figure 9.6.

This principle was verified by spectrophotometric determination of Fe^{2+} ions in the presence of Fe^{3+} ions in synthetic samples and natural water samples [2]. The obtained analytical results are summarized in Table 9.1. A clear improvement in the accuracy of the results after compensating for the effect of Fe^{2+} ions on the determination of Fe^{2+} ions in synthetic samples can be seen. It is important to note how large a systematic error can be made when determining an analyte using the external

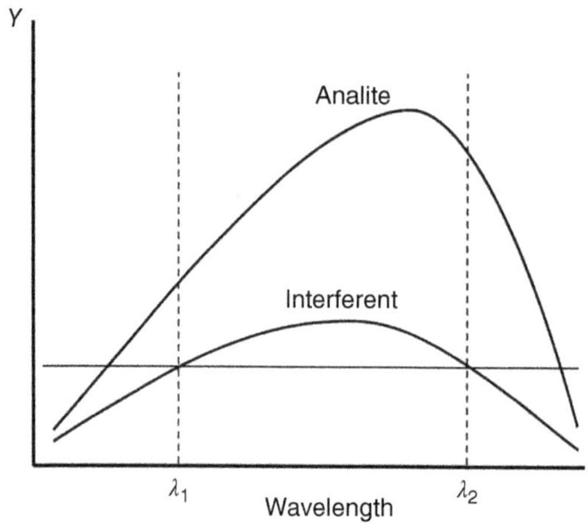

Figure 9.6 Two wavelengths, λ_1 and λ_2, selected at the analyte and interferent spectra in accordance with the calibration compensation for the additive interference effect.

Table 9.1 Calibration compensation for the additive effect caused by Fe^{3+} ions on the spectrophotometric determination of Fe^{2+} ions in synthetic and water samples.

				Concentration found (mg l^{-1})		
		c_0 (mg l^{-1})	c_{m1} (mg l^{-1})	Before compensation		After compensation
Sample		**Fe(II)**	**Fe(III)**	c_x'	c_x''	c_x
Synthetic	I	1.00	5.00	2.90 ± 0.02	3.77 ± 0.03	1.00 ± 0.07
	II	5.00	20.00	11.24 ± 0.01	14.17 ± 0.00	4.71 ± 0.01
Spring water	Jagielloński	–	–	4.75 ± 0.18	4.70 ± 0.25	4.66 ± 0.22
	Lajkonik	–	–	4.66 ± 0.30	4.62 ± 0.55	4.60 ± 0.07
River water	Ruczaj	–	–	12.92 ± 0.07	17.41 ± 0.18	5.73 ± 0.32
	Wisła	–	–	2.19 ± 0.02	3.17 ± 0.13	0.55 ± 0.18
	Wilga	–	–	3.34 ± 0.09	5.06 ± 0.22	0.97 ± 0.18
	Drwinka	–	–	3.10 ± 0.15	4.51 ± 0.25	0.91 ± 0.18

Source: Wieczorek et al. [2], table 1 / Springer Nature / Licensed under CC BY 4.0.

standard method when the analyst is not aware of the additive effect in the sample or does not take other steps to eliminate or compensate for it. Based on the results obtained for synthetic samples, it can be thought that the accuracy of the determined values of Fe^{3+} concentration in water samples (especially river waters) has also been significantly improved in this way.

Thus, it has been proved that this type of calibration method for the compensation of additive effects is conceptually correct and can find its important place among the few methods used for this purpose in single-component analysis. However, interest in this method among analysts is negligible and there are no further reports in the literature demonstrating the use of this method in analytical practice.

9.3.2 Extrapolative Procedure

An **extrapolation version of the MCM method** was first developed and applied in 1988 by Bosh-Reig and Campíns-Falcó [3] under the term H-point standard addition method (HPSAM). As the name suggests, it uses calibration by the conventional standard addition method, the general idea is however analogous to the interpolation calibration-based version described above.

Assume as before that a sample containing an analyte at concentration c_0, as well as components causing additive and proportional interference effects, is analyzed twice under two different conditions P_1 and P_2, but this time based on the E-SAM method. This situation is shown in Figure 9.7.

In the linear range of the real function, the analytical signals measured for the sample, $Y_x^{(1)}$, $Y_x^{(2)}$, and for the sample with added standard, $Y_2^{(1)}$, $Y_2^{(2)}$, containing the analyte at a concentration of Δc_{st}, are given by the formulas:

$$Y_x^{(1)} = A^{(1)} + B_1^{(1)} \cdot k_v c_x' + B_2^{(1)} H \cdot k_v c_x' \tag{9.17}$$

$$Y_x^{(2)} = A^{(2)} + B_1^{(2)} \cdot k_v c_x'' + B_2^{(2)} H \cdot k_v c_x'' \tag{9.18}$$

$$Y_2^{(1)} = A^{(1)} + B_1^{(1)} \cdot \left[k_v c_x' + (1 - k_v)\Delta c_{st}\right] + B_2^{(1)} H \cdot \left[k_v c_x' + (1 - k_v)\Delta c_{st}\right] \tag{9.19}$$

$$Y_2^{(2)} = A^{(2)} + B_1^{(2)} \cdot \left[k_v c_x'' + (1 - k_v)\Delta c_{st}\right] + B_2^{(2)} H \cdot \left[k_v c_x'' + (1 - k_v)\Delta c_{st}\right] \tag{9.20}$$

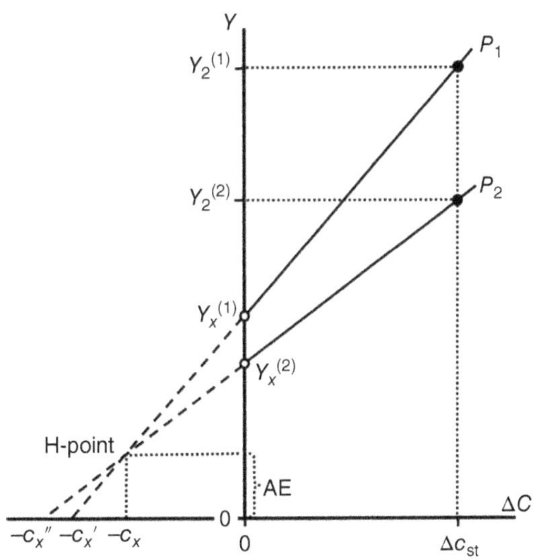

Figure 9.7 HPSAM method: concentration of an analyte in a sample is estimated by two different values, c_x' and c_x'', and by single value, c_x, before (a) and after (b) correction of additive effect, AE, respectively (for details see text).

where k_v is the degree of dilution of the sample with the standard solution added to it. If the interferents produce equal additive effects under both experimental conditions, $\mathrm{AE} = A^{(1)} = A^{(2)}$, then the apparent concentrations, c_x' and c_x'', are calculated from the formulas:

$$c_x' = \frac{Y_x^{(1)} + \mathrm{AE}}{Y_2^{(1)} - Y_x^{(1)}} \cdot \frac{1 - k_v}{k_v} \cdot \Delta c_{\mathrm{st}} \tag{9.21}$$

$$c_x'' = \frac{Y_x^{(2)} + \mathrm{AE}}{Y_2^{(2)} - Y_x^{(2)}} \cdot \frac{1 - k_v}{k_v} \cdot \Delta c_{\mathrm{st}} \tag{9.22}$$

It follows from the system of Eqs. (9.21) and (9.22) that the analytical result, c_x, free of additive effect and the value of the additive effect itself, AE, are calculated in this case from the formulas:

$$c_x = \frac{Y_x^{(1)} - Y_x^{(2)}}{\left(Y_2^{(1)} - Y_2^{(2)}\right) - \left(Y_x^{(1)} - Y_x^{(2)}\right)} \cdot c_{\mathrm{st}} \tag{9.23}$$

$$\mathrm{AE} = \frac{Y_x^{(2)} \cdot Y_2^{(1)} - Y_x^{(1)} \cdot Y_2^{(2)}}{\left(Y_2^{(1)} - Y_2^{(2)}\right) - \left(Y_x^{(1)} - Y_x^{(2)}\right)} \tag{9.24}$$

As can be seen in Figure 9.7, when an additive effect is present, the two graphs produced in calibration by the E-SAM method intersect at a single point (H-point) that indicates both the value of this effect (AE) and the corrected analytical result (c_x).

Thus, it can be definitively concluded that **if the test sample contains components that produce only an additive and proportional interference effect, this effect can be determined by performing the E-SAM calibration at least twice under different experimental conditions that ensure a change in the measurement sensitivity of the analyte, provided that the change does not alter the value of this effect**.

It should be noted that the discussed extrapolation version accepts – unlike the interpolation version – the occurrence of the proportional interference effect in the analytical system under study due to the fact that extrapolation methods – unlike interpolation methods – are resistant to an interference effect of this character. This is very important from a practical point of view because, as will be shown, the proportional effect can be induced intentionally thus satisfying the conditions necessary for the HPSAM method to effectively compensate for the additive effect.

On the other hand, it is obvious that the HPSAM method is subject to all the limitations characteristic of the E-SAM method, resulting primarily from the process of extrapolation of the model function. Nevertheless, in the author's opinion, it represents a very important step towards the possibility of controlling additive effects in chemical analysis. It should be emphasized that – similarly to interpolation methods – the HPSAM method can be realized not only on the basis of E-SAM calibration but also by means of other extrapolation methods (in particular, extrapolation indirect method and extrapolation internal standard method).

The first experiments with the HPSAM method were performed by UV/VIS spectrophotometry using the wavelength of radiation absorbed by the analyte and interferent as parameter P. The corresponding wavelength values, λ_1 and λ_2, were chosen

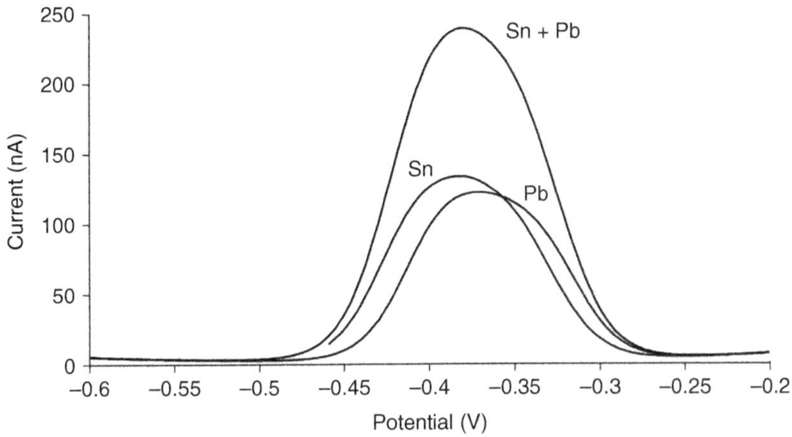

Figure 9.8 Differential pulse anodic stripping voltammograms of lead, tin, and their mixture; the conditions selected for the lead determination by HPSAM: −0.4026 and −0.3663 V. Source: Shams et al. [4], fig 3 (p. 362)/with permission of Elsevier.

in the same way as shown in Figure 9.6. In this basic version, the method has been successfully applied many times, most often in spectrophotometric analysis, but also in electrochemical and chromatographic analysis at different instrumental parameter values characteristic of these methods. Figure 9.8 shows an example of selecting conditions for calibration by differential pulse anodic stripping voltammetry with the HPSAM method [4].

Due to limitations associated with the selection of appropriate instrumental parameters different attempts to modify the basic HPSAM procedure have been made. For instance, it was revealed that the HPSAM could be successively used for determination of an analyte in the presence of two interferents as long as the signals produced by them at two wavelengths represent the same ratio, which is constant and independent of concentration of species [5]. It was also suggested to avoid the restrictive conditions mentioned above by dividing a spectrum of the analyte/interferent mixture by a standard normalized spectrum of the interferent [6]. In more sophisticated approaches it is proposed to isolate mathematically the unknown interferent spectrum from a sample spectrum prior to application of HPSAM using either the curve resolution method [7] or the concept of the net analyte signal [8].

In 1995 the so-called **generalized H-point standard addition method (G-HPSAM)** was developed by Campíns-Falcó et al. [9]. In this approach, the measurements need to be performed at three selected wavelengths (third wavelength is selected in most cases as an average of two preselected wavelengths). Two calibration graphs are then constructed based on the difference between the signals measured at the middle wavelength and at two extreme wavelengths. This method enables to compensate additive effects even in situations when interference effect is not the same at the three selected wavelengths. Then, the interfering signal should vary linearly for these wavelengths, while the analytical signal may change nonlinearly in the selected range. The wavelength selection conditions of the G-HPSAM method are less restrictive than those of the HPSAM method, making it applicable in such cases where the type of interferent is unknown.

Kinetic analysis also offers great potential for applying the HPSAM method to detect additive effects. In the **kinetic HPSAM method**, a reaction between the analyte and a reactant added to the calibration solutions (sample alone and with added standards) is involved and the reaction rate is measured using the fixed time method (see Figure 7.21a), i.e. the signal for the reaction product is measured at two well-defined times after the reaction initiation. In most often cases, the measurements are made over a time interval in which the concentration of the analyte increases markedly as a result of the reaction, while the concentration of the interferent produced under the same conditions remains constant. An example of this situation is shown in Figure 9.9 [10].

From the calibration point of view, it is interesting that, due to the involvement of chemical reaction in the kinetic version of the HPSAM method, calibration is performed with the E-SAM method in an indirect version (i.e. extrapolation indirect method), measuring signals not directly for the analyte and interferent, but for the products of their reaction with a specific reagent. The principle of the method also does not preclude its use when the type of interferent causing the additive effect is unknown.

Recently, one more modification of the HPSAM method has been developed [11]. It was decided to take advantage of the applicability of this method when proportional interference is present in the sample. It was then proposed to perform the E-SAM method in two (or more) different chemical conditions causing proportional

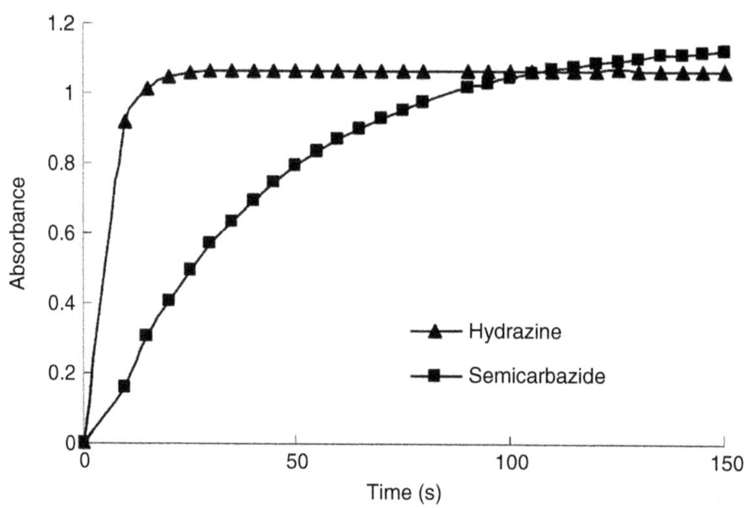

Figure 9.9 Spectrophotometric kinetic curves obtained for semicarbazide and hydrazine as the result of reduction of Cu^{2+} to Cu^{+} by both species in the presence of 2,9-dimethyl-1,10-phenantheroline (Nc) and formation of complex between Cu^{+} and Nc; the conditions selected for the semicarbazide determination by HPSAM: 60 and 210 seconds. Source: Safavi et al. [10], fig 2 (p. 150)/Elsevier.

interference effects of different magnitude. The conditions must be chosen still in the same way: so that this affects only the analyte and does not cause a change in the signal of the interferent being responsible for the additive effect. Consequently, in accordance with the principle of the method, the calibration graphs of different slopes can be constructed which are crossed in the H-point indicating additive effect. As the proportional effects are compensated for in the E-SAM calibration and the additive effect is detected, the analytical result calculated at the H-point can be expected to be accurate. The method has been named the **chemical H-point standard addition method (C-HPSAM)**.

The usefulness of C-HPSAM for identification of additive effects caused by unknown interferents was verified in spectrophotometric determination of ascorbic acid in juices and soft drinks. The samples contained unidentified additive interferents, mainly colored components. The extrapolative indirect method was used based on the reaction of ascorbic acid with Fe(III) followed by formation of a complex of Fe(II) with *o*-phenantroline. Reaction equilibrium was controlled by slight variation of pH values in the ranges carefully optimized, hence the calibrations graphs of different slopes were able to be constructed. To verify constancy of the additive effect in different conditions, a procedure encompassing construction of three calibration graphs was tested, as presented in Figure 9.10.

A similar implementation of the HPSAM method was used for the determination of total acidity in semisweet wines of different colors [11]. The results obtained (Table 9.2) compared with those obtained with the E-SAM method under pH conditions ensuring the highest sensitivity of measurements prove the possibility of

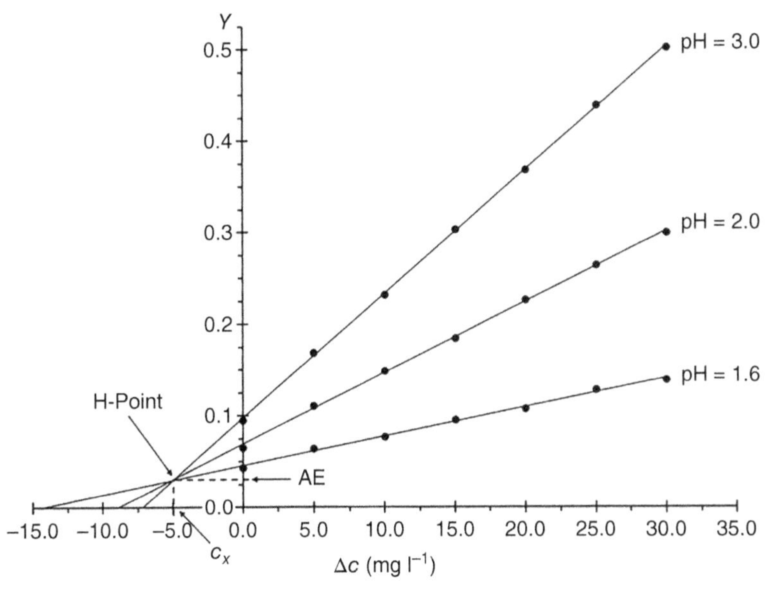

Figure 9.10 Calibration according to the C-HPSAM procedure: seven-point calibration graphs constructed at three different pH values used for the spectrophotometric determination of ascorbic acid in a juice; AE – value of the additive interference effect.

Table 9.2 Results of the spectrophotometric determination of total acidity in wines by E-SAM and C-HPSAM with the use of two-point calibration graphs constructed at pH = 7.0, 7.5, and 8.0; reference method – potentiometric titration.

	Concentration found, c_x, (mmol l^{-1})		
Sample	**Reference method**	**E-SAM**	**C-HPSAM**
Fresco rose	45.14 ± 1.16	54.84 ± 7.42	45.69 ± 2.34
Carlo Rossi rose	36.82 ± 1.18	47.67 ± 6.92	37.57 ± 1.84
El Sol rose	35.36 ± 1.13	55.08 ± 8.20	36.24 ± 2.28
Fresco red	46.12 ± 1.77	102.14 ± 12.16	45.11 ± 2.56
Carlo Rossi red	38.94 ± 1.04	97.13 ± 8.96	39.4 ± 2.10
El Sol red	42.58 ± 2.07	113.25 ± 15.44	43.26 ± 2.40

Source: Adapted from Wieczorek et al. [11].

significant improvement in the accuracy of determinations after compensation of the additive effect of unknown origin. It was also found that changing the version of the E-SAM method from basic (two measurement points) to multipoint (seven points) does not improve the accuracy and precision of the results if only three calibration graphs intersect exactly at one point (i.e. the additive effect does not change under different chemical conditions).

Guided by the conclusions of the above experiments, the C-HPSAM procedure implemented to flow analysis with the use of gradient ratio technique was proposed [12]. The idea is to inject a sample prepared in a chemical condition (e.g. pH) defined by parameter P_1 into the stream of the same sample prepared at quite different parameter P_2. Due to the difference between conditions P_1 and P_2 the resulting peak represents the gradient of signals induced by local concentrations of the analyte (and possibly the additive interferent) under conditions determined by the gradually changing parameter P from the value of P_1 to P_2 and vice versa. This situation is illustrated in Figure 9.11.

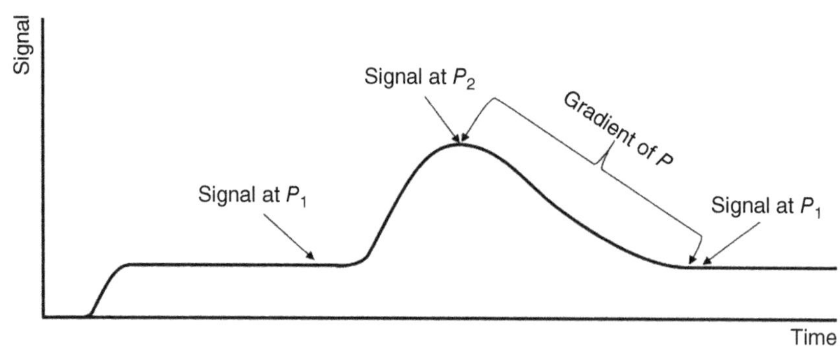

Figure 9.11 The way of generating a gradient of chemical conditions defined by parameter P in a sample zone in flow injection analysis.

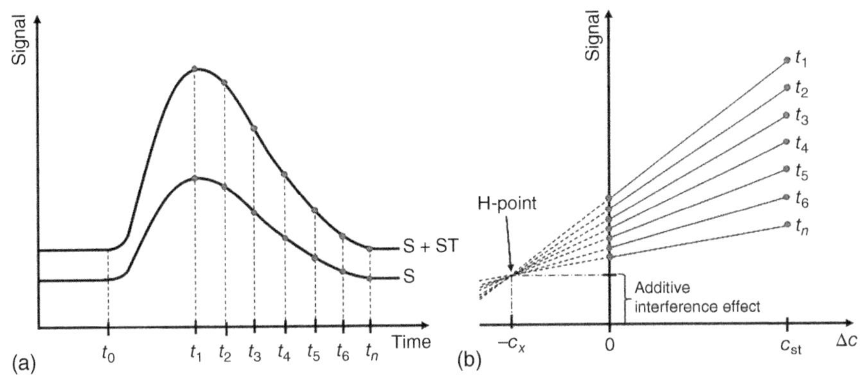

Figure 9.12 The C-HPSAM procedure at the measurement (a) and transformation (b) stages in flow injection analysis; S and ST – peaks obtained for a sample and sample with standard of the analyte concentration c_{st}, respectively, $t_1, \ldots t_n$ – moments chosen for measurements, c_x – analytical result obtained. Source: Wieczorek et al. [12]/with permission of Elsevier.

Further, maintaining the same measurement conditions, the procedure is repeated for the sample with standard addition, and the two peaks generated are overlaid as shown in Figure 9.12a. A series of two-point calibration graphs, shown in Figure 9.12b, are produced from the dozen pairs of measurement points obtained for the sample and the sample with the standard additive after equal time since injection of both solutions. If there is an additive effect in the sample and the chemical conditions have been chosen appropriately, all graphs intersect at the point indicating the value of this effect.

The data in Table 9.3, for the determination of ascorbic acid in juices and beverages, indicate that the use of the described procedure yields very accurate and precise results. Due to the use of a flow-injection manifold automatically generating a gradient of chemical conditions the analysis is not laborious, low costly, and fast (time of a single peak registration is about 100 seconds). The possibility of obtaining as

Table 9.3 Results of the spectrophotometric determination of ascorbic acid in drinks and juices by E-SAM and C-HPSAM with the use of a set of two-point calibration graphs produced by the gradient ratio flow technique; reference method – capillary electrophoresis.

	Concentration found, c_x, (mmol l^{-1})		
Sample	**Reference method**	**E-SAM**	**C-HPSAM**
Schweppes orange	41.88 ± 6.72	50.34 ± 8.06	40.60 ± 3.55
Hortex blackcurrant	147.00 ± 21.60	229.18 ± 36.93	166.00 ± 9.60
Fanta orange	71.32 ± 9.65	76.18 ± 2.28	68.95 ± 4.70
Cherry syrup	610.00 ± 86.60	849.08 ± 110.37	644.00 ± 55.00

Source: Wieczorek et al. [12], table 1 (p. 14)/with permission of Elsevier.

many as a dozen calibration graphs in a short time allows the stability of the additive effect to be excellently controlled and the chemical conditions to be changed quickly if insufficient stability is found. The total volume of the sample, standard solution, and reagents was about 40 ml, definitely meeting the principles of "green analytical chemistry."

Thus, it appears that a version of the HPSAM method with chemical control of measurement conditions and the use of a flow-through sample processing mode is particularly recommendable. The strict applicability conditions of the HPSAM method are relatively easy to meet in this case by taking advantage of the natural resistance of the extrapolation method to proportional interference effects. By using the flow mode of analysis, the preparation of even a few calibration graphs is quick and easy. As a result, the HPSAM method becomes a unique, efficient, and "green" way to solve such a difficult analytical problem as detecting and compensating for additive effects coming from unknown sample components.

The HPSAM method is often used for two-component analysis (more components are determined simultaneously in occasional cases). This is basically only possible if the interferent present in the sample is known. The calibration procedure looks like this: first, the method is used (as shown above) to determine one component (analyte) while compensating for the additive AE effect caused by the second component (see Figure 9.7). If it is certain that only the second component (interferent) is responsible for the AE effect, it is sufficient under conditions P_1 or P_2 to make a single calibration graph for it, determine its measurement sensitivity, and calculate its concentration from the AE signal value. If an additive effect of the first component on the second component is suspected, the HPSAM method is again applied by determining the second component (now acting as an analyte) under the appropriate conditions required by this method.

9.3.3 Integrated Procedure

The interpolation and extrapolation versions of the MCM method can be integrated according, for example, to the complementary dilution method (CDM) principle presented earlier and implemented in a single calibration procedure. The process of integration allows, as previously demonstrated, to obtain several estimates of the analyte concentration in the sample, which in turn allows to diagnose the analytical system for interference effects (except for additive effects) and to obtain an analytical result with increased accuracy. The MDM method allows compensation for additive interference effects. The combination of both calibration approaches is thus undoubtedly an attractive analytical proposition.

At the basis of the **CDM–MCM method** is a special rule of the MCM method whereby the difference of calibration graphs obtained under two conditions satisfying the MCM method produces a graph that allows the analytical result free of additive effects to be determined by interpolation or extrapolation. This is shown graphically in Figure 9.13. This rule was originally called the "absorbance increment" [13], although when generalizing it to calibration graphs obtained by different measurement methods, it should rather be called "signal increment."

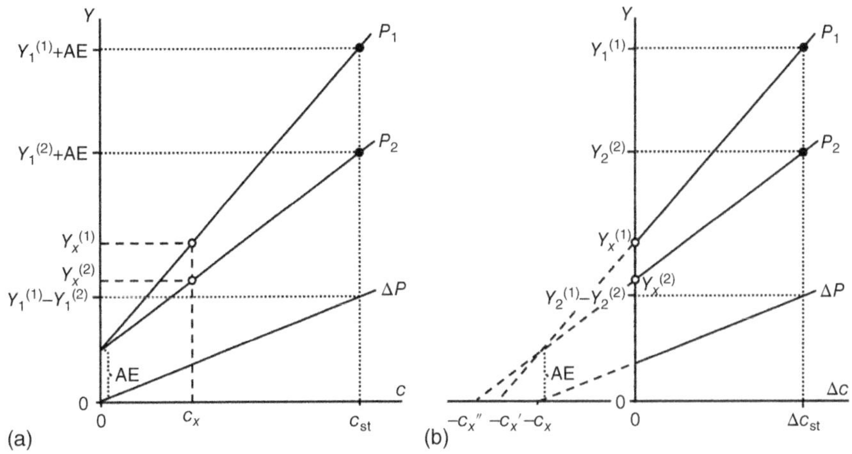

Figure 9.13 Creation of calibration graphs released from the additive effect in the interpolation (a) and extrapolation (b) versions of the MCM method based on the "signal increment" property.

The signal increment rule is also confirmed by formulas (9.15) and (9.23), which can be written as follows:

$$c_x = \frac{\Delta Y_x}{\Delta Y_{st}} \cdot c_{st} \tag{9.25}$$

$$c_x = \frac{\Delta Y_x}{\Delta Y_{st} - \Delta Y_x} \cdot c_{st} \tag{9.26}$$

where the symbols ΔY_x and ΔY_{st} denote the difference in signals obtained for the sample and standard (or sample with standard), respectively, under the two measurement conditions required by the MCM method.

The complete procedure of the CDM–MCM method is thus as follows: in the preparative step, six calibration solutions are prepared as shown in Figure 8.21, that is, according to the procedure of the CDM method. These solutions contain diluent, sample, standard solution, and sample with standard solution added in such volumes that sample and standard are diluted in these solutions in mutually complementary degrees P and Q ($P + Q = 1$).

The six prepared calibration solutions are then subjected to analyte signal intensity measurements under two different conditions, P_1 and P_2, which satisfy the assumptions of the MCM method, i.e. provide a change in the analyte signal without changing the interferent signal. Thus, under both conditions, signals are obtained for the analyte compact in the sample, $Y_{xP}{}^{(1)}$, $Y_{xQ}{}^{(1)}$, $Y_{xP}{}^{(2)}$, $Y_{xQ}{}^{(2)}$, the standard, $Y_{1P}{}^{(1)}$, $Y_{1Q}{}^{(1)}$, $Y_{1P}{}^{(2)}$, $Y_{1Q}{}^{(2)}$, and the sample with the addition of the standard, $Y_{2P}{}^{(1)}$, $Y_{2Q}{}^{(1)}$, $Y_{2P}{}^{(2)}$, $Y_{2Q}{}^{(2)}$. Based on the "signal increment" rule, the differences of the signals obtained under the two measurement conditions are then calculated: $\Delta Y_{xP} = Y_{xP}{}^{(1)} - Y_{xP}{}^{(1)}$, $\Delta Y_{xQ} = Y_{xQ}{}^{(1)} - Y_{xQ}{}^{(1)}$, $\Delta Y_{1P} = Y_{1P}{}^{(1)} - Y_{1P}{}^{(1)}$, $\Delta Y_{1Q} = Y_{1Q}{}^{(1)} - Y_{1Q}{}^{(1)}$, $\Delta Y_{2P} = Y_{2P}{}^{(1)} - Y_{2P}{}^{(1)}$, $\Delta Y_{2Q} = Y_{2Q}{}^{(1)} - Y_{2Q}{}^{(1)}$. These results allow us to construct four calibration graphs, which are shown in Figure 9.14 (compare Figure 8.22).

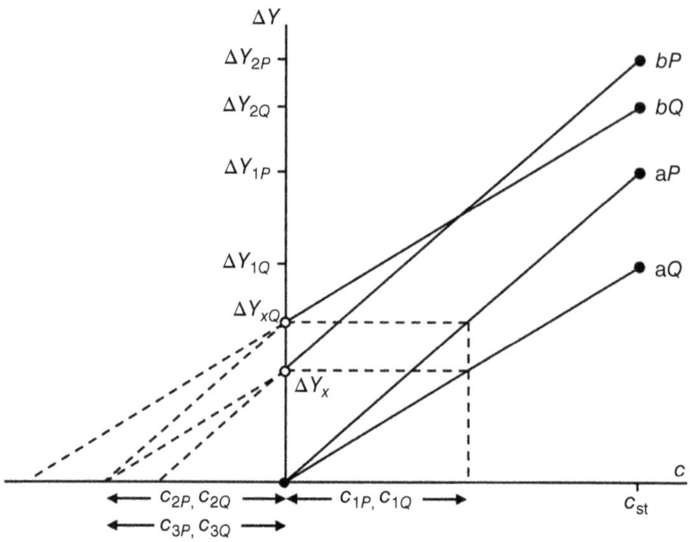

Figure 9.14 Calibration graphs obtained in the CDM–MCM method as the dependencies between the signal difference, ΔY, and analyte concentration, c (compare with Figure 8.22 in Chapter 8).

From the graphs and measurement results shown in Figure 9.14, six apparent concentration values can be obtained by interpolation, c_{1P}, c_{1Q}, quasi-extrapolation, c_{2P}, c_{2Q}, and extrapolation, c_{3P}, c_{3Q}, which are estimates of the analyte concentration in the sample after compensation for additive effect:

$$c_{1P} = \frac{\Delta Y_{xQ}}{\Delta Y_{1P}} \cdot c_{st} \tag{9.27}$$

$$c_{1Q} = \frac{\Delta Y_{xP}}{\Delta Y_{1Q}} \cdot c_{st} \tag{9.28}$$

$$c_{2P} = \frac{\Delta Y_{xQ}}{\Delta Y_{2P} - \Delta Y_{xP}} \cdot c_{st} \tag{9.29}$$

$$c_{2Q} = \frac{\Delta Y_{xP}}{\Delta Y_{2Q} - \Delta Y_{xQ}} \cdot c_{st} \tag{9.30}$$

$$c_{3P} = \frac{\Delta Y_{xP}}{\Delta Y_{2P} - \Delta Y_{xP}} \cdot \frac{\Delta Y_{1P}}{\Delta Y_{1Q}} \cdot c_{st} \tag{9.31}$$

$$c_{3Q} = \frac{\Delta Y_{xQ}}{\Delta Y_{2Q} - \Delta Y_{xQ}} \cdot \frac{\Delta Y_{1Q}}{\Delta Y_{1P}} \cdot c_{st} \tag{9.32}$$

These concentrations are interpreted as in the CDM method, i.e. the uncontrolled effects are diagnosed according to the rules in Table 8.18, and the analytical result is calculated from this diagnosis. In Table 9.4, the results obtained by the CDM–MCM method are included and compared with the results obtained by the CDM method (without compensation for additive interference effects) under two experimental conditions defined by two wavelengths selected respectively [14]. The implementation of the MCM method brings a visible improvement in the accuracy of the

Table 9.4 Results of spectrophotometric determination of calcium in water samples (tap water and two mineral waters) and the errors caused by proportional and additive interference effects reduced due to application of the CDM-MCM method (complexometric titration was used as the reference method).

Sample	Concentration ($ng\,l^{-1}$)				Relative errors (%)		
	Reference method	CDM λ_1	CDM λ_2	CDM–MCM $\Delta\lambda$	Proportional	Additive *P*	Additive *Q*
Tap water	125.0	197.8	175.1	126.2	+6.2	+34.4	+61.1
Bronchales	4.91	7.06	5.82	4.93	+51.0	+46.9	+15.3
Lanjaron	31.0	36.8	32.9	26.5	−6.5	−10.5	+55.1

Source: Adapted from Świt et al. [14].

analytical results, which become very close to the expected results. Based on the obtained measurement results, it is also possible to evaluate the errors that would have been made if the CDM–MCM method had not been applied. The evaluation concerns the errors caused by both the proportional interference effect, compensated by the application of the extrapolation method, and the additive effects compensated after their disclosure in the sample diluted in the calibration solutions in *P* and *Q*.

We described above, the MCM method can be combined with the CDM method integrating various other interpolation and extrapolation approaches than the CDM and E-SAM methods. There is also nothing to prevent one from using a calibration procedure consistent with the GRCM–MCM method, that is, calibration solutions of the composition shown in Figure 8.21 are gradually diluted and, as they are diluted, subjected to measurements at two appropriately chosen wavelengths. In any case, it is possible to perform calibrations not only under different instrumental conditions but also – it is easier – under selected chemical conditions that meet the requirements of the MCM method.

Thus, the MCM method – quite rarely seen today – can be developed in the future in various forms, each of which provides specific analytical advantages. The critical stage, however, still remains the choice of appropriate conditions ensuring the efficiency of elimination of additive effects, especially when the source of these effects is unknown interferents present in the analyzed sample. Further efforts to develop clear and relatively simple criteria for this selection using different measurement methods is, it seems, a basic condition for the MDM method to develop and become more popular among analysts.

References

1 Thompson, M., Ellison, S.L.R., and Wood, R. (2002). Harmonized guidelines for single-laboratory validation of methods of analysis. *Pure and Applied Chemistry* 74 (5): 835–855.

2 Wieczorek, M., Dębosz, M., and Kościelniak, P. (2019). Elimination of additive interference effects by H-point calibration curve method. *Monatshefte für Chemie* 150 (9): 1563–1567.
3 Bosh-Reig, F. and Campíns-Falcó, P. (1988). H-point standard addition method. Part 1 Fundamentals and application to analytical spectroscopy. *Analyst* 113 (7): 1011–1016.
4 Shams, E., Abdollahi, H., Yekehtaz, M. et al. (2004). H-point standard addition method in the analysis by differential pulse anodic stripping voltammetry. Simultaneous determination of lead and tin. *Talanta* 63 (2): 359–364.
5 Verdú-Andrés, J., Bosch-Reig, J., and Campíns-Falcó, P. (1995). H-point standard additions method for analyte determination in ternary mixtures. *Analyst* 120 (2): 299–304.
6 Yehia, A.M. (2013). Development and validation of new spectrophotometric ratio H-point standard addition method and application to gastrointestinal acting drugs mixture. *Spectrochimica Acta A* 109: 193–200.
7 Blasco-Gómez, F., Campíns-Falcó, P., Bosch-Reig, F. et al. (1998). Curve resolution procedure for isolating the spectra of unknown interferences from the sample spectrum in analyte determinations. *Analyst* 123 (12): 2857–2863.
8 Shariati-Rad, M., Irandoust, M., and Sheikhi, S. (2015). Net analyte signal based interferent modelling (NAS-IM) for solving matrix effects and unknown spectral interferents. *Analytical Methods* 7 (20): 8901–8910.
9 Campíns-Falcó, P., Verdú-Andrés, J., Bosch-Reig, F. et al. (1995). Generalized H-point standard additions method for analyte determinations in unknown samples. *Analytica Chimica Acta* 302 (2, 3): 323–333.
10 Safavi, A., Abdollahi, H., Sedaghatpour, F. et al. (2003). Indirect simultaneous kinetic determination of semicarbazide and hydrazine in micellar media by H-point standard addition method. *Talanta* 59 (1): 147–153.
11 Wieczorek, M., Rengevicova, S., Świt, P. et al. (2017). New approach to H-point standard addition method for detection and elimination of unspecific interferences in samples with unknown matrix. *Talanta* 170: 165–172.
12 Wieczorek, M., Dębosz, M., Świt, P. et al. (2018). Application of gradient ratio flow-injection technique to implementation of the chemical H-point standard addition method. *Talanta* 186: 215–220.
13 Campíns-Falcó, P., Bosch-Reig, F., and Verdú-Andrés, J. (1992). Application of the H-point standard additions method by using absorbance increment values as analytical signals. *Talanta* 39 (1): 1–7.
14 Świt, P., Verdú-Andrés, J., Wieczorek, M. et al. (2018). New calibration model: combining integrated calibration method and H-point standard addition method to detect and avoid interference effects. *Analytical Letters* 51 (8): 1194–1207.

10

Calibration-Based Procedures for Correction of Preparative Effects

10.1 Introduction

The calibration methods presented in Chapters 6–9 were mainly used to obtain analytical results free of uncontrolled effects, especially interference effects (IEs). However, they were concerned with the sample in a state ready for signal measurement. Thus, they did not account for those changes in the analytical signal that were caused by factors that manifested themselves from the time sample was collected for analysis until the time the measurement was made. In other words, they did not include those preparative effects (PEs) that result from the change (usually loss) of some amount of analyte that occurs during the physicochemical processing of the sample.

The determination of the change in the amount of an analyte in a sample as a result of its processing is one of the most important problems in analytical chemistry. The change is also characterized by the term **recovery** (implicitly: recovery of analyte from the analyzed sample), which is defined as "yield of a preconcentration or extraction stage of an analytical process for an analyte divided by amount of analyte in the original sample" [1]. In a slightly extended approach, recovery, measured by the recovery factor RV, should be understood as the ratio of the concentration of that part of the analyte which, at the time of measurement, remained in the sample after all preparative treatments necessary in a given case (e.g. sampling, dilution, digestion, extraction, saponification, derivatization, heating, etc.), i.e. real value, c_0, to the initial, true concentration of the analyte, c_{true}, not yet changed as a result of these treatments:

$$\text{RV} = \frac{c_0}{c_{\text{true}}} \tag{10.1}$$

If one assumes that the value of c_0 is equal to the analytical result, c_x (i.e. the real function is accurately represented by the model function), then the RV value allows one to calculate the true concentration of the analyte in the sample from the formula:

$$c_{\text{true}} = \frac{c_x}{\text{RV}} \tag{10.2}$$

The RV, therefore, represents a correction to the analytical result that needs to be taken into account when it is feared or certain that there has been a change in the amount of analyte naturally present in the sample during the preparative step.

Calibration in Analytical Science: Methods and Procedures, First Edition. Paweł Kościelniak.

The higher the accuracy of the recovery factor determination, the smaller the number of preparative treatments and the smaller the changes in the amount of analyte these treatments cause. Furthermore, this accuracy depends on the accuracy of the determination of the analytical result, c_x, on the calibration route. A prerequisite for this is the absence of other – besides preparative – statistically significant uncontrolled effects, or their complete elimination or compensation, which is of course very difficult to achieve. Because of this "calibration" nature of recovery, it is even recommended to use the name "apparent recovery" [1]. For the sake of simplicity, the term "recovery" will continue to be used in this paper, although its association with the calibration process and all the resulting consequences should be kept in mind.

As mentioned earlier, analytical calibration offers the possibility to compensate for all uncontrolled effects as long as the standard used, especially the chemical standard, and the sample are physically similar and are treated the same within the analytical method. Thus, if the standard is subjected to the same preparative treatment that the sample is subjected to prior to measurement, the preparative effect should be compensated for and the analytical result obtained by, for example, the external standard method should be free of error due to this effect. However, this is very rarely the case, mainly because subjecting standard and sample to the same treatment is not equivalent to the analyte concentration in sample and standard changing to the corresponding degree. This is especially true for the analyte isolation process, which can occur with a completely different efficiency from the natural sample matrix than from the standard, and even when it contains components that mimic the sample matrix. Another reason is that compensation for preparative effect does not allow to determine the value of this effect, and such information (e.g. extraction efficiency) is usually very useful for the analyst.

So how to calculate, even approximately, this correction so that, according to Eq. (10.2), it is accounted for in the final analytical result? There are several ways, which are outlined below.

10.2 Specific Procedures

According to IUPAC recommendations [2], the recovery factor is best estimated using certified reference materials appropriately matched in chemical composition to the sample under analysis. It is then assumed that the certified analyte content of such a material, c_{cert}, is a good approximation of its true content ($c_{cert} \approx c_{true}$). The certified material is subjected to all preparative treatments to which the sample is subjected, the analyte is determined in it and the recovery factor is calculated from Eq. (10.1) assuming that $c_{true} \approx c_{cert}$ and $c_0 \approx c_x$. The calculated value is taken to correspond to the recovery factor in the test sample.

The advantage of this procedure is that the analyte present in the reference material is similarly bound to the sample matrix as the analyte in the test sample, creating the chance that in both cases the analyte will be subject to all preparative treatments to a similar degree. The problem, however, is the limited availability of suitable reference materials. It is clear that the accuracy of determining the recovery factor in this way is greater than the mutual similarity of the reference material and the test sample in terms of chemical composition, and achieving a sufficiently high similarity is very difficult in practice.

Some sample processing techniques are so well recognized on the theoretical side that their mechanism of operation is described mathematically, which can be the basis for calculating the recovery factor. An example of such a technique is solid-phase microextraction (SPME) technique invented in 1989 by R.P. Belardi and J.B. Pawliszyn [3]. This technique uses micro solid-phase extractors in the form of fused silica fibers, which are modified by either chemically bonding or polymerizing an organic layer. An analyte to be extracted from the polar aqueous solution is absorbed into the less polar stationary phase. By this means, different kinds of analytes (including both volatile and nonvolatile) are able to be extracted from different kinds of samples that can be in liquid or gas phase.

The most well-established and widely used quantification **method using SPME** is the equilibrium extraction method. When SPME fiber coating is exposed to a sample matrix for sufficiently long time an equilibrium between both phases is reached. According to the law of mass conservation, this condition is described by [4]:

$$n_e = \frac{K_{\mathrm{fs}} \cdot V_{\mathrm{f}} \cdot V_{\mathrm{s}}}{V_{\mathrm{s}} + K_{\mathrm{fs}} \cdot V_{\mathrm{f}}} \cdot c_{\mathrm{in}} \tag{10.3}$$

where c_{in} and n_e is the initial and extracted concentration of the analyte, respectively, K_{fs} is the distribution coefficient of the analyte between the fiber coating and the sample, V_{f} is the volume of the fiber coating, and V_{s} is the volume of sample matrix. Distribution coefficient is experimentally determined based on extraction performed with the use a standard solution of the analyte extracted and in some cases can be taken from the literature.

Thus, if c_{in} and n_e are assumed to be measures of the true concentration of the analyte in the sample and the analytical result, respectively, then the expression linking these quantities can be viewed as a measure of the recovery factor. The analyte recovery so determined is obviously subject to errors if only because the above assumptions are never completely fulfilled. Besides, the value of the distribution coefficient depends on the experimental conditions (e.g. temperature) and is determined with a certain approximation. Nevertheless, this theoretical–experimental approach allows a relatively quick estimation of analyte recovery without additional calibration efforts.

The amount of analyte, extracted by the SPME technique from a sample can also be determined in various kinetic ways. Especially interesting procedure is that announced as "standard-free kinetic calibration method" [5]. It is based on the equation describing the kinetic process for absorption in SPME technique [6]:

$$\frac{n}{n_e} = 1 - e^{-at} \tag{10.4}$$

where n is the amount of analyte in the extraction phase at time t, n_e the amount of analyte in the extraction phase at equilibrium, and a is a constant that is dependent

Table 10.1 Validation of the standard free kinetic calibration method on the example of four PAHs (each calculated n_e is the average value from results obtained for n_1 and n_2).

	n (ng)		n_e (ng)	
Analyte	n_1 **(after 5 s)**	n_2 **(after 10 s)**	**Determined**	**Calculated**
Benzene	1.83	2.11	5.6	5.9
Toluene	8.11	14.15	31.1	31.6
Ethylbenzene	5.17	9.34	25.9	26.5
o-Xylene	7.49	13.73	42.7	42.1

Source: Adapted from Ouyang et al. [5].

on the volumes of the extraction phase, mass-transfer coefficients, distribution coefficients, and the surface area of the extraction phase. When extraction is perfumed for different periods of time, t_1 and t_2, at constant conditions, even if the conditions of sampling are unknown and the equilibrium is not reached, the kinetic process for the absorption can be described by:

$$\frac{t_2}{t_1} \ln\left(1 - \frac{n_1}{n_e}\right) = \ln\left(1 - \frac{n_2}{n_e}\right) \tag{10.5}$$

Thus, concentration n_e can be calculated (e.g. by iteration) without a need to estimate coefficient a experimentally.

Table 10.1 shows the results obtained by the described method for polycyclic aromatic hydrocarbons (PAHs). As can be seen, the calculated results were comparable to the experimental data, demonstrating that this method is suitable for accurate determination of analytes even after very rapid sampling (well before equilibrium is reached).

Various methods allowing the recovery to be estimated using the SPME technique are summarized and discussed in paper [7].

10.3 Surrogate Recovery Method

The previously discussed approaches to recovery determination can be used to a limited extent, i.e. either with a properly selected certified reference material or when using a sample processing technique with a well-understood and mathematically described mechanism of action. An approach with much greater versatility is the so-called **surrogate recovery method (SRM)**. The name of the method is derived from the term "surrogate recovery," which is defined as "recovery of a pure compound or element specifically added to the test portion or test material as a spike" [1].

The principle of the SRM method is illustrated in Figure 10.1. The procedure is that, as early as possible in the analytical procedure, the test sample is divided into two (at least) portions of equal volume. A standard solution of known analyte concentration, c_{st}, is added to one portion, while the other portion is supplemented with diluent and both portions are subjected to laboratory treatment and measurements.

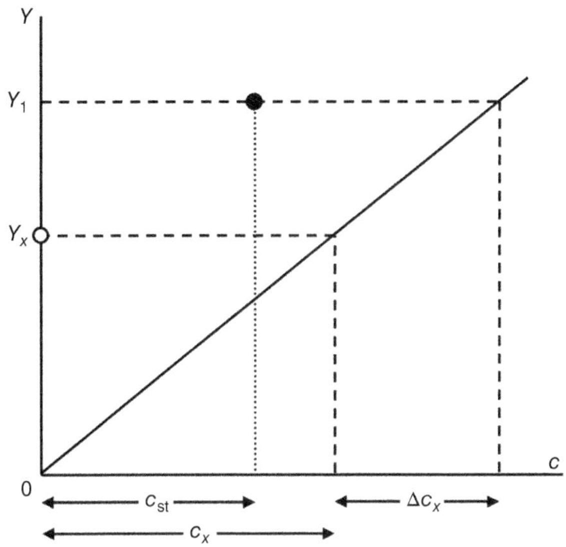

Figure 10.1 Principle of the SRM method: full line – ESM calibration graph, Y_x, Y_1 – signals obtained for the sample and for the sample spiked with analyte, respectively, measured after processing of both solutions, c_{st} – concentration of the analyte added to a sample; Δc_x – the concentration difference found on the basis of signals Y_x (empty point) and Y_1 (full point).

The two signals are then related to the calibration graph made with pure analyte according to the ESM method and the difference Δc_x in analyte concentrations corresponding to these signals is determined. Recovery factor, RV, is calculated from the equation:

$$\mathrm{RV} = \frac{\Delta c_x}{c_{st}} \tag{10.6}$$

Note that the sample handling in the SRM method is the same as in the conventional extrapolative standard addition method (E-SAM). If the standard is added to the sample in such a way that the addition of the standard does not change the concentration of the native constituents of the sample, then the only thing that distinguishes this treatment from the E-SAM method is that the standard is added to the sample before it is prepared for measurement, rather than immediately before measurement. Therefore, it can be said that by introducing the analyte to the sample according to the E-SAM method, one ensures the similarity of the chemical composition of the standard and the sample by substituting a suitable reference material in this role.

The SRM method can be performed on a sample that does not contain the analyte (or actually contains it in a concentration below the limit of quantification). It is then sufficient to add a known amount of analyte, c_{st}, to this sample, process it together with the standard added, and determine the analyte, c_x, remaining after processing. The recovery factor is then calculated using the simplified formula:

$$\mathrm{RV} = \frac{c_x}{c_{st}} \tag{10.7}$$

The role of surrogate may be played by an isotopically modified version of the analyte. In such an approach the recovery of the surrogate can be estimated separately by the radiometric or mass spectrometry measurements and then related to the native analyte. Regardless of its form, however, the surrogate is more or less a "foreign" component of the sample and may therefore be subject to preparative treatments to a different extent than the native analyte. Therefore, it is very important to bring the native analyte and the analyte added to the sample, if possible, to some equilibrium state that ensures similar behavior of both forms in the preparative step.

Because of its versatility and simplicity, the SRM method has become the primary means of determining analyte recovery from a sample to determine the efficiency of the sample processing technique used. According to Eq. (10.2), the determined RV value also allows, at least theoretically, to evaluate the discrepancy between the obtained and the true concentration of analyte in the sample that arose during the preparative step of the analytical procedure. Since 1999, IUPAC has officially recommended that "recovery values should always be established as part of method validation ..." and allows the SRM method to be used to determine recovery values as a measure of the accuracy of an analytical result when a suitable reference material is not available [2].

In this situation, it is worthwhile to take a closer look at the SRM method from a calibration point of view and, on this basis, determine the additional conditions to which it should be subjected to lead to reliable results.

10.3.1 Reliability of the Method

The mathematical relationship between the accuracy of the analytical result, relative error (RE), and the recovery factor, RV, is very simple. If one assumes that the analyte is recovered from the added form to the same extent as from the native form, then the equation occurs:

$$\frac{\Delta c_x}{c_{st}} = \frac{c_x}{c_{true}} \tag{10.8}$$

where c_x is the portion of the analyte recovered from the sample that is determined. If the RE of an analytical result is expressed in the form:

$$RE = \frac{c_x - c_{true}}{c_{true}} \cdot 100\% \tag{10.9}$$

it follows from Eqs. (10.7) and (10.8) that:

$$RE = RV - 100\% \tag{10.10}$$

This interpretation of the SRM method is the basis for treating recovery as a measure of the accuracy of the analytical result. However, the question arises: can SRM perform this role correctly regardless of the nature of the preparative effect? Let us base our considerations from the calibration point of view.

First of all, the SRM method should be presented as a composite of two calibration methods. E-SAM and ESM with their characteristic calibration graphs. In the absence of preparative and other uncontrolled effects, the graph obtained by the E-SAM method is then simultaneously an image of the true function, a, the real function, b, and the model function, c, as illustrated in Figure 10.2A. It is apparent that the recovery value, RV, determined in this situation is 100%, and the analytical

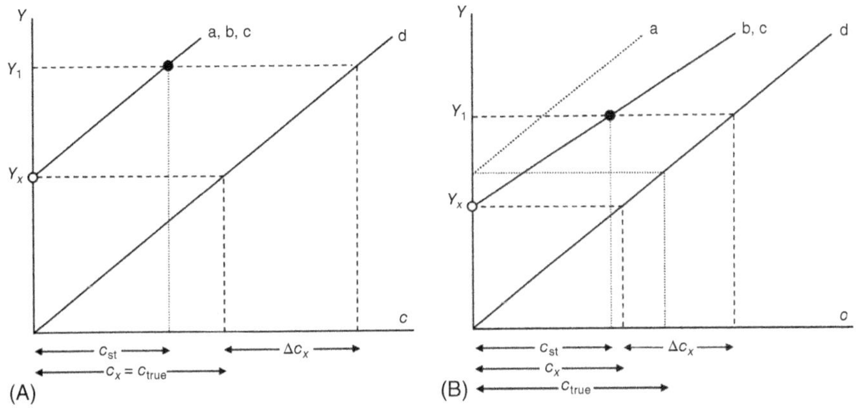

Figure 10.2 The SRM graphs in cases of absence of preparative effect (A) and of presence of proportional preparative effect (B): (a) true function, (b), real function, (c) E-SAM calibration graph, (d) ESM calibration graph; the meaning of the other symbols as in Figure 10.1.

result, c_x, obtained from the signal intensity obtained for the sample itself, is equal to the true concentration of analyte in the sample ($c_x = c_{true}$), i.e. the error of this result, RE, is 0%. Equation (10.10) is therefore satisfied.

The occurrence of a preparative effect is manifested by a change in the position of the graphs of the true and model functions relative to the position of the true function. Preparative treatments generally involve the loss of some analyte from both solutions, so there is a decrease in the b and c graphs relative to the a graph. If this change is proportional to the concentration of analyte in both portions of the sample (without and with addition of standard), i.e. as shown in Figure 10.2B, it manifests itself as a decrease in the difference in analyte concentration, c_x, to the same extent that the value of the analytical result, c_x relative to the true value, c_{true}. Thus, in this case, Eq. (10.10) is also satisfied.

Both situations shown in Figure 10.2 are very likely to occur in practice. Nevertheless, the occurrence of effects of a different nature cannot be ruled out, especially since the native and added analyte may be bound differently to the sample matrix and be released from it to different degrees. For example, they may be lost in equal amounts, producing a preparative effect of an additive nature. In another extreme case, which may be called a selective effect, the sample processing conditions may not favor the release of the native analyte from the sample as opposed to the added analyte.

As can be seen in Figure 10.3, in both cases the SRM method is unable to detect the change in the recovery value of the analyte from the sample caused by the additive effect, while at the same time it allows the determination of the reduced amount of the analyte due to its loss in the sample. When a selective effect is present, the performance of the SRM method is reversed: the detected reduced recovery value is not adequate for the unchanged concentration of the analyte in the sample. In both situations, the obtained recovery value does not allow the determination of an

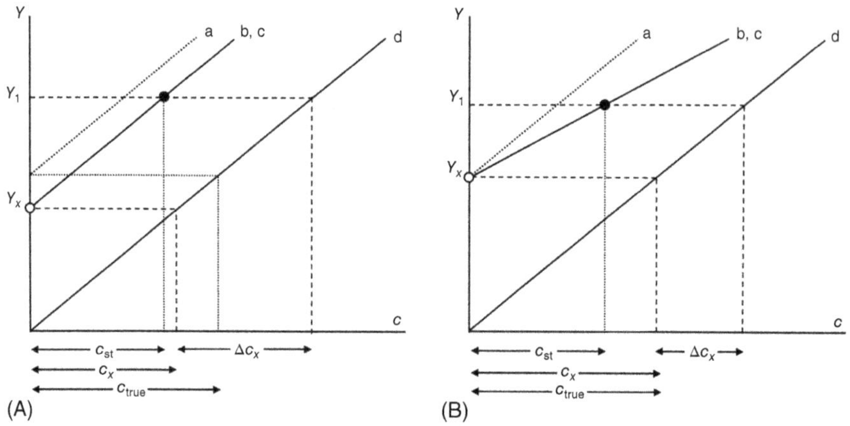

Figure 10.3 The SRM graphs in cases of presence of additive (A) sand selective (B) effect: (a) true function, (b), real function, (c) E-SAM calibration graph, (d) ESM calibration graph; the meaning of the other symbols as in Figure 10.1.

Table 10.2 Reliability of the SRM method depending on the nature of the preparative effect.

Preparative effect	RV (%)	c_x vs. c_{true}	RE (%)	RE vs. RV
No	=100	$c_x = c_{true}$	=0	RE = RV – 100%
Proportional	≠100	$c_x \neq c_{true}$	≠0	RE = RV – 100%
Additive	=100	$c_x \neq c_{true}$	≠0	RE ≠ RV – 100%
Selective	≠100	$c_x = c_{true}$	=0	RE ≠ RV – 100%
Other	≠100	$c_x \neq c_{true}$	≠0	RE ≠ RV – 100%

analytical result with the correct value (RE ≠ 100% – RV). The same conclusion can be made if the preparative effect is intermediate between additive and selective.

Table 10.2 summarizes the relations that exist between the analytical parameters determined by the SRM method in the various cases considered above.

Therefore, it follows from the above considerations **that the SRM method can be used to evaluate the accuracy of an analytical result only if the handling of the sample and standard at the preparative stage does not induce changes in the analytical signal (i.e. the preparative effect does not exist) or the changes are proportional to the total concentration of analyte in the sample with the addition of the standard (i.e. the preparative effect is proportional).**

The relation between recovery and trueness was studied more thoroughly for selenium in the synthetic samples analyzed by atomic fluorescence spectrometry with hydride generation (HG-AFS) [8]. To obtain the effects of required character and degree, they were simulated at the preparative stage of the analytical procedure. Preparative effect of both the proportional (PE1) and additive (PE2) natures was created by preparation of the sample and standard solutions containing "native" and

Table 10.3 Relation between recovery, RV, and accuracy, RE, estimated by the SRM method when preparative effect of proportional, PE1, and additive, PE2, characters occur.

	Theoretical values		Experimental values	
Effect/Effects combined	**RV (%)**	**RE (%)**	**RV (%)**	**RE (%)**
PE1 (−20%)	80	−20	80.7	−19.9
PE1 (−60%)	40	−60	40.8	−57.6
PE2 (+40%)	100	+40	100.8	+41.5
PE1 (−20%) + PE2 (+40%)	80	+20	80.4	+19.3
PE2 (+40%) + PE1 (−20%)	80	+12	79.8	+12.6
PE1 (−60%) + PE2 (+40%)	40	−20	40.5	−21.8
PE2 (+40%) + PE1 (−60%)	40	−44	40.3	−44.9

Source: Adapted from Stafiński et al. [8].

added analyte in decreased or increased analyte concentrations in comparison to the initial ones (20 and 60 mmol l^{-1}) adequately to the kind and value of a given effect. For instance, when PE1 of −20% was considered, the analyte concentrations in solutions were decreased to 16 and 48 mmol l^{-1}, while in case of PE2 of 40% the concentrations were increased to 28 and 60 mmol l^{-1}, respectively.

A mathematical model has been developed, including the contribution of different effects, which may potentially occur during sample preparation and measurements. The relationship between recovery and accuracy predicted by the model was verified experimentally using HG-AFS technique. The results obtained are shown in Table 10.3.

The results obtained confirmed the observations made earlier: proportional effect influences both recovery, RV, and accuracy, RE, to the same extent, which is equal to the level of these effects (e.g. if PE1 = −20% then RV = 80% and RE = −20%), however, the additive effect provides recovery equal to 100% and accuracy different from 0%. When proportional and additive effects occur together, RV is influenced only by proportional effects, but RE is affected by both proportional and additive effects. If both kinds of effects are of different course of action (e.g. negative and positive), then RE ≠ 100% and RV ≠ 0% but RE ≠ RV − 100% (particularly, RE may be equal to 0% in case of total compensation of both effects). It is also interesting that when the two effects occur one after the other in a different order, this difference does not affect the recovery of the analyte, but it does affect the accuracy of its determination.

10.3.2 Interpretations of the Method

The recovery of an analyte from a sample can be expressed not only by the change in analyte concentration relative to the expected concentration (see Eq. (10.7)) but also by the change in analytical signals responsible for the change in concentrations. This interpretation of the results obtained by the SRM method is shown in Figure 10.4.

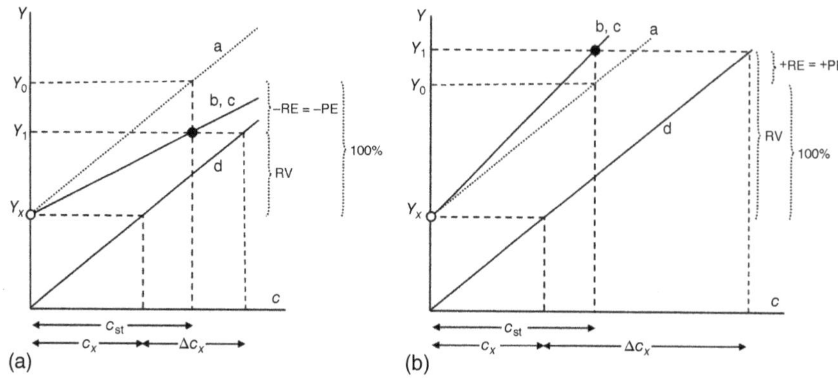

Figure 10.4 Interpretation of the SRM method basing exclusively on the signal values in the case negative (a) and positive (b) preparative effect: Y_0 – signal corresponding to concentration c_s on the graph of true function, RV, RE, PE – recovery, relative error, and preparative effect, respectively; the meaning of the other symbols as in the previous figures.

It is easily demonstrated that in this approach the recovery factor, RV, and the relative error of the analytical result, RE, expressed as percentages are given by the formulas:

$$\mathrm{RV} = \frac{Y_1 - Y_x}{Y_x} \cdot 100\% \tag{10.11}$$

$$\mathrm{RE} = \frac{Y_1 - Y_0}{Y_x} \cdot 100\% \tag{10.12}$$

where Y_0 is the signal, whose value can be calculated by knowing coefficient B of the calibration graph obtained for the analyte alone (d in Figure 10.4) and the c_{st} concentration of the analyte added to the sample:

$$Y_0 = \mathrm{B} \cdot \frac{Y_1}{Y_1 - Y_x} \cdot c_{st} \tag{10.13}$$

The presentation of the measurement results illustrated in Figure 10.4 provides an even better understanding of the correct relationship between analyte recovery and the relative error of its determination (RE = RV – 100%). If the value of 100% is taken as a measure of the expected signal, that is, the signal that would have been measured if there had been no loss (Figure 10.4a) or gain (Figure 10.4b) due to a preparative effect, then this change (loss or gain) is a measure of the relative error, RE, and the remainder of the expected signal expresses the recovery, RV, of the analyte from the sample. If the signal change is a representation of the preparative effect only (PE), then the relative error can be fully equated with the magnitude of that effect (RE = PE).

However, it is worth noting that in this otherwise very illustrative approach, the RV and RE values depend on the changes recorded for the sample with the added standard after treatment, and ignore the varying degree of signal change that may, as shown earlier, affect the sample itself. The question then arises: how to account for the fact that this interpretation of the measurement results in the SRM method is only valid if the preparative effect is proportional?

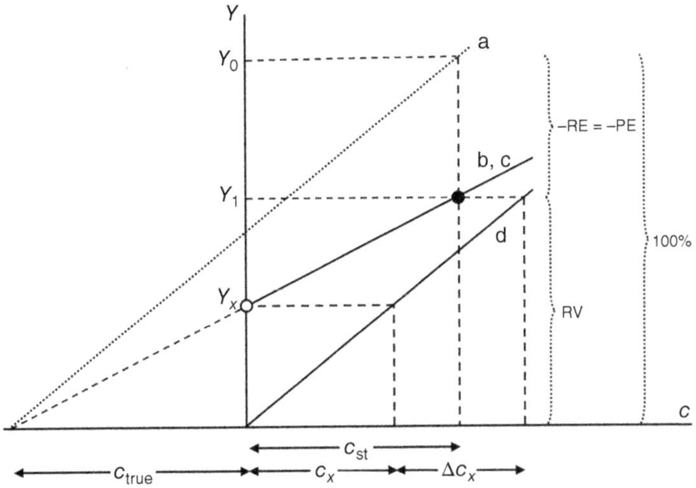

Figure 10.5 Extended interpretation of the SRM method based on both the signal values and calibration by the E-SAM method; the meaning of symbols as in the previous figures.

The answer to this question is easily found when the measurement results obtained in the SRM method for the sample without and with the addition of the standard are interpreted in a manner characteristic of the E-SAM method: to extrapolate the calibration graph fitted to these points to the zero analytical signal and thus to determine the concentration of analyte in the sample before processing. This way of interpreting the SRM method is illustrated in Figure 10.5.

As is well known, the E-SAM method is capable of compensating for analytical effects provided that they are proportional in nature. In other words, calibration graphs produced according to the E-SAM method indicate the true analytical result, c_{true}, as long as the effects present (including preparative effects) are proportional. After combining the graphs a, b, and c (see Figure 10.5) at a single point indicating the c_{true} concentration, the values of recovery, RV, relative error, RE, and c_{true} concentration can be calculated using the following formulas with confidence that they simultaneously express both the proportional character of the preparative effect occurring, PE, and the correct relationship between RV and RE (RE = RV – 100%):

$$\mathrm{RV} = \frac{Y_1}{Y_0} \cdot 100\% \tag{10.14}$$

$$\mathrm{RE} = \frac{Y_1 - Y_0}{Y_0} \cdot 100\% \tag{10.15}$$

$$c_{true} = \frac{Y_x}{Y_1 - Y_x} \cdot 100\% \tag{10.16}$$

The above "calibration" interpretation of the results obtained by the SRM method was checked on the example of the determination of arsenic and selenium in blood by HG-AFS [9]. Before measurement, a sample was digested in nitric acid using the microwave system, and after this process, the gas was removed from above the sample and digested for 10 minutes by a stream of argon. The sample thus processed

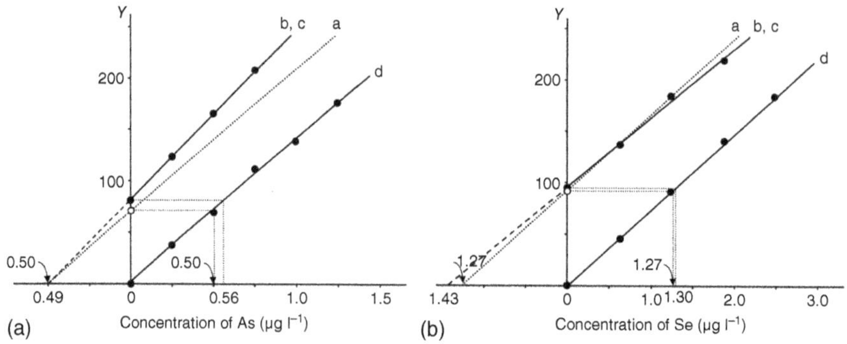

Figure 10.6 Results obtained using SRM method in case of the HG-AFS determination of arsenic (a) and selenium (b) in certified material of blood; the meaning of symbols – as in the previous figures, analytical results obtained by the ESM and E-SAM methods are denoted by arrows. Source: Adapted from Kościelniak et al. [9].

was prepared for measurement according to the SRM method and introduced into the measuring system. To confirm the linear character of the graph produced by the E-SAM method, the sample was processed with several standard additions of different analyte concentrations. The results obtained for the blood reference material are shown in Figure 10.6 (the expected concentrations of arsenic and selenium were 0.50 and 1.27 $\mu g\,l^{-1}$, respectively).

Based on the tests performed, arsenic was recovered from the sample in excess (RV = 111.9%) and with an accuracy corresponding (including random error) to this recovery value (RE = +12.8%). The proportional nature of the effect responsible for this error was confirmed by interpreting the results obtained for the sample and the sample with standard additions in a manner characteristic of the E-SAM method (see Figure 10.6a). In the case of selenium determination, the recovery rate obtained (RV = 87.3%) indicated that the analyte was lost during sample processing, although this was not confirmed by the accuracy of its determination (RE = +1.9%). It was concluded that a selective effect was the cause of this inconsistency (see Figure 10.6b).

A situation in which the determined value of the recovery factor is greater than 100% requires separate attention. If this is attributed solely to the occurrence of a preparative effect, such an "excessive" result is usually explained by random errors made at the preparative stage. When this result is clearly caused by systematic factors (as in the example above for arsenic), it is explained by accidental or intentional addition of the analyte to the sample during sample processing. It is worth noting, however, that the amount of this "contaminant" introduced into two (or more) portions of a sample prepared for SRM testing should be equal, producing an additive effect and therefore not changing the recovery factor value. The most likely reason for the appearance of RV values significantly greater than 100%, which is not always remembered, will be explained in Section 10.3.3.

10.3.3 Recovery vs. Interference Effect

Underlying the definition of recovery and the SRM method is the assumption that the final analytical result, c_x, is determined only by the preparative step and the resulting analytical result, c_x, is treated as the true concentration, c_0, and the model function (calibration graph) with the true function (as seen in Figures 10.3–10.5). It is thus assumed that the accuracy of the analytical result depends only on the determined value of the recovery factor and no further calibration procedure is actually needed. This ignores the possibility that there may be other analytical effects manifested by changes in the analytical signal and potentially affecting the accuracy of the signal transformation to the analytical result. Among the effects manifested during this calibration step of the analytical procedure is the interference effect (IE).

Unlike the preparative effect, the interference effect is not caused by a change in the amount of analyte in the sample, but both effects are manifested by a change in the intensity of the analytical signal. For the SRM method, which "handles" only the measurement results, both effects are therefore indistinguishable [10]. It follows that the recovery determined by the SRM method can be a measure of either the amount of analyte that remains in the sample after processing, a measure of the signal "recovered" after the interference effect, or a joint measure of both of these "recoveries." When the two effects occur side by side and both are proportional in nature, they can be represented as shown in Figure 10.7, that is, as effects jointly responsible for the present value of analyte recovery from the sample.

The presence of the interference effect means that the real concentration of analyte in the sample obtained after the preparative step (c_0) cannot be treated as the final analytical result ($c_x \neq c_0$), and the real function, b, is not accurately represented by the model function, c (as also seen in Figure 10.7). However, if the preparative and interference effects are proportional, the model function c obtained by the E-SAM

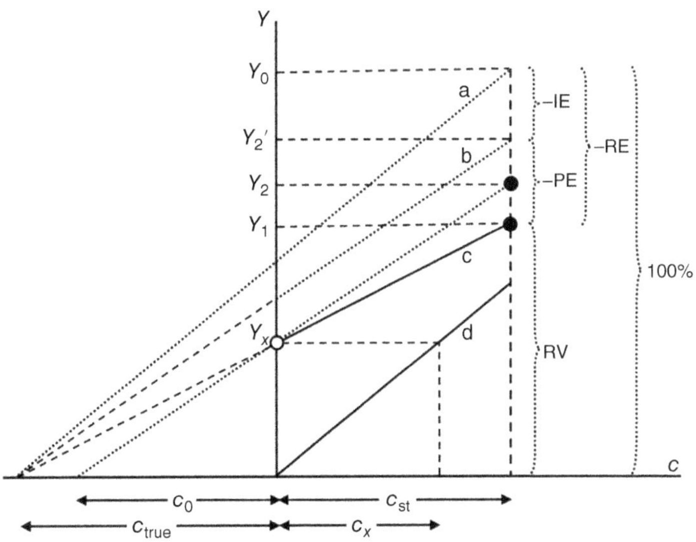

Figure 10.7 The SRM method incorporating the interference effect, IE, into the balance RE vs. RV: Y_2 – signal obtained for the processed sample spiked with unprocessed standard solution, Y_2' – signal Y_2 corrected, a – true function, b – real function, c – E-SAM model function, d – ESM calibration graph, c_0 – analyte concentration not affected by the interference effect (the meaning of symbols – as in the previous figures).

method allows one to hold an analytical result that is an accurate measure of the true concentration, c_{true}, of the analyte in the sample. It follows that the RV recovery value obtained by the SRM method is a potential indirect measure of c_{true} concentration regardless of whether the analyte is recovered due to the presence of preparative or interference effects, or both effects simultaneously.

Following this line of reasoning, one arrives at a general conclusion that is important from a calibration point of view: **in the presence of uncontrolled effects of a proportional nature in the analytical system under investigation, an accurate approximation of the real function by means of the model function can be achieved through calibration by the E-SAM provided that the calibration solutions are prepared before these effects become apparent.**

To determine the interference effect, IE, one has, as is known, to perform the E-SAM calibration in the conventional manner, that is, to add the standard solution of analyte concentration c_{st} to the sample after its processing and to measure signal Y_2 for this solution, as seen in Figure 10.7. Any difference between the Y_1 and Y_2 signals indicates the presence of the interference effect. To determine the contribution of the IE effect to the total effect, IE + PE, the resulting graph must be shifted along the Y axis (i.e. keeping the magnitude of the IE effect) to a position that accounts for its proportional nature and the possibility of determining the c_{true} concentration after extrapolation of this graph (see Figure 10.7). The sum of the IE + PE effects is then also a measure of the RE error, which can be determined from the relationship RE = RV − 100%.

In this approach the effects of PE and IE can be expressed by the following formulas:

$$\mathrm{PE} = \frac{Y_1 - Y_2'}{Y_0} \cdot 100\% \tag{10.17}$$

$$\mathrm{IE} = \frac{Y_2' - Y_0}{Y_0} \cdot 100\% \tag{10.18}$$

where:

$$Y_2' = \frac{Y_2 - Y_x}{Y_1 - Y_x} \cdot Y_1 \tag{10.19}$$

and the recovery factor, RV, and relative error, RE, are calculated from Eqs. (10.11) and (10.12). Note that the graph connecting the points Y_x and Y_2 in Figure 10.7 indicates, after extrapolation, the actual concentration, c_0, of the analyte in the sample, and the difference ($c_{true} - c_0$) is a measure of the loss of analyte concentration from the sample due to preparative effects.

The presence of an interference effect further limits the utility of the SRM method. This is because the effect can take on a character different from the proportional one and this is, it seems, more likely than in the case of the preparative effect. This problem was noted during the determination of selenium by the HG-AFS technique [11, 12]. As mentioned, the blood sample was digested in nitric acid using the microwave system. During digestion of a sample with nitric acid nitric oxide was produced that was then oxidized to nitric dioxide in liquid phase. When the solution is chemically treated for the hydride generation purpose, NO_2 can prevent

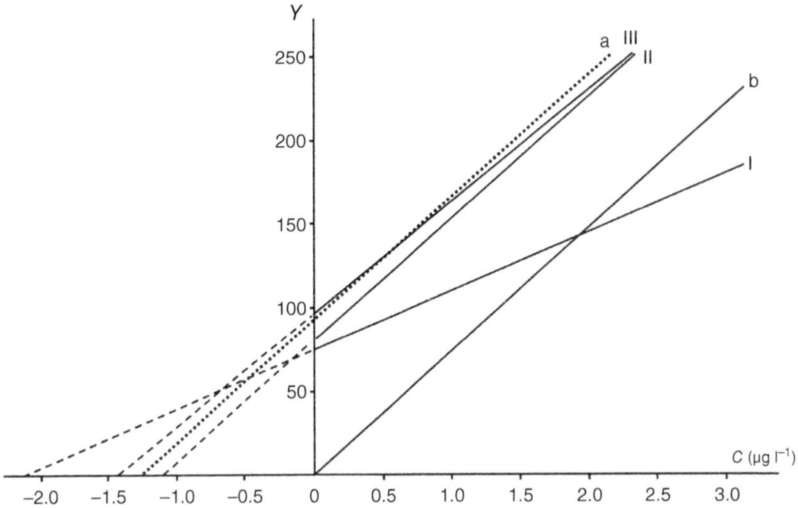

Figure 10.8 Results obtained using SRM method in case of the HG-AFS determination of selenium in certified material of blood: a, b – true function and ESM calibration graph, respectively, I, II, III – E-SAM calibration graphs obtained when nitrogen oxides were removed for 0, 5, and 10 minutes, respectively (experimental points have been removed for clarity). Source: Adapted from Kościelniak et al. [11].

reduction of selenium by $NaBH_4$ from Se^{6+} to Se^{4+} (which is the only form for generation of selenium hydride) and, as a consequence, the analytical signal was weakening. To eliminate this interference effect the gas was removed from above the sample digest. The significance of the timing of this process in light of the SRM method is shown in Figure 10.8.

It can be seen that the unremoved nitrogen oxides cause a decrease in the analytical signal (compare graphs a and I), with the change becoming greater as more of the added standard is present in the sample. Apparently, nitrogen oxides interact more readily with selenium added to the sample than with native selenium, resulting in a strong selective effect. When exposed to argon, this effect disappears, turning first (after 5 minutes) into a small additive effect and then (after 10 minutes) into an equally small selective effect. Although at subsequent stages selenium is determined by the ESM method with increasing accuracy, this cannot be revealed from the calculated recovery values because the effects occurring are not proportional.

It is worth noting that the signal changes shown in Figure 10.8 are caused by changes in the preparative parameter (gas removal time). However, they are not due to changes in the amount of analyte in the sample, but to changes in the amount of its form that causes the analytical signal as a result of the presence of a sample component. Thus, they are not, as one might think, preparative but interference effects. Although this fact is unimportant from the point of view of the SRM method, it is a good illustration of the close influence that the sample processing procedure can have not only on preparative but also on interference effect.

The importance of the interference effect in the context of the SRM method is also expressed by the fact that it can be much more frequently revealed by an increase in the signal measured for the analyte alone than the preparative effect. If a positive interference effect is present without the presence of a preparative effect or the sum of these effects is positive, the determined recovery factor is greater than 100% (see Figure 10.4b). Thus, one can venture to say that the most common cause of a statistically significant positive recovery value is a positive interference effect, which is not taken into account if the recovery is attributed to a preparative effect alone. This is also most likely to be attributed to the result shown in Figure 10.6a for arsenic since this analyte can undergo a positive interference effect when the blood sample is prepared for analysis by the HG-AFS technique as described above in relation to selenium determination [11].

The influence of the preparative and interference effects, separated and combined with each other, on the relation between recovery and trueness was also theoretically and experimentally studied in the example of the HG-AFS determination of selenium in synthetic samples [8]. As previously, the effects were simulated by preparation of the sample and standard solutions containing "native" and surrogate analyte in decreased or increased concentrations in comparison to the initial ones (20 and 60 $mmol\,l^{-1}$). The results obtained are collected in Table 10.4.

In general, the interference effect works as the preparative one if the character of both effects, considered separately or together, is the same. The correct relationship between recovery and relative error, RE = RV − 100%, occurs only if the interference effect itself is proportional or combined with a preparative effect of the same nature. When the interference effect is proportional and positive, the recovery value is greater than 100% and does not depend on whether the effect is combined with an additive effect or not. If both effects are additive and act together, the resultant recovery value may be greater or less than 100% but is not equal to the sum of the values of the two effects. If the two effects are proportional in nature, then combined they affect the RV vs. RE relationship in the same way as when they act individually:

Table 10.4 Relation between recovery, RV, and accuracy, RE, estimated by the SRM method when preparative (PE) and interference effect (IE) of proportional, PE1, IE1, and additive, PE2, IE2 characters occur.

	Theoretical values		Experimental values	
Effect/Effects combined	**RV (%)**	**RE (%)**	**RV (%)**	**RE (%)**
IE1 (+50%)	150	+50	148.8	+53.6
IE2 (+40%)	100	+40	100.8	+41.5
IE1 (+50%) + IE2 (+40%)	150	+90	149.4	+89.5
PE1 (−20%) + IE1 (+50%)	120	+20	122.0	+19.9
PE1 (−60%) + IE1 (+50%)	60	−40	62.0	−40.6
PE2 (+40%) + IE1 (+50%)	150	+110	155.5	+107.8
PE2 (+40%) + IE2 (+40%)	100	+80	99.7	+79.8

Source: Adapted from Stafiński et al. [8].

the recovery is always equal to 100%, and the RE value results from the sum of the values of the two effects.

10.3.4 Recovery vs. Speciation Effect

Another analytical effect potentially present in qualitative and quantitative analysis is the speciation effect. Like the interference effect, the speciation effect can be eliminated by appropriate preparative treatments, but in many cases, this is neglected or the elimination does not occur completely. As already demonstrated, calibration methods are not able to help eliminate the speciation effect, resulting in inaccurate analytical results. It is therefore interesting to ask whether the systematic error committed in this situation can be controlled by the value of the recovery factor of the analyte in the sample determined by the SRM method.

Considering the issue from the theoretical side, it is difficult to expect a positive answer. As a rule, the analyte is added to the sample in the same form in which it was used to prepare the calibration graph by the ESM method. If the analyte is present in the sample in a different form exhibiting a different measurement sensitivity, the value of the true concentration of the analyte in the sample can in principle never be predicted and determined. Even under such favorable conditions, when there are no or proportional effects in the analytical system, the speciation effect generally fails to correctly determine the systematic error of the analytical result, RE, from the formula: RE = RV – 100% based on the determined value of the recovery factor RV. This is shown in Figure 10.9.

It may accidentally happen that due to a selective analytical effect (preparative and/or interference), the measurement sensitivity of the surrogate analyte becomes equal to that of the native analyte, and extrapolation of the standard addition method (SAM) graph leads to the correct analytical result. Such a situation is depicted in Figure 10.10a. This is quite likely because the surrogate analyte may, for example, release from the sample at a different rate than the native analyte due to different

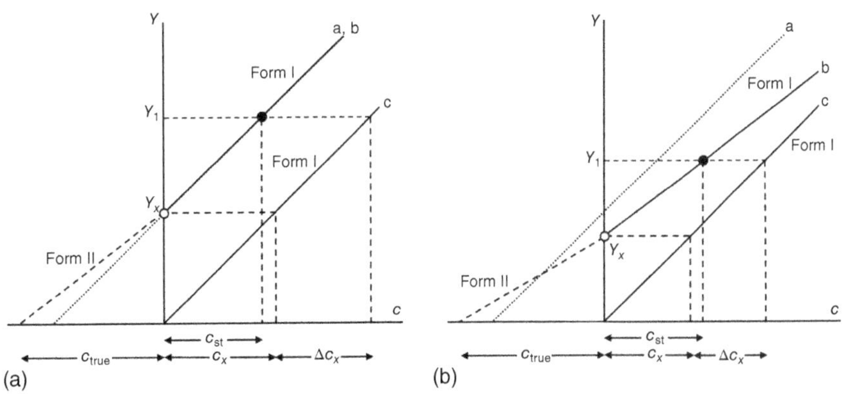

Figure 10.9 Inconsistency between recovery, RV, and relative error, RE (Eq. (10.8)), due to the speciation effect occurring in the absence of other effects (a) and in the presence of proportional effects (b).

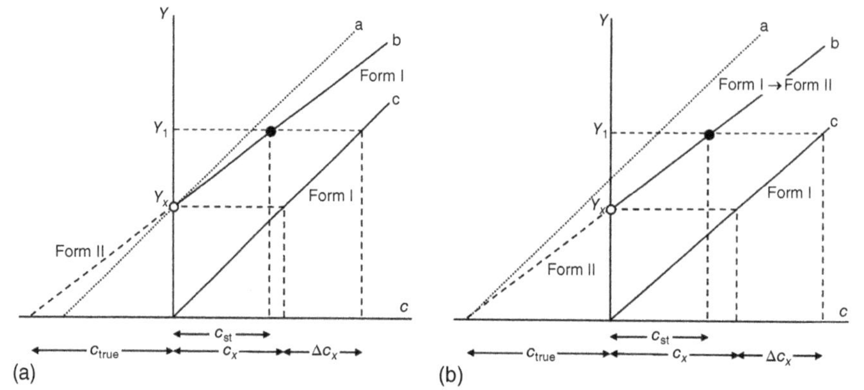

Figure 10.10 Illusive lack of the speciation effect due to release in different degree (a) or change in chemical form (b) of native and surrogate analyte of different forms, which may lead to misinterpretation of the SRM results and to erroneous analytical results (for details see text).

physicochemical properties. However, due to the selective nature of the effect, the determined recovery factor will also not be a valid measure of the accuracy of the analytical result.

The most "dangerous" situation can occur when the analyte added to the sample has the same chemical form as the native analyte but is different from the form of the analyte used to produce the ESM calibration graph. As can be seen in Figure 10.10b, there will then be a proportional effect caused by the different measurement sensitivity values of the two forms of analyte, which will occur regardless of whether there is no accompanying analytical effect or whether the effect is proportional. In any such case, the RE of the analytical result will be correctly determined from the relationship RE = RV − 100%, although on the other hand, this result will not be correct because it is determined from a calibration graph with an incorrect slope.

The above theses have been confirmed by experimental study of selenium recovery from synthetic and natural samples using the HG-AFS technique [13]. First, selenium in organic (selenomethionine, Se-Met) and inorganic forms played a role of both the native and surrogate analyte. The speciation effect was controlled with the use of the UV lamp: when it was turned on, Se-Met was totally digested to inorganic form, otherwise incomplete digestion of Se-Met (caused by high temperature in strongly acidic environment) was observed at the level of c. 50%. In addition, negative interference effect of proportional character was provoked by spiking the sample with copper in the concentration producing c. 50% decrease in fluorescence intensity. The results obtained are shown in Table 10.5.

Incomplete digestion of a sample provoked the speciation effect and, consequently, a risk of incorrect estimation of the RE value based on the RV value. Only in the specific case, predicted theoretically, when both the native and the surrogate analyte were in organic form, relation RE = RV − 100% was valid. However, in all cases, including the presence of interferent, the speciation effect has been effectively eliminated due to digestion process.

Table 10.5 Relation between recovery, RV, and accuracy, RE, estimated experimentally by the SRM method when speciation effect is not eliminated or eliminated due to the UV lamp operation [13].

Analyte			UV lamp doesn't work		UV lamp works	
Native	**Surrogate**	**In ESM standards**	**RV (%)**	**RE (%)**	**RV (%)**	**RE (%)**
Se-Met	Se	Se	100.1	−48.7	97.8	−0.1
Se-Met	Se-Met	Se	51.3	−48.3	102.2	−2.0
Se	Se-Met	Se	55.7	−1.0	98.2	−1.0
Se-Met + Se	Se-Met	Se	53.7	−25.8	103.7	0.8
Se-Met + Cu	Se	Se	54.2	−84.3	57.7	−41.9

In the same way the speciation effect in natural samples of thermal spring water and vitamin drink were examined. The selenium concentrations in both samples were declared by produces, hence these values were assumed as the true ones" ones. Surrogate analyte was added to the samples in inorganic or organic form. Recovery values obtained after one- and twofold addition of surrogate were taken into account [13].

As seen in Table 10.6, the analyte was found in much lower concentration than the declared values due to the strong additive preparative effect diagnosed. As selenium is particularly unstable in samples with stored at room temperature, the most likely reason of this systematic error was a loss of selenium as a result of inappropriate storage of the samples for a too long period of time (i.e. from production to selling). The additive nature of the effect is indicated by the RV values close to 100% and inconsistent with the RE values that were obtained after mineralization of the samples (i.e. after the probable complete elimination of the speciation effect). It was therefore assumed that the selenium concentrations determined in this situation in both samples (approximately 14 and 11 $\mu g\, l^{-1}$, respectively) are the actual selenium concentrations remaining in these samples after the loss caused by the preparative effect.

The form of selenium in the sample of thermal spring water was expected to be inorganic only. This is confirmed by the obtained results. If the form of surrogate analyte was inorganic (originally or due to complete digestion), RE′ value was close to 0% and RV1 values were adequately close to 100%. However, if the surrogate analyte was organic and the sample was not digested completely, RE′ value (still close to 0%) could not be predicted by the recovery values. This evidently indicated the speciation effect.

In the case of the sample of vitamin drink, the situation appeared to be even more complicated. When the sample was spiked with selenium in an inorganic form and exposed to UV radiation, it was digested totally and the analyte was recovered completely. However, if selenium was added in an organic form, the RV2 value was different from 100% and different from RV1. Apparently, the sample enriched with

Table 10.6 Results of the speciation effect examination in real samples using the SRM method: c_{true}, c_0, c_x – true (declared), real and obtained concentrations of selenium, respectively; RV1, RV2 – recovery coefficients obtained for one- and twofold addition of surrogate analyte to the sample, RE, RE′ – relative errors biased and not biased by the additive preparative effect, respectively (inorganic selenium was used to prepare the calibration graph).

				Results obtained					
Sample	**Surrogate analyte**	c_{true} (μg l^{-1})	**UV lamp**	c_0 (μg l^{-1})	c_x (μg l^{-1})	**RV1 (%)**	**RV2 (%)**	**RE (%)**	**RE′ (%)**
Thermal spring water	Se	60	Off	~14	13.96	87.1	95.3	−76.7	−0.3
Thermal spring water	Se-Met	60	Off	−14	13.96	47.4	54.0	−76.7	−0.3
Vitamin drink	Se	200	Off	~11	2.16	99.5	118.2	−98.9	−80.3
Vitamin drink	Se-Met	200	Off	−11	2.16	47.2	53.9	−98.9	−80.3
Thermal spring water	Se	60	On	~14	14.76	96.9	97.6	−75.4	5.4
Thermal spring water	Se-Met	60	On	−14	14.76	96.2	104.6	−75.3	5.4
Vitamin drink	Se	200	On	~11	10.92	92.9	101.4	−94.5	−0.7
Vitamin drink	Se-Met	200	On	−11	10.92	104.5	64.3	−94.5	−0.7

Source: Adapted from Stafiński et al. [13].

greater addition of the analyte contained too much organic selenium to be digested completely. This assumption was confirmed by the observed speciation effect when the sample spiked with both inorganic and inorganic selenium was not exposed to UV radiation. Namely, the following results: RE $\approx$ RE′ $\neq$ 0%, RV1 = RV1 $\neq$ 100%, but RE $\neq$ RV − 100% may indicate the presence of different forms of selenium in the sample.

It is clear that this type of analysis of results obtained by the SRM method as presented in this chapter is only possible if the concentration of the analyte in the sample being analyzed is known. Thus, the studies presented here do not provide much helpful information on how to proceed in real analytical situations to obtain the best possible results with this method.

Certainly, the SRM method is very simple and can be used in any case, as long as the condition of the sample before processing allows the analyte to be added to it and the sample to be fully homogenized with the addition. It is also important that the added analyte has properties conditioning a behavior similar to that manifested by the native analyte at the preparative stage. If, in addition, this step is not too complex and there are no or very little analytical effects, then the SRM method has a good chance of correctly assessing the magnitude of biases made in the preparative step and accurately determining the analytical result.

However, it has also been proved that the SRM method may be unreliable. From calibration point of view there are some additional conditions critical to be of its working correctly:

- analyte recovery cannot be considered only as the part of the analyte remaining after the preparative effect, but also as the effect of other analytical effects, including interference and speciation effects,
- preparative and interference effects should be considered as two independent and equal factors determining the amount of analyte recovery (consequently, a recovery value significantly higher than 100%, indicating an increase in the amount of analyte in the sample, should also be considered as a consequence of a positive interference effect),
- the method can only be used to assess the accuracy of the analytical result if it is known (based on e.g. literature data or preliminary experiments performed) that preparative and/or interference effects are proportional in nature,
- in case there is certainty or suspicion that the native and surrogate analytes have a different chemical form, additional preparative procedures (chemical or physicochemical) leading to the elimination of this difference should be applied,
- to increase the reliability of the interpretation of the measurement results, the surrogate analyte should be added to the sample strictly according to the rules of the E-SAM method and the function connecting the two measurement points should be treated as a calibration graph leading to an estimation of the analytical result by extrapolation,
- at least two analyte additions should be made to the sample so that it can be verified that the analyte recovery from the sample is equal in each case (i.e. that the calibration graph is linear) [10].

As already mentioned, there is a tendency to treat the SRM method as the primary approach to assess the accuracy of analytical results. However, the arguments presented above argue that this method should be used with much more caution and criticism. Thus, one should try to combine it as often as possible with the use of reference materials, analyte-free or surrogate isotopically modified matrices, and when this is not possible, try to verify the results obtained by alternative approaches (e.g. reference analytical methods).

References

1 Burns, D.T., Danzer, K., and Townshend, A. (2002). Use of the terms "recovery" and "apparent recovery" in analytical procedures. *Pure and Applied Chemistry* 74 (11): 2201–2205.

2 Thompson, M., Ellison, S.L.R., Fajgelj, A. et al. (1999). Harmonised guidelines for the use of recovery information in analytical measurement. *Pure and Applied Chemistry* 71 (2): 337–348.

3 Belardi, R.P. and Pawliszyn, J. (1989). The application of chemically modified fused silica fibers in the extraction of organics from water matrix samples and their rapid transfer to capillary columns. *Water Quality Research Journal* 24 (1): 179–191.

4 Pawliszyn, J. (1997). *Solid-Phase Microextraction-Theory and Practice*. New York: Wiley-VCH.

5 Ouyang, G., Cai, J., Zhang, X. et al. (2008). Standard-free kinetic calibration for rapid on-site analysis by solid-phase microextraction. *Journal of Separation Science* 31 (6, 7): 1167–1172.

6 Ai, J. (1997). Solid phase microextraction for quantitative analysis in nonequilibrium situations. *Analytical Chemistry* 69 (6): 1230–1236.

7 Ouyang, G. and Pawliszyn, J. (2008). A critical review in calibration methods for solid-phase microextraction. *Analytica Chimica Acta* 627 (2): 184–197.

8 Stafiński, M., Wieczorek, M., Janicki, P. et al. (2012). Theoretical and experimental examination of recovery in the context of trueness. *Talanta* 96: 39–43.

9 Kościelniak, P. (2004). On analytical usefulness of the recovery method. *Analytical Letters* 37 (12): 2097–2112.

10 Kościelniak, P. (1999). Some remarks on the recovery method. *Analusis* 27 (4): 381–382.

11 Kościelniak, P. and Wietecha, R. (2003). A critical look at the recovery test as a way for evaluation of analytical accuracy. *Analytical Letters* 36 (4): 861–870.

12 Cutter, G.A. (1983). Elimination of nitrite interference in the determination of selenium by hydride generation. *Analytica Chimica Acta* 149: 391–394.

13 Stafiński, M., Wieczorek, M., and Kościelniak, P. (2013). Influence of the species effect on trueness of analytical results estimated by the recovery test when determining selenium by HG-AFS. *Talanta* 117: 64–69.

11

Calibration-Related Applications of Experimental Plans

11.1 Introduction

Theory of the design of experiments is a widely used methodology [1–3] that originated in the 1930s. It was initially used in agronomy and found technical applications during World War II. The real development of this one came in the 1960s with the development of space exploration. One of the first papers on its use in the field of analytical chemistry was published in 1974 [4]. Since then, it has become one of the basic research in this field.

The outline of the methodology of the design of experiments given below, based mainly on the results obtained by the author, is intended to give an idea of the various possibilities of its use in research related to the issue of analytical calibration. Therefore, the problem of optimization of experimental conditions in analysis will be omitted, although it is worth remembering that in this field experimental plans also find great application.

In general, the method considered takes into account various factors that affect the final outcome of each experiment performed. Unlike the most common research approach, which involves changing a single factor while keeping other factors at a fixed level, the method provides for a simultaneous change in the values of all the factors under study, which allows information to be obtained not only about the influence of individual factors, but also about their interaction. By appropriately planning of experiments, the efficiency of testing is also increased, which in chemistry translates into savings in reagent consumption and waste production. Thus, this methodology harmonizes with the increasingly respected idea of "green chemistry" (including "green analytical chemistry").

In the literature one can find many types of experimental plans, of different structures and sizes, allowing to obtain the required amount and type of information in arbitrarily complex analytical systems. This is aided by computer programs that free the analyst from the need to perform sometimes very tedious calculations. It should be remembered, however, that the more information that is expected about the analytical system under study, the number of experiments required to be performed often increases dramatically. Therefore, a properly adopted research strategy is crucial, consisting in the proper selection of a plan that allows to achieve a specific goal at a given stage with the minimum possible expenditure of time and effort.

Calibration in Analytical Science: Methods and Procedures, First Edition. Paweł Kościelniak.

Proper interpretation of the obtained results is also very important. All this requires from the analyst not only the ability to use appropriate computer programs but also considerable knowledge and experience in analytical chemistry.

Calibration-related experiments based on the chosen plan first consist in preparing a series of standard solutions containing a number n of test substances (factors) at known concentrations fixed at least at two levels: lower, $c_{i,l}$, and higher, $c_{i,u}$ ($i = 1, 2, \ldots, n$). The manner in which the concentrations at each level are determined must be consistent with the so-called planning matrix, i.e. the arrangement of concentrations, $\tilde{c}_i$, expressed in coded units, −1 and +1, corresponding to concentrations $c_{i,l}$ and $c_{i,u}$.

In this study, special attention will be focused on the use of so-called orthogonal factorial experimental plans for analytical calibration. The matrices of these plans, indicating the composition of the standard solutions, adjusted to test two, three, or four components with concentrations fixed at two levels, are shown in Table 11.1. Other orthogonal plans will be presented later in this chapter.

After preparing standard solutions according to a given experimental plan, each solution is subjected to measurements of the intensity of the analytical signal, Y, under conditions characteristic of a particular component, and from this, a polynomial model of general formula is formulated:

$$\hat{Y} = B_0 + \sum_{i=1}^{n} B_i \cdot \tilde{c}_i + \sum_{i,j=1}^{n} B_{ij} \cdot \tilde{c}_i \cdot \tilde{c}_j + \sum_{i,j,k=1}^{n} B_{ijk} \cdot \tilde{c}_i \cdot \tilde{c}_j \cdot \tilde{c}_k + \ldots \quad (11.1)$$

Table 11.1 Preparation of the standard solutions in accordance with experimental plans: 2^2 (2 factors, experiments 1–4), 2^3 (3 factors, experiments 1–8), 2^4 (4 factors, experiments 1–16), and 2^{4-1} (4 factors, experiments 1–8).

Experiment	$\tilde{c}_1$	$\tilde{c}_2$	$\tilde{c}_3$	$\tilde{c}_4$	c_1	c_2	c_3	c_4
1	+1	+1	+1	+1	$c_{1,u}$	$c_{2,u}$	$c_{3,u}$	$c_{4,u}$
2	−1	−1	+1	+1	$c_{1,l}$	$c_{2,l}$	$c_{3,u}$	$c_{4,u}$
3	−1	+1	+1	−1	$c_{1,l}$	$c_{2,u}$	$c_{3,u}$	$c_{4,l}$
4	+1	−1	+1	−1	$c_{1,u}$	$c_{2,l}$	$c_{3,u}$	$c_{4,l}$
5	+1	+1	−1	−1	$c_{1,u}$	$c_{2,u}$	$c_{3,l}$	$c_{4,l}$
6	−1	−1	−1	−1	$c_{1,u}$	$c_{2,l}$	$c_{3,l}$	$c_{4,l}$
7	−1	+1	−1	+1	$c_{1,l}$	$c_{2,u}$	$c_{3,l}$	$c_{4,u}$
8	+1	−1	−1	+1	$c_{1,u}$	$c_{2,l}$	$c_{3,l}$	$c_{4,u}$
9	+1	+1	+1	−1	$c_{1,u}$	$c_{2,u}$	$c_{3,u}$	$c_{4,l}$
10	−1	−1	+1	−1	$c_{1,l}$	$c_{2,l}$	$c_{3,u}$	$c_{4,l}$
11	−1	+1	+1	+1	$c_{1,l}$	$c_{2,u}$	$c_{3,u}$	$c_{4,u}$
12	+1	−1	+1	+1	$c_{1,u}$	$c_{2,l}$	$c_{3,u}$	$c_{4,u}$
13	+1	+1	−1	+1	$c_{1,u}$	$c_{2,u}$	$c_{3,l}$	$c_{4,u}$
14	−1	−1	−1	+1	$c_{1,u}$	$c_{2,l}$	$c_{3,l}$	$c_{4,u}$
15	−1	+1	−1	−1	$c_{1,l}$	$c_{2,u}$	$c_{3,l}$	$c_{4,l}$
16	+1	−1	−1	−1	$c_{1,u}$	$c_{2,l}$	$c_{3,l}$	$c_{4,l}$

where $i \leq j \leq k$, $i < k$. Coefficient B_0 is the average value of the signals $Y_1, \ldots, Y_{n+1}$, and coefficients B_1, B_{ij}, B_{ijk}, ... are calculated as scalar products of the coded concentration vectors, correspondingly $\tilde{c}_i$, $\tilde{c}_i \times \tilde{c}_j$, $\tilde{c}_i \times \tilde{c}_j \times \tilde{c}_k$, and vectors $Y/(n+1)$. In each experiment, the measurement of the analytical signals is repeated several times to estimate the measurement random errors. The equivalence of the variance of the measurement results obtained in successive experiments and the significance of the coefficients B_1, B_{ij}, B_{ijk}, ... in model (11.1) are tested statistically (e.g. using the F-Snedecor test and Student's t-test, respectively).

Coefficients B_1, B_{ij}, B_{ijk}, ... express individual (B_1) and complex (B_{ij}, B_{ijk}) effects of the components on the analytical signal, and their sign indicates the positive or negative direction of a particular interaction. The coded form of the factors simplifies the statistical analysis of these effects, as well as normalizes them (makes them independent of natural concentration values), facilitating their comparison with effects obtained, for example, for the same analyte but determined by a different analytical or measurement method. A model of the form (1) is therefore useful, for example, for studying interference effects.

Model (1) can be transformed to a model that relates the analytical signal to concentrations c expressed in natural units:

$$\hat{Y} = \hat{Y} = B_0 + \sum_{i=1}^{n} B_i \cdot f(c_i) + \sum_{i,j=1}^{n} B_{ij} \cdot f(c_i) \cdot f(c_j) + \sum_{i,j,k=1}^{n} B_{ijk} \cdot f(c_i) \cdot f(c_j) \cdot f(c_k) + \ldots \tag{11.2}$$

where $i \leq j \leq k$, $i < k$, and $f(c)$ is for each value i, j, i, and k a linear function that satisfies the concentration-coding conditions, i.e.: $f(c_l) = -1$ and $f(c_u) = +1$:

$$f(c) = \frac{2c - c_u - c_l}{c_u - c_l} \tag{11.3}$$

Equation (11.2) can be regarded as a calibration model function mapping the real function taking into account the interference effects potentially present in the analytical system under study. If a series of such models is formulated from measurements made under conditions characteristic of each component included in the experimental plan, the multidimensional calibration space thus formulated can be used to perform multicomponent analysis.

All three fields of utilizing experimental plans – the study of interference effects, real function modeling, and multicomponent analysis – will be presented and discussed in the following sections.

11.2 Examination of Interference Effects

The idea of how to design experiments and how to use them is best illustrated by the example of the simplest plan, plan 2^2, shown in Table 11.1. As can be seen, this plan provides for fixing the concentrations of two components at two levels of value in all possible combinations of components and levels. In such cases, the plan is called the total factorial plan, and the number of experiments required is then implied by its name (e.g. $2^2 = 4$). Model (1) obtained from the results of experiments performed according to plan 2^2 takes the form:

$$\hat{Y} = B_0 + B_1 \cdot \tilde{c}_1 + B_2 \cdot \tilde{c}_2 + B_{12} \cdot \tilde{c}_1 \cdot \tilde{c}_2 \tag{11.4}$$

The question arises: how should model (11.4) be interpreted when it is to be applied to the study of interference effects? This interpretation is very well reflected in its graphic illustration presented in two-dimensional space in Figure 11.1. If we assume that the component c_1 is an analyte and c_2 is an interferent, then it can be seen that the coefficient B_0 expresses the average value of the analytical signals measured in all experiments, B_1 – analyte measurement sensitivity, B_2 and B_3 – the interference effect expressed by additive (B_2) and proportional (B_3) change of the analytical signal under the influence of aluminum in the planned concentration range of both components.

As an example of the application of plan 2^2, the study of the influence of aluminum on the determination of calcium by FAAS method, i.e. the interference effect already discussed several times in this book, can be shown. In the work [5], this effect was examined within the limits from 4 to 10 mg l^{-1} of calcium and from 0 to 160 mg l^{-1} of aluminum. The results are presented in Table 11.2.

Based on the obtained measurement results, a model was formulated that provides measurable information about the interference effect in the assumed range of calcium and aluminum concentrations:

$$\hat{Y} = 52.61 + 23.76 \cdot \tilde{c}_1 - 15.88 \cdot \tilde{c}_2 - 6.20 \cdot \tilde{c}_1 \cdot \tilde{c}_2 \tag{11.5}$$

All coefficients in model (11.5) were found to be statistically significant (on 5% significance level). Thus, model (11.5) confirms that aluminum strongly decreases

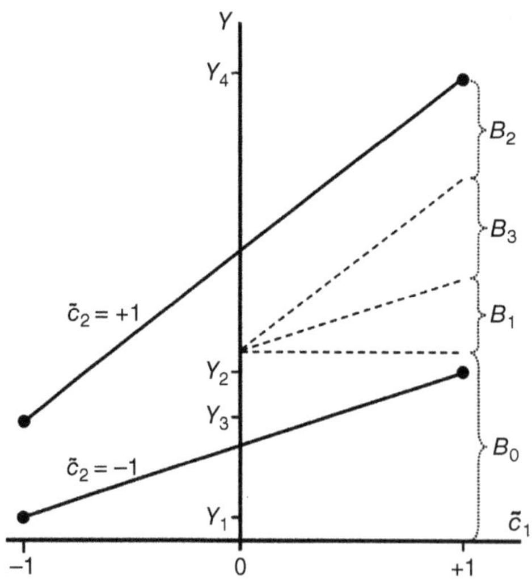

Figure 11.1 Graphic interpretation of model (11.4) as a tool for interference examination; the symbols are as in Table 11.1 and model (11.4).

Table 11.2 Application of 2^2 experimental plan to examination of the effect caused by aluminum (c_2) on the signals measured for calcium (c_1) using FAAS.

Experiment	Coded unites				Natural units ($mg\,l^{-1}$)		Signals Y
	$\tilde{c}_0$	$\tilde{c}_1$	$\tilde{c}_2$	$\tilde{c}_i \times \tilde{c}_j$,	c_1	c_2	
1	+1	+1	+1	+1	10	160	54.29
2	+1	+1	−1	−1	10	0	98.44
3	+1	−1	+1	−1	4	160	19.18
4	+1	−1	−1	+1	4	0	38.53

Source: Kościelniak et al. [5], table 1 (p. 115)/with permission of Elsevier.

the analytical signal measured for calcium and further provides an opportunity to evaluate the additive and proportional component of this effect.

It is very important to interpret the determined effects only with respect to the range of concentrations of the test components assumed in the experimental plan. In particular, this applies to the additive component of the interference effect. Thus, the detection of this component in the concentration range of 4–10 $mg\,l^{-1}$ calcium only means that aluminum additively reduces the analytical signal only in this range of analyte concentrations. To find out whether this effect of aluminum on calcium is additive in general it would be necessary to examine it over the full linear range of analyte concentration including the zero concentration.

Using an appropriately chosen type 2^k factorial plan, the effect of a larger number of potential interferents on the signal measured for the analyte can be investigated in an analogous manner as shown in the example of plan 2^2. The plans shown in Table 11.1 offer a reasonable number of experiments sufficient to thoroughly investigate the effect of one, two, or three interferents.

When there is a need to determine the effects of a larger number of interferents on the analyte, the use of a 2^k-type plan may be too long and tedious (e.g. to study a system with seven components, 27 = 128 experiments are needed). In this situation, a different type of two-level plan requiring fewer experiments can be used. An example is the so-called Plackett–Burman plan [6], which for detecting the effect of, e.g. six interferents on a selected analyte provides for performing only eight experiments. The planning matrix applicable in this case is shown in Table 11.3.

The model formulated from this plan has the form:

$$\hat{Y} = B_0 + B_1 \cdot \tilde{c}_1 + B_2 \cdot \tilde{c}_2 + B_3 \cdot \tilde{c}_3 + B_4 \cdot \tilde{c}_4 + B_5 \cdot \tilde{c}_5 + B_6 \cdot \tilde{c}_6 + B_7 \cdot \tilde{c}_7 \quad (11.6)$$

As can be seen, the plan does not provide a way to separately assess the additive and proportional nature of the effects. It should be emphasized that each coefficient in model (11.6) is a combined measure of both types of effects. Thus, tests made using this plan provide only a general detection and evaluation of interference effects, without providing detailed information about their nature.

Table 11.3 Matrix of the seven-factor Plackett–Burman plan.

Experiment	$\tilde{c}_1$	$\tilde{c}_2$	$\tilde{c}_3$	$\tilde{c}_4$	$\tilde{c}_5$	$\tilde{c}_6$	$\tilde{c}_7$
1	+1	+1	+1	+1	+1	+1	+1
2	+1	+1	−1	−1	−1	−1	+1
3	+1	−1	+1	+1	−1	−1	−1
4	+1	−1	−1	−1	+1	+1	−1
5	−1	+1	+1	−1	+1	−1	−1
6	−1	+1	−1	+1	−1	+1	−1
7	−1	−1	+1	−1	−1	+1	+1
8	−1	−1	−1	+1	+1	−1	+1

Table 11.4 Application of the Plackett–Burman plan to examination of mutual interferences between seven main components of dolomite basing on measurements performed by FAAS; the effects significant on 5% level are bolded.

Kind of effect	Effect value detected at conditions specific for the individual analytes					
	Fe	Mg	Ca	Na	K	Mn
$\tilde{c}_{Fe}$	**5.010**	−0.009	0.013	0.010	−0.177	−0.042
$\tilde{c}_{Mg}$	−0.013	**1.468**	**0.058**	−0.001	**1.244**	0.022
$\tilde{c}_{Ca}$	0.009	**0.018**	**3.980**	−0.009	**0.358**	−0.005
$\tilde{c}_{Na}$	−0,039	**0.012**	−0.020	**1.950**	**0.250**	0.049
$\tilde{c}_{K}$	0.032	−0.001	0.008	**0.019**	**10.276**	0.098
$\tilde{c}_{Mn}$	0.019	0.001	−0.007	−0.005	0.016	**4.976**
$\tilde{c}_{Al}$	−0.018	**−0.009**	**−0.059**	−0.002	0.023	0.015

Source: Adapted from Kościelniak et al. [7].

The plan is so efficient and economical that measurements using once prepared solutions can be made not only under conditions characteristic of one selected component but also for other components under various conditions strictly defined for them. In other words, under certain conditions, each component can be treated as an analyte and the others as interferents. An example of the results obtained in this way for the study of mutual interference of seven components is presented in Table 11.4 [7]. The collected data indicate that in the predetermined concentration ranges, iron and manganese are the least susceptible to interference from other elements, and magnesium is the most influenced by calcium, sodium, and aluminum.

The results obtained using the Plackett–Burman plan can be treated as preliminary data on the effects occurring in a given analytic system. From these, a more detailed investigation of the effects of one, two, or three selected components on a particular analyte can be made using plans 2^2, 2^3, or 2^4, respectively, shown in Table 11.1.

Thus, to determine the type of effect of calcium, sodium, and aluminum detected earlier on the analytical signal of magnesium (i.e. to find out what type of interference is hidden by the coefficients B_{Ca}, B_{Na}, and B_{Al}), the factorial plan 2^4 was used (see Table 11.1) and the following model (where only statistically significant coefficients were considered) was formulated based on the measured results obtained for these solutions:

$$\hat{Y} = B_0 + 1.550 \cdot \tilde{c}_{Mg} + 0.077 \cdot \tilde{c}_{Ca} + 0.056 \cdot \tilde{c}_{Al} + 0.035 \cdot \tilde{c}_{Mg}\tilde{c}_{Na} + 0.032 \cdot \tilde{c}_{Mg}\tilde{c}_{Ca} - 0.082 \cdot \tilde{c}_{Na}\tilde{c}_{Al} + 0.039 \cdot \tilde{c}_{Mg}\tilde{c}_{Ca}\tilde{c}_{Al} \quad (11.7)$$

The use of plan 2^4 thus confirmed the partially additive effect of calcium and aluminum, although it also showed their proportional effect. Furthermore, both metals produce a common synergistic effect on the signal measured for magnesium. In contrast, sodium shows only a compound effect and not only in the presence of the analyte but also in the presence of aluminum alone.

If there is a need to investigate nonlinear effects in a given analytical system, experiments should be performed using a three-level factorial plan. Examples of the simplest plans of this type are shown in Table 11.5. The concentration of each component in successive standard solutions should then be fixed at a third, so-called central level, $c_{i,c}$, whose value should be the mean value of the extreme concentrations ($c_{i,c} = (c_{i,l} + c_{i,u})/2$; $i = 1, 2$).

Based on the measurement results, a second-order polynomial model is then formed, which for plan 3^2 takes the form:

$$\hat{Y} = B_0 + B_1 \cdot \tilde{c}_1 + B_2 \cdot \tilde{c}_2 + B_3 \cdot \tilde{c}_1 \cdot \tilde{c}_2 + B_4 \cdot \tilde{c}_1^{\,2} + B_5 \cdot \tilde{c}_2^{\,2} + B_6 \cdot \tilde{c}_1^{\,2} \cdot \tilde{c}_2 + B_7 \cdot \tilde{c}_1 \cdot \tilde{c}_2^{\,2} + B_8 \cdot \tilde{c}_1^{\,2} \cdot \tilde{c}_2^{\,2} \quad (11.8)$$

Thus, after nine performed experiments, it is possible to distinguish and evaluate the mutual influence of the two components expressed not only by additive and

Table 11.5 Experimental plans: 2 × 3 (experiments 1–6) and 3^2 (experiments 1–9).

	Coded units		Natural unites	
Experiment	$\tilde{c}_1$	$\tilde{c}_2$	c_1	c_2
1	−1	−1	$c_{1,l}$	$c_{2,l}$
2	+1	−1	$c_{1,u}$	$c_{2,l}$
3	−1	+1	$c_{1,l}$	$c_{2,l}$
4	+1	+1	$c_{1,u}$	$c_{2,u}$
5	−1	0	$c_{1,l}$	$c_{2,u}$
6	+1	0	$c_{1,u}$	$c_{2,c}$
7	0	−1	$c_{1,c}$	$c_{2,c}$
8	0	+1	$c_{1,c}$	$c_{2,u}$
9	0	0	$c_{1,u}$	$c_{2,u}$

proportional effects but also by all nonlinear effects (approximated by a parabolic function) that may take place in the analytical system under study.

11.3 Modeling of Real Functions

As mentioned, experimental plans can be used to formulate model functions that relate analytical signals to the concentrations of the test components that act as analytes and interferents. Models of the general formula (11.2) are used for this purpose. If one of the components is treated as an analyte and the others as interferents, the model can be interpreted as a set of multidimensional model functions (calibration surfaces) representing the dependence of the analytical signal on the concentrations of the individual components within the limits of the values determined in the experimental plan. If these surfaces are to be further used to determine analytical results in multicomponent analysis, they should of course be the best possible representation of the true function.

Consideration of this question is again best begun with the simplest plan 2^2 allowing us to formulate a model of the following form:

$$\hat{Y} = B_0 + B_1 \cdot f\left(c_1\right) + B_2 \cdot f\left(c_2\right) + B_{12} \cdot f\left(c_1\right) \cdot f\left(c_2\right) \tag{11.9}$$

where $f(c_i)$ $(i = 1, 2)$ is a linear function of the form (11.3). If the signals are measured under conditions characteristic of the component c_1, the function (11.9) is a model function approximating the true dependence of the analytical signal on the concentration of the analyte, c_1, and the interferent, c_2, in their concentration intervals $[c_{1,l}, c_{1u}]$ and $[c_{2,l}, c_{2u}]$. Because the two components are only at two concentration levels, this function is linear with respect to both components. Therefore, it can only be used to model the real function if the determinations are performed in a linear concentration range and the predicted interferences are also linear and relatively small (because then there is a chance of maintaining this linearity over the analyte concentration range considered).

In order for the model function to account for the nonlinearity of the dependence of the analytical signal on both analyte and interferent concentrations, both components should be at least three concentration levels. Thus, the tests should be performed using a factorial plan 3^2, whereby the concentrations of the components are set at levels c_1, c_c, and c_u and a second-order polynomial model of the form (11.8) is formulated. Transformation of this model to a model expressed in real units by the linear formula (3) satisfying the coding conditions: $(f(c_1) = -1, f\left(c_c\right) = 0$, and $f(c_u) = +1)$ requires that the concentration c_c be set equal to the mean value of the extreme concentrations $(c_c = (c_l + c_u/2))$. After such a transformation, the model maintains its mathematical character of a second-degree polynomial (parabolic function).

However, it is obvious that a parabolic function passing through three points corresponding to equally distant concentrations will not always be a sufficiently accurate approximation of the real function, especially with respect to modeling strong and complex interference effects. This is evidenced by the experimental results shown

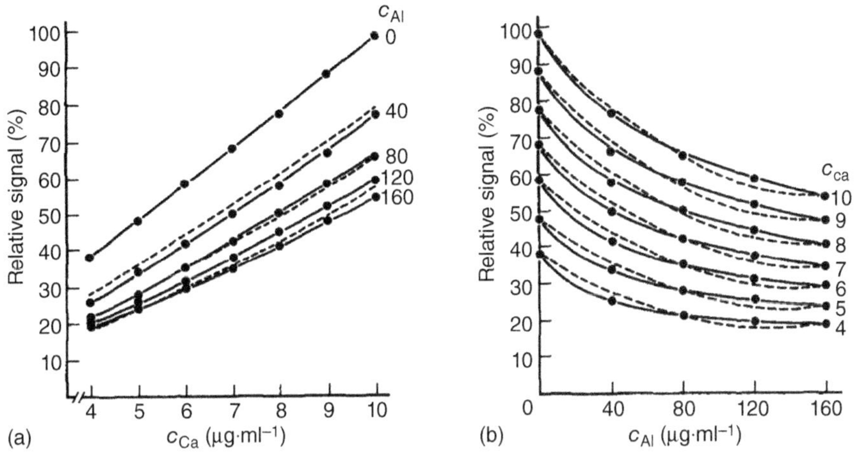

Figure 11.2 Approximation of real function (points) in the ranges from 4 to 10 mg ml^{-1} of Ca (a) and from 0 to 160 mg ml^{-1} of Al (b) using model functions formulated on the base of 3^2 plan with linear (dashed lines) and logarithmic (solid lines) transformation of concentrations. Source: Kościelniak et al. [5], fig 3 (p. 114)/with permission of Elsevier.

in Figure 11.2 obtained in the calcium–aluminum system during measurements performed by the FAAS method [5]. Thus, while plan 3^2 is helpful in evaluating nonlinear interference effects, its use for modeling the real function may be unreliable.

In work [8] this problem was solved by formulating nonlinear functions, $f(c, q)$, of various forms that allow the transformation of coded to real concentrations according to the basic coding conditions: $\tilde{c}(c_l) = -1$, $\tilde{c}(c_c) = 0$, and $\tilde{c}(c_u) = +1$. They have the primary advantage of providing an arbitrary choice of concentration values c_c between c_l and c_u levels. Formulas for several such nonlinear transformation functions are given in Table 11.6.

Nonlinear transformation functions are used such that a central concentration level, c_c, within the limits of the extreme levels, c_l and c_u, is selected and the value

Table 11.6 Formulas for nonlinear transformations of concentrations $\tilde{c} = f(c, q)$ satisfying the coding conditions required in the theory of experiments.

Function f	$\tilde{c} =$	$q =$
Parabolic	$\frac{q \cdot (c_u - c_l) \cdot (c - c_u) \cdot (c - c_l) + 2c - c_u - c_l}{c_u - c_l}$	$\frac{c_u + c_l - 2c_c}{(c_u - c_l) \cdot (c_c - c_u) \cdot (c_c - c_l)}$
Exponential	$\frac{2e^{q \cdot c} - e^{q \cdot c_u} - e^{q \cdot c_l}}{e^{q \cdot c_u} - e^{q \cdot c_l}}$	$\frac{1}{c_c} \cdot \ln \frac{e^{q \cdot c_u} + e^{q \cdot c_l}}{2}$
Hyperbolic	$\frac{2c - c_u - c_l + q \cdot (c_u - c_l)}{c_u - c_l + q \cdot (2c - c_u - c_l)}$	$\frac{c_u + c_l - 2c_c}{c_u - c_l}$
Logarithmic	$\log_q \frac{(q^2 - 1) \cdot (c - c_l) + c_u - c_l}{q \cdot (c_u - c_l)}$	$\frac{c_u - c_c}{c_c - c_l}$

of the parameter q is calculated from this. A function with this parameter is used to transform model (1) to model (2), which then takes the general form:

$$\hat{Y} = B_0 + \sum_{i=1}^{n} B_i \cdot f_i\left(c_i, q_i\right) + \sum_{i,j=1}^{n} B_{ij} \cdot f_i\left(c_i, q_i\right) \cdot f_2\left(c_j, q_j\right)$$
$$+ \sum_{i,j,k=1}^{n} B_{ijk} \cdot f_1\left(c_i, q_i\right) \cdot f_2\left(c_j, q_j\right) \cdot f_3\left(c_k, q_k\right) + \ldots \tag{11.10}$$

where $i \leq j \leq k, i < k$. In case when $c_c = (c_l + c_u)/2$, parameter q takes the value 0, and any nonlinear transformation function becomes a linear function.

Thanks to nonlinear transformations, the modeling of the real function becomes much more flexible than in the situation of having only a linear transformation. This is due not only to the possibility of arbitrary choice of both the value of central concentration and the type of nonlinear function in relation to each considered component. It should also be emphasized that the use of this type of concentration transformation does not have to involve the performance of additional experiments beyond those provided by the chosen experimental plan.

However, the next problem remains: how to choose the concentration $c_{i,c}$ so that the model (6) after a nonlinear transformation describes the real function, and in particular the effect of interferents on the analytical signal, better than after a linear transformation? Investigations of interference effects in the FAAS method in this respect have shown that the criterion is very simple: if the transformation is done with an exponential, hyperbolic, or logarithmic function, it is best if the concentration of the interferent at c_c corresponds approximately to the mean value of the signals obtained for the extreme concentrations, c_l and c_u, of that interferent [9].

The results of this study were verified, among others, by the example of the effect of aluminum on calcium, shown in Figure 11.2 [5]. It can be seen that the nonlinear (logarithmic) transformation of the concentrations in model 3^2, combined with the selection of the central concentration level according to the above criterion, allows a significant improvement in the fit of the model function to the real function in this analytical system.

When it is certain that the dependence of the analytical signal on the concentration of analytes is linear in the investigated concentration range of interferents, and only the interference effects are presumably nonlinear, the number of experiments necessary for modeling can be reduced by using the so-called mixed plans. These allow selected factors to be set at two and others at three concentration levels, thus taking into account the linear and nonlinear nature of the relationships under study, respectively. If two components are involved, plan 2×3 requiring six rather than nine experiments can be used in place of plan 3^2. A schematic of this plan (as part of plan 3^2) is shown in Table 11.5, and an example of its use for successful modeling of the real function is shown in Figure 11.3 [10].

Interestingly, in some cases, modeling of the relationship analytical signal vs. interferent concentration on the basis of a linear function (i.e. resulting from assuming two levels of interferogram concentration) with a nonlinear transformation of the variables may yield better results than when a transformation of this type is

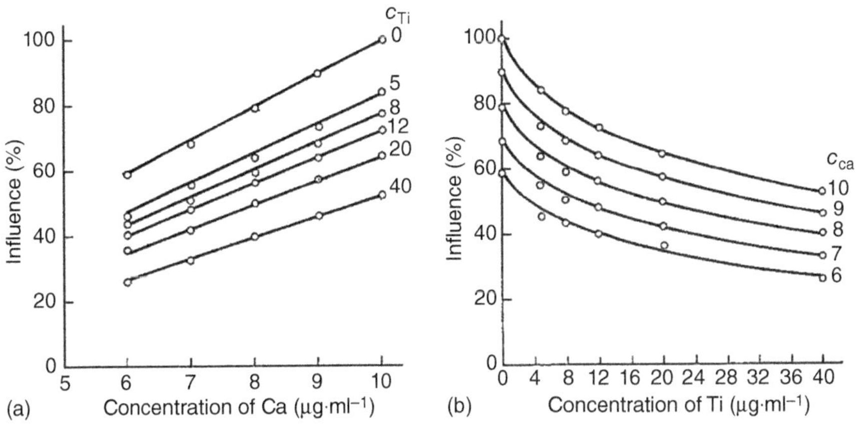

Figure 11.3 Modelling (solid lines) of real function (points obtained by FAAS) in the ranges from 6 to 10 mg ml^{-1} of Ca (analyte) (a) and from 0 to 40 mg ml^{-1} of Ti (b) using 2×3 plan with linear and logarithmic transformations of the analyte and interferent concentrations, respectively. Source: Kościelniak et al. [10], fig 2 (p. 105)/with permission of Elsevier.

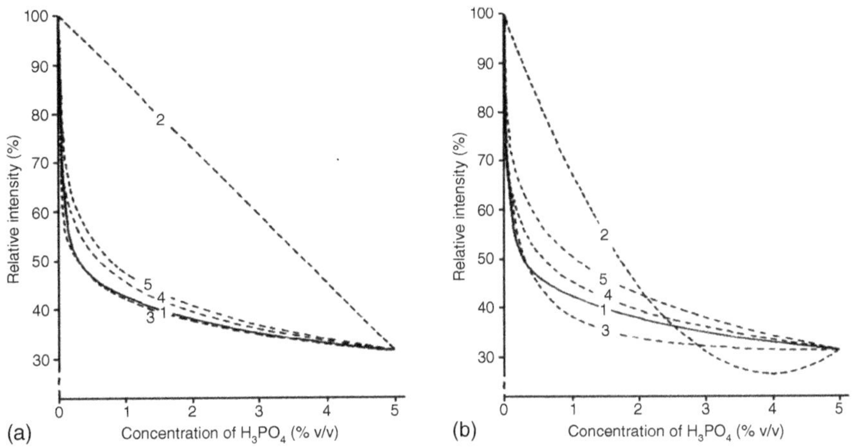

Figure 11.4 Modeling of real function (solid line 1) expressing effect of H_3PO_4 on the Ca signal (measured by FAAS) using first-order (a) and second-order (b) models with linear (2) and logarithmic (3, 4, 5) transformation of concentrations and with central concentration c_c equal to 0.375 (3), 0.025 (4), and 0.1 mg l^{-1} (5) of H_3PO_4 [8].

applied to a second-order model transformation (i.e. formulated for three levels of interferogram concentration). In other words, overlaying the parabolic nature of the function (11.8) with a transformation function of another type (e.g. logarithmic) may over interpret the form of the real function. An example of such a situation is presented by Figure 11.4.

Figure 11.4 also shows that the selection tolerance of the central concentration, c_c, within the limits of the extreme interferent concentrations, c_l and c_u, in the nonlinear transform function is larger for linear than for nonlinear modeling. Thus, all

Table 11.7 Extended 2^2 plan used for nonlinear (hyperbolic) modeling of the real function in two-component analytical system.

No. of experiment	Coded unites		Natural units		Signals
	$\tilde{c}_1$	$\tilde{c}_1$	c_1	c_2	
1	−1	−1	$c_{1,l}$	$c_{2,l}$	Y_1
2	+1	+1	$c_{1,u}$	$c_{2,l}$	Y_3
3	−1	−1	$c_{1,l}$	$c_{2,u}$	Y_2
4	+1	+1	$c_{1,u}$	$c_{2,u}$	Y_4
5	q_1'	q_1'	$c_{1,c}$	$c_{2,u}$	Y_5
6	q_1''	q_1''	$c_{1,c}$	$c_{2,u}$	Y_6
7	−1	−1	$c_{1,l}$	$c_{2,c}$	Y_7
8	+1	+1	$c_{1,u}$	$c_{2,c}$	Y_8

indications are that real functions can be efficiently approximated by models formulated using two-level experimental plans with a nonlinear transformation (preferably exponential, hyperbolic, or logarithmic).

To avoid having to choose a central concentration to formulate the optimal value of the parameter q of the nonlinear transformations, another way of formulating the nonlinear model functions can be used. It requires performing additional experiments to the number required in the experimental plan. For example, plan 2^2 can be extended to include an additional four experiments in which one of the components is at its minimum, c_l, or maximum, c_u, and the other component is at c_c, which is the average value of its extreme concentrations ($c_c = (c_l + c_u)/2$). Thus, the entire study requires the performance of eight experiments according to the plan presented in Table 11.7.

The measurement results, Y_1, Y_2, Y_3, and Y_4, obtained in the experiments according to plan 2^2 allow the determination of the values of the coefficients B_0, B_1, B_2, and B_{12} in the model:

$$\hat{Y} = B_0 + B_1 \cdot f\left(c_1, q_1\right) + B_2 \cdot f\left(c_2, q_2\right) + B_{12} \cdot f\left(c_1, q_1\right) \cdot f\left(c_2, q_2\right) \tag{11.11}$$

Values Y_5, Y_6, Y_7, and Y_8 are used to determine the parameter values of the selected nonlinear transformation function (Table 11.6). In this case, the hyperbolic transformation is recommended because, as can be seen that this function for the value $c = c_c = (c_l + c_u)/2$ takes the simple form $\tilde{c} = q$, and for $c = c_l$ and $c = c_u$ takes the values $\tilde{c} = -1$ and $\tilde{c} = +1$, respectively.

$$\hat{Y}_5 = B_0 - B_1 + B_2 \cdot q_2' - B_{12} \cdot q_2' \tag{11.12}$$

$$\hat{Y}_6 = B_0 + B_1 + B_2 \cdot q_2'' + B_{12} \cdot q_2'' \tag{11.13}$$

$$\hat{Y}_7 = B_0 + B_1 \cdot q_1' - B_2 - B_{12} \cdot q_1' \tag{11.14}$$

$$\hat{Y}_8 = B_0 + B_1 \cdot q_1'' + B_2 + B_{12} \cdot q_1'' \tag{11.15}$$

From equations pairs (11.12), (11.13), and (11.14), (11.15) the values of $q_2 = (q_2' + q_2'')/2$ and $q_1 = (q_1' + q_1'')/2$ can be calculated, respectively. The final form of the model is thus as follows:

$$\begin{aligned}\hat{Y} = B_0 + B_1 \cdot & \frac{2 \cdot c_1 - c_{1,u} - c_{1,l} + q_1 \cdot (c_{1,u} - c_{1,l})}{c_{1,u} - c_{1,l} + q_1 \cdot (2 \cdot c_1 - c_{1,u} - c_{1,l})} \\ + B_2 \cdot & \frac{2 \cdot c_2 - c_{2,u} - c_{2,l} + q_1 \cdot (c_{2,u} - c_{2,l})}{c_{2,u} - c_{2,l} + q_2 \cdot (2 \cdot c_2 - c_{2,u} - c_{2,l})} \\ + B_{12} \cdot & \frac{2 \cdot c_1 - c_{1,u} - c_{1,l} + q_1 \cdot (c_{1,u} - c_{1,l})}{c_{1,u} - c_{1,l} + q_1 \cdot (2 \cdot c_1 - c_{1,u} - c_{1,l})} \cdot \frac{2 \cdot c_2 - c_{2,u} - c_{2,l} + q_1 \cdot (c_{2,u} - c_{2,l})}{c_{2,u} - c_{2,l} + q_2 \cdot (2 \cdot c_2 - c_{2,u} - c_{2,l})}\end{aligned} \tag{11.16}$$

It is worth noting that the interpretation of the parameter q in the hyperbolic transformation function is the same as that of the parameter Q defined by Eq. (6.11): it is a measure of the nonlinearity of the function $f(c, q)$ and takes on a positive or negative value depending on whether the function $f(c, q)$ is of convex or concave shape, respectively, with respect to the linear function connecting the points corresponding to concentrations c_l and c_u. The above method of nonlinear modeling thus consists in determining the average value of the degree of nonlinearity and the shapes (convex or concave) of the dependence of the analytical signal on the concentration of the analyte (q_1) and the interferent (q_2) on the basis of the degrees of nonlinearity of these dependences determined for the extreme values of the concentrations, c_l and c_u, of the two components.

The above modeling method is particularly convenient for use in multicomponent analysis because it provides a good representation of the real function, while at the same time it offers the possibility of determining the concentration of components in the analyzed sample using a relatively simple mathematical way.

11.4 Multicomponent Analysis

Mathematical models formulated from experimental plans can be the basis for performing multicomponent analyses. If the analysis is to concern n components of the sample under study, a series of m models (where $m \geq n$) of the type (11.2) should be formulated using standard solutions, performing measurements of the signal intensities Y under m conditions characteristic of each component under consideration. In analytical practice, multicomponent analysis procedures are usually limited to multicomponent analysis. The examples presented below, therefore, concern the use of experimental plans to implement this type of analysis. The general principle of the procedure will be presented on the example of a two-component analysis realized by means of a complete and an extended plan 2^2.

If the dependences of the signals $Y^{(j)}$ on the concentrations c_i $(i, j = 1, 2)$ is assumed to be linear, a series of four standard solutions is prepared according to plan 2^2, and then, under two conditions characteristic of both analytes, measurements are made

Table 11.8 Scheme of two-component analysis performed on the basis of 2^2 and extended 2^2 experimental plans.

Solution	Coded units		Natural units		Signals	
	$\tilde{c}_1$	$\tilde{c}_2$	c_1	c_2		
Standard solutions	−1	−1	$c_{1,l}$	$c_{2,l}$	$Y_1^{(1)}$	$Y_1^{(2)}$
	+1	−1	$c_{1,u}$	$c_{2,l}$	$Y_2^{(1)}$	$Y_2^{(2)}$
	−1	+1	$c_{1,l}$	$c_{2,u}$	$Y_3^{(1)}$	$Y_3^{(2)}$
	+1	+1	$c_{1,u}$	$c_{2,u}$	$Y_4^{(1)}$	$Y_4^{(2)}$
Standard solutions	$q_1^{(1)\prime}$	−1	$c_{1,c}$	$c_{2,l}$	$Y_5^{(1)}$	–
	$q_1^{(2)\prime}$	−1	$c_{1,c}$	$c_{2,l}$	–	$Y_5^{(2)}$
	$q_1^{(1)\prime\prime}$	+1	$c_{1,c}$	$c_{2,u}$	$Y_6^{(1)}$	–
	$q_1^{(2)\prime\prime}$	+1	$c_{1,c}$	$c_{2,u}$	–	$Y_6^{(2)}$
	−1	$q_2^{(1)\prime}$	$c_{1,l}$	$c_{2,c}$	$Y_7^{(1)}$	–
	−1	$q_2^{(2)\prime}$	$c_{1,l}$	$c_{2,c}$	–	$Y_7^{(2)}$
	+1	$q_2^{(1)\prime\prime}$	$c_{1,u}$	$c_{2,c}$	$Y_8^{(1)}$	–
	+1	$q_2^{(2)\prime\prime}$	$c_{1,u}$	$c_{2,c}$	–	$Y_8^{(2)}$
Sample	–	–	c_{x1}	c_{x2}	$Y_x^{(1)}$	$Y_x^{(2)}$

for each standard solution $Y_1^{(1)}, \ldots, Y_4^{(1)}, Y_1^{(2)}, \ldots, Y_4^{(2)}$ and for the analyzed sample $Y_x^{(1)}, Y_x^{(2)}$ (see Table 11.8). The analyte concentration ranges in the standard solutions, $[c_{1,l}, c_{1,u}]$ and $[c_{2,l}, c_{2,u}]$, should encompass the presumed concentrations of these components, c_{x1} and c_{x2}, in the analyzed sample.

The formulated calibration model functions then have the general form:

$$\hat{Y}^{(1)}\hat{Y}_x\hat{Y}^{(1)} = B_0^{(1)} + B_1^{(1)} \cdot f\left(c_1, q_1^{(1)}\right) + B_2^{(1)} \cdot f\left(c_2, q_2^{(1)}\right) + B_{12}^{(1)} \cdot f\left(c_1, q_1^{(1)}\right) \cdot f\left(c_2, q_2^{(1)}\right) \tag{11.17}$$

$$\hat{Y}^{(2)} = \hat{Y}^{(2)} = B_1^{(2)} \cdot f\left(c_1, q_1^{(2)}\right) + B_2^{(2)} \cdot f\left(c_2, q_2^{(2)}\right) + B_{12}^{(2)} \cdot f\left(c_1, q_1^{(2)}\right) \cdot f\left(c_2, q_2^{(2)}\right) \hat{Y}_x^{(2)} \tag{11.18}$$

where coefficients $B_0^{(j)}$, $B_i^{(j)}$, and $B_{ij}^{(j)}$ $(i, j = 1, 2)$ are calculated by means of the values of the signals obtained for the standard solutions, and the transformation function $f(c_i, q_i^{(j)})$ is linear (Eq. (11.3)), so the parameters $q_i(j)$ $(i, j = 1, 2)$ take the zero value. For signals $Y_x^{(1)}$ and $Y_x^{(2)}$ measured for both analytes present in the sample at concentrations c_{x1} and c_{x2}, models (11.17) and (11.18) take the form of equations:

$$Y_x^{(1)} = B_0^{(1)} + B_1^{(1)} \cdot f\left(c_{x1}\right) + B_2^{(1)} \cdot f\left(c_{x2}\right) + B_{12}^{(1)} \cdot f\left(c_{x1}\right) \cdot f\left(c_{x2}\right) \tag{11.19}$$

$$\hat{Y}_x^{(2)} = B_0^{(2)} + B_1^{(2)} \cdot f\left(c_{x1}\right) + B_2^{(2)} \cdot f\left(c_{x2}\right) + B_{12}^{(2)} \cdot f\left(c_{x1}\right) \cdot f\left(c_{x2}\right) \tag{11.20}$$

From the equation systems (11.19) and (11.20), the concentration values of c_{x1} and c_{x2} can be calculated.

If the dependencies of the signals $Y^{(j)}$ on the concentrations c_i $(i, j = 1, 2)$ are or can be nonlinear, then the values of the parameters $q_i^{(j)}$ in formulas (11.17) and (11.18) must be determined. As described in previous sections, this can be done in two ways. One is to choose a nonlinear transformation function $f(c_i, q_i^{(j)})$ (Table 11.8) with a form suitable for the presumed nonlinear nature of the relationship (in the extreme case, four different nonlinear functions $f(c_i, q_i^{(j)})$ can be used in models (11.17) and (11.18)). The parameters $q_i^{(j)}$ are determined from the values of the central concentrations corresponding to the mean values of the signals measured for the extreme concentrations of each component in the standard solutions. As mentioned, the estimation of each value of $q_i^{(j)}$ should be approached individually, preferably in the form of additional preliminary experiments.

The second way of determination of parameters $q_i(j)$ in formulas (11.17) and (11.18) requires preparation of additional four standard solutions according to the extended plan 2^2, in which concentrations of both analytes are set at levels $c_{i,c} = (c_{i,l} + c_{i,u})/2$ $(i = 1, 2)$ and subject these solutions to measurements of signals $Y_5^{(1)}, \ldots, Y_8^{(1)}, Y_5^{(2)}, \ldots, Y_8^{(2)}$ (see Table 11.6). If one assumes that all the dependencies of the signals $Y^{(j)}$ on the concentrations c_i can be described by a hyperbolic function, then the signals $Y_5^{(1)}, \ldots, Y_8^{(1)}, Y_5^{(2)}, \ldots, Y_8^{(2)}$ allow the determination of the parameters $q_i^{(j)\prime}$ and $q_i^{(j)\prime\prime}$ $(i, j = 1, 2)$ along the path described by formulas (11.12)–(11.15). Finally, the nonlinearity of the relationship $Y^{(j)}$ vs. c_i is estimated by $q_i^{(j)} = (q_i^{(j)\prime} + q_i^{(j)\prime\prime})/2$.

Regardless of how the parameters $q_i^{(j)}$ are determined, the analytical results c_{x1} and c_{x2} are determined from a system of equations:

$$\begin{aligned} Y_x^{(1)} = {} & B_0^{(1)} + B_1^{(1)} \cdot f\left(c_{x1}, q_1^{(1)}\right) + B_2^{(1)} \cdot f\left(c_{x2}, q_2^{(1)}\right) \\ & + B_{12}^{(1)} \cdot f\left(c_{x1}, q_1^{(1)}\right) \cdot f\left(c_{x2}, q_2^{(1)}\right) \end{aligned} \tag{11.21}$$

$$\begin{aligned} Y_x^{(2)} = {} & B_0^{(2)} + B_1^{(2)} \cdot f\left(c_{x1}, q_1^{(2)}\right) + B_2^{(2)} \cdot f\left(c_{x2}, q_2^{(2)}\right) \\ & + B_{12}^{(2)} \cdot f\left(c_{x1}, q_1^{(2)}\right) \cdot f\left(c_{x2}, q_2^{(2)}\right) \end{aligned} \tag{11.22}$$

A graphical representation of this calibration procedure is shown in Figure 11.5.

What analytical advantage results from the above calibration procedure? The answer is obvious: Eqs. (11.19) and (11.20) allow for the determination of two analytical results taking into account the interference effects caused by each other's two components. The components thus appear both as analytes and as potential interferents. Analysis of this type, therefore, leads to results of theoretically higher accuracy than those obtained by single-component analysis.

Experimental examples for the realization of two-component analysis in the manner described are numerous in the literature. Quite unusual and worth special attention is the use of plan 2^2 for calibration in titration analysis. It has been found, for example, that the results of complexometric titration (with EDTA) of Fe^{3+} and Fe^{2+} ions with potentiometric detection are subject to large systematic errors when both analytes are determined from a single titration curve and reactions proceed through nonequilibrium states [11]. The data in Table 11.9 shows that the two-component calibration allows the occurring mutual interferences of the two analytes to be accounted for and determined with greater accuracy.

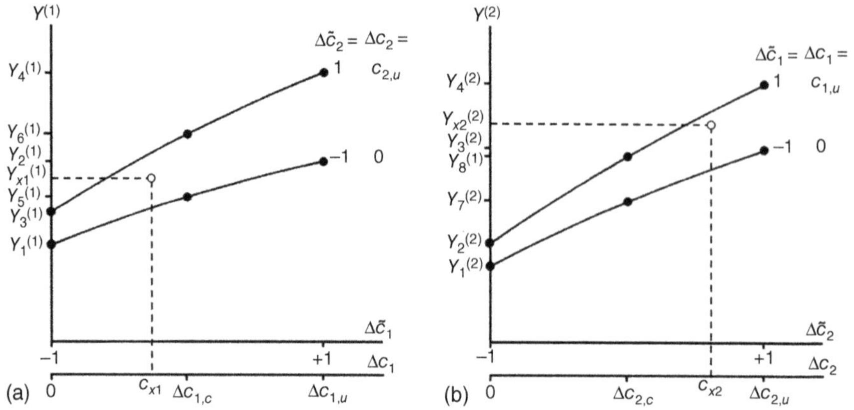

Figure 11.5 Determination of first (a) and second (b) analyte in two-component analysis performed by the external standard method in accordance to the extended 2^2 experimental plan.

Table 11.9 Results of simultaneous determination of Fe^{3+} and Fe^{2+} by complexometric titration in synthetic samples using 2^2 experimental plan; c_0 – expected results, c_x – found results obtained with no interference effects and all effects are considered in calculations .

c_0 (μg)		c_x (μg)			
		No effects considered		Mutual effects	
Fe^{3+}	Fe^{2+}	Fe^{3+}	Fe^{2+}	Fe^{3+}	Fe^{2+}
199	422	228	369	198	421
597	422	616	388	593	436
398	633	437	560	394	636
300	319	310	275	299	310
499	319	512	283	497	321
300	530	332	498	304	549

Source: Adapted from Kochana et al. [11].

The use of experimental plans to compensate for interference effects in multicomponent analysis is also possible when sample components are determined by different measurement methods. Some models can then be formulated on the basis of signals measured by a method suitable for one group of components, while other models are formulated by a method more suitable for other components. This division must of course be maintained when measurements are made for standard solutions and samples.

This type of analysis was used to compensate for the effects of zinc, nickel, and iron on the results of copper determination by the titration (complexometric) method [12]. A four-component calibration was performed using plan 2^{4-1} with the matrix shown in Table 11.1. Copper was determined by titration while zinc,

Table 11.10 Results of complexometric titration of Cu in the synthetic samples containing Zn, Ni, and Fe and of the simultaneous determination of all four metals (Zn, Ni, and Fe by FAAS) in the same samples using 2^{4-1} experimental plan; c_0, c_x – expected and found results, respectively [12].

c_0 (μg)				c_x (μg)				
				Titration	Multicomponent analysis			
Cu	Zn	Ni	Fe	Cu	Cu	Zn	Ni	Fe
90.0	6.0	8.0	3.6	91.3	89.9	10.3	8.2	3.6
60.0	6.0	5.0	1.2	60.7	60.3	5.9	5.1	1.2
120.0	10.0	5.0	2.4	121.3	120.6	6.1	5.1	2.3
90.0	14.0	3.0	3.6	91.5	90.0	4.1	3.1	3.5
120.0	4.0	8.0	1.8	121.1	120.6	13.2	7.6	1.7
60.0	12.0	3.0	3.0	61.0	59.7	4.0	3.1	2.9

nickel, and iron were determined by FAAS. As can be seen from the data in Table 11.10, this procedure had the expected effect of improving the accuracy of copper determination in a number of samples analyzed and "by the way" allowed the determination of the other components in these samples also with very good accuracy.

Noteworthy here is the use of the 2^{4-1} plan, which is an example of a so-called fractional plan. Each 2^k plan consists of two 2^{k-1} fractional plan, which are twinned in the sense that they are identical in terms of the arrangement of $k-1$ factor levels, with the remaining factor appearing in the two plans mutually at different levels (-1 or $+1$). Thus, the 2^{k-1} plan allows us to examine k components in experiments with twice the number of experiments that result from the full 2^k plan. Reducing the number of experiments results in a loss of information in the sense of overlapping effects occurring in the system. However, a fractional plan can be generated from the 2^k plan that provides more relevant information in a given case. In the research described above, a 2^{4-1} plan was used resulting in the following overlapping effects: Cu with $\tilde{c}_{Zn}\tilde{c}_{Ni}\tilde{c}_{Fe}$, $\tilde{c}_{Zn}$ with $\tilde{c}_{Cu}\tilde{c}_{Ni}\tilde{c}_{Fe}$, $\tilde{c}_{Ni}$ with $\tilde{c}_{Cu}\tilde{c}_{Zn}\tilde{c}_{Fe}$ and $\tilde{c}_{Fe}$ with $\tilde{c}_{Cu}\tilde{c}_{Zn}\tilde{c}_{Ni}$ as well as $\tilde{c}_{Cu}\tilde{c}_{Zn}$ with $\tilde{c}_{Ni}\tilde{c}_{Fe}$, $\tilde{c}_{Cu}\tilde{c}_{Ni}$, with $\tilde{c}_{Zn}\tilde{c}_{Fe}$ and $\tilde{c}_{Cu}\tilde{c}_{Fe}$, with $\tilde{c}_{Zn}\tilde{c}_{Ni}$, assuming beforehand, that all ternary and binary interactions between components different from copper are so insignificant that they do not play a major role in modeling the real function.

A separate interesting aspect is the possibility to prepare standard solutions of composition according to the chosen experimental plan in the flow mode in a mechanized way, which obviously speeds up and facilitates the calibration procedure at the laboratory stage. An example of a suitable instrumental solution is the already described (see Figure 7.4) fully rotary valve (FRV) with the configuration shown in Figure 11.6. It is readily seen that if a component of concentration c_1 is supplied to two channels of the stator, a component of concentration c_2 to two

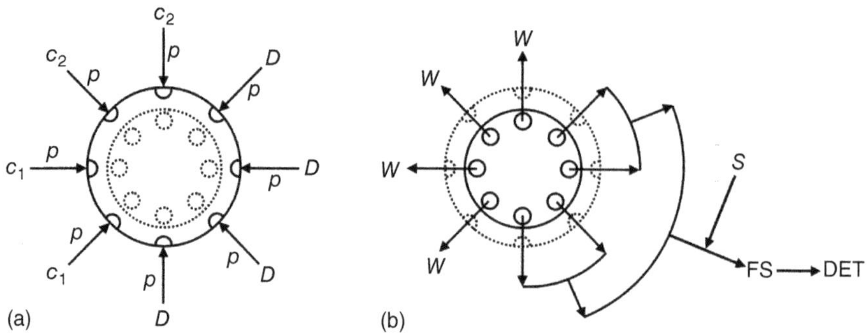

Figure 11.6 Fully rotary valve with stator (a) and rotor (b) used for preparation of a set of standard solutions in accordance with the incomplete 3^2 experimental plan: c_1, c_2 – concentrations of two components in standard solutions, S – sample, D – diluent, W – waste, FS – flow system, DET – detection system, p – flow rate. Source: Kościelniak et al. [13], fig 1 (p. 230)/with permission of John Wiley & Sons.

adjacent channels, and diluent D to the remaining channels (Figure 11.6a) and all solutions flow at a uniform velocity p, then after a further eight rotations of the rotor (Figure 11.6b) in a clockwise direction, standard solutions with compositions according to the extended experimental plan 2^2 shown in Table 11.8 will be introduced into the flow system. After the measurements for the standard solutions have been made, the sample, S, can be introduced into the system through a separate tube (as indicated in Figure 11.6b).

The results of the spectrophotometric determination of Cr^{6+} and Co^{2+} ions with the use of the FRV valve are presented in Table 11.11 [13]. Models (11.17) and (11.18) were created at two wavelengths, and the predicted nonlinear nature of the mutual interferences of both analytes was taken into account by hyperbolic concentration transformation and determination of parameters $q_i(j)$ based on the results of experiments supplementing plan 2^2 (as described in the previous chapter). Not only is a significant improvement in the accuracy of the analytical results obtained after applying the model function formulated in this way deserving attention, but also a very good precision of these results, which was certainly due to the automatic mode of preparing standard solutions.

The calibration procedures presented so far follow the external standard method because the standard solutions are prepared separately from the samples and the analytical results, as shown in Figure 11.5, are determined by interpolation. This begs the question: Is it possible to use experimental plans to calibrate with the standard addition method (E-SAM)? From a theoretical point of view, the intended purpose of including the E-SAM method in this type of procedure would be to enhance the efficiency of compensation for interference effects exhibited by the components being determined, as well as to compensate for potential interference effects caused by other components of the sample being analyzed.

The idea of creating multivariate polynomial models to compensate for interference effects by the standard addition method appeared in 1979 under the name "generalized standard addition method" (GSAM) [14]. Several years later, the theory

Table 11.11 Results of simultaneous spectrophotometric determination of Co^{2+} and Cr^{6+} in synthetic samples using extended 2^2 plan; c_0 – expected results, c_x – found results obtained with no interference effects, and all effects are considered in calculations; in parenthesis, the RSD values (%) are given.

c_0 (mmol l^{-1})		c_x (mmol l^{-1})			
		No effects		Total effects	
Co^{2+}	Cr^{6+}	Co^{2+}	Cr^{6+}	Co^{2+}	Cr^{6+}
50	20	59.8 (1.1)	19.5 (0.9)	50.7 (1.4)	19.5 (0.6)
100	20	107.3 (0.4)	22.7 (0.6)	98.8 (0.9)	20.1 (1.0)
50	30	70.3 (0.4)	30.8 (2.0)	49.1 (1.5)	30.4 (2.1)
100	30	119.4 (0.1)	33.1 (0.4)	100.0 (0.4)	30.3 (0.7)
50	40	79.5 (0.6)	35.5 (0.1)	49.7 (0.7)	39.0 (0.1)
100	40	130.3 (0.6)	43.5 (0.1)	102.0 (0.4)	40.6 (0.4)

Source: Adapted from Kościelniak et al. [13].

of the design of experiments was adapted to implement this method [15], and a calibration procedure for the GSAM method in multicomponent analysis with linear and nonlinear concentration transformations was developed [16]. This procedure will be described by using the total and extended plan 2^2 as an example.

In this case, the planning matrix indicates the composition of the standard solutions added to the sample. The lower concentration level of the analyte in the calibration solution, $c_{i,l}$, should therefore be equal to 0, and the possible central level, $c_{i,c}$, considered according to the extended plan 2^2 should be half the concentration value determined at the higher level, $c_{i,u}$, $(i = 1, 2)$. The complete scheme of how to prepare the calibration solutions is shown in Table 11.12.

On the basis of measurements carried out in conditions characteristic of both analytes, the models are formulated of the form:

$$\hat{Y}^{(1)} = B_0^{(1)} + B_1^{(1)} \cdot f\left(\Delta c_1, q_1^{(1)}\right) + B_2^{(1)} \cdot f\left(\Delta c_2, q_2^{(1)}\right) + B_{12}^{(1)} \cdot f\left(\Delta c_1, q_1^{(1)}\right) \cdot f\left(\Delta c_2, q_2^{(1)}\right) \tag{11.23}$$

$$\hat{Y}^{(2)} = B_0^{(2)} + B_1^{(2)} \cdot f\left(\Delta c_1, q_1^{(2)}\right) + B_2^{(1)} \cdot f\left(\Delta c_2, q_2^{(2)}\right) + B_{12}^{(2)} \cdot f\left(\Delta c_1, q_1^{(2)}\right) \cdot f\left(\Delta c_2, q_2^{(2)}\right) \tag{11.24}$$

where the coefficients $B_0^{(j)}$, $B_i^{(j)}$, and $B_{ij}^{(j)}$ $(i, j = 1, 2)$ are calculated using the values of the signals obtained for the calibration solutions prepared according to plan 2^2. The transformation functions $f(c_i, q_i^{(j)})$ $(i, j = 1, 2)$ can be linear (see Eq. (11.3)) or nonlinear. In the latter case, the parameters $q_i^{(j)}$ are determined, as before, from the values of the signals obtained for the solutions prepared according to an extension of Plan 2^2 or the values of the central concentrations corresponding to the average values of the signals measured for the sample and the sample with additions c_i of both analytes. Finally, the concentrations of the analytes in the sample, c_{x1} and c_{x2}

Table 11.12 Scheme of two-component analysis performed on the basis of 2^2 experimental plan in accordance with the generalized standard addition method.

	Coded unites of additions		Natural units of additions			
Solution	$\tilde{c}_1$	$\tilde{c}_1$	Δc_1	Δc_1	**Signals**	
Sample + additions	−1	−1	0	0	$Y_x^{(1)}$	$Y_x^{(2)}$
	+1	−1	$\Delta c_{1,u}$	0	$Y_2^{(1)}$	$Y_2^{(2)}$
	−1	+1	0	$\Delta c_{2,u}$	$Y_3^{(1)}$	$Y_3^{(2)}$
	+1	+1	$\Delta c_{1,u}$	$\Delta c_{2,u}$	$Y_4^{(1)}$	$Y_4^{(2)}$
	$q_1^{(1)'}$	−1	$\Delta c_{1,u}/2$	0	$Y_5^{(1)}$	–
	$q_1^{(2)'}$	−1	$\Delta c_{1,u}/2$	0	–	$Y_5^{(2)}$
	$q_1^{(1)''}$	+1	$\Delta c_{1,u}/2$	0	$Y_6^{(1)}$	–
	$q_1^{(2)''}$	+1	$\Delta c_{1,u}/2$	0	–	$Y_6^{(2)}$
	−1	$q_2^{(1)'}$	0	$\Delta c_{2,u}/2$	$Y_7^{(1)}$	–
	−1	$q_2^{(2)'}$	0	$\Delta c_{2,u}/2$	–	$Y_7^{(2)}$
	+1	$q_2^{(1)''}$	0	$\Delta c_{2,u}/2$	$Y_8^{(1)}$	–
	+1	$q_2^{(2)''}$	0	$\Delta c_{2,u}/2$	–	$Y_8^{(2)}$

are determined by extrapolation from a system of equations:

$$0 = B_0^{(1)} + B_1^{(1)} \cdot f\left(-c_{x1}, q_1^{(1)}\right) + B_2^{(1)} \cdot f\left(-c_{x2}, q_2^{(1)}\right) + B_{12}^{(1)} \cdot f\left(-c_{x1}, q_1^{(1)}\right) \cdot f\left(-c_{x2}, q_2^{(1)}\right) \tag{11.25}$$

$$0 = B_0^{(2)} + B_1^{(2)} \cdot f\left(-c_{x1}, q_1^{(2)}\right) + B_2^{(1)} \cdot f\left(-c_{x2}, q_2^{(2)}\right) + B_{12}^{(2)} \cdot f\left(-c_{x1}, q_1^{(2)}\right) \cdot f\left(-c_{x2}, q_2^{(2)}\right) \tag{11.26}$$

A graphical representation of the measurement and transformation step of this calibration procedure is shown in Figure 11.7.

The GSAM method in this version was applied, among others, to the determination of sodium and calcium by FAES [16]. This is a particularly difficult analytical system to master, because in the adopted concentration ranges the dependence of analytical signals on the concentration of both analytes is strongly nonlinear, and additionally they exhibit mutual interference effects. This is seen in Figure 11.8. Under these difficult conditions the results obtained with the two-component GSAM method proved to be more accurate than the determinations of both analytes performed for each of them separately with the one-component E-SAM method. This is demonstrated by the results summarized in Table 11.13.

The above-described procedures of the external standard method and the standard addition method implemented according to the theory of the design of experiments can, of course, be adapted to the simultaneous determination of any number of analytes on the basis of any chosen experimental plan. It should be kept in mind, however, that as the analytical system is extended, not only does the number of

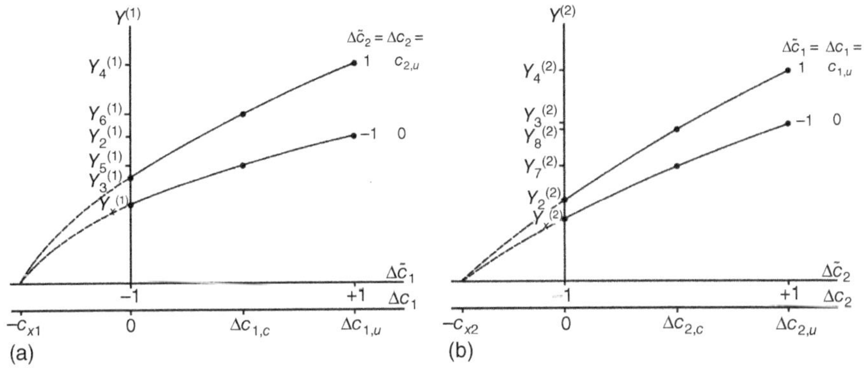

Figure 11.7 Determination of first (a) and second (b) analyte in two-component analysis performed by the generalized standard addition method in accordance to the extended 2^2 experimental plan.

Table 11.13 Results of the determination of Na and Ca by flame emission spectrometry in the synthetic samples using calibration by the E-SAM method and two-component GSAM method based on the extended 2^2 experimental plan.

Concentrations added, Δc_u (mg l^{-1})		Expected concentrations, c_0 (mg l^{-1})		Obtained concentrations, c_x (mg l^{-1})			
				E-SAM		GSAM	
Na	Ca	Na	Ca	Na	Ca	Na	Ca
8	200	4	200	4.63	97.2	4.11	97.6
16	200	4	200	4.37	97.2	3.80	101.3
8	400	4	200	4.63	86.6	4.03	88.1
16	400	4	200	4.37	86.6	3.91	87.2

Source: Adapted from Kościelniak [16].

experiments required increase, but also the difficulty of accurately representing a multivariate real function by means of formulated model functions increases, and the advantage of multicomponent analysis over single- or multicomponent analysis in terms of the ability to obtain accurate analytical results decreases.

It is also clear that the experimental plans can be used as the basis for performing analytical calibrations with the various methods described in this book (e.g. the indirect method or the internal standard method). Nor is there any obstacle to their being adapted to determinations made by a variety of measurement methods. It is therefore a fully versatile analytical tool. Certainly, the growth of interest in these issues is conditioned, among other things, by the availability of commercial mathematical programs to facilitate the necessary calculations. Freeing the analyst from this step of the calibration procedure will allow greater focus on the selection of an appropriate experimental plan and the adaptation of a suitable calibration method to the analytical system under study.

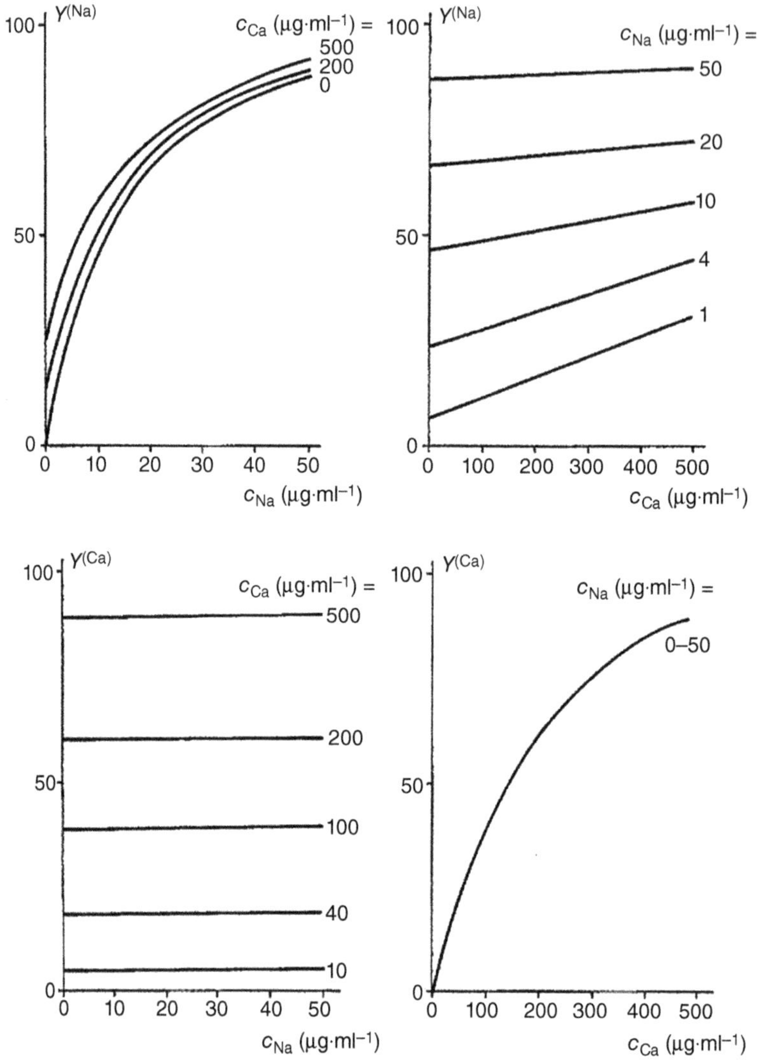

Figure 11.8 Relationships between the signal intensities measured by FAES for sodium, $Y^{(Na)}$, and calcium, $Y^{(Ca)}$, and the concentration of both analytes, c_{Na} and c_{Ca}. Source: Kościelniak et al. [16], fig 2 (p. 374)/with permission of Springer Nature.

References

1 Cochran, W.C. and Cox, G.M. (1950). *Experimental Designs*. New York: Wiley.
2 Gupta, A.K. and Kabe, D.G. (2013). *Design and Analysis of Experiments*. World Scientific Publishing Co Pte Ltd.
3 Montgomery, D.C. (2019). *Design and Analysis of Experiments*, 10e. Wiley.

4 Pravčeva, C.A., Delijska, A., and Ninova, V. (1974). Rapid spectrographic determination of calcium fluoride in slags (in German). *Fresenius' Zeitschrift für analytische Chemie* 272 (2): 120–123.

5 Kościelniak, P. and Parczewski, A. (1983). Empirical modelling of the matrix effect in atomic absorption spectrometry. *Analytica Chimica Acta* 153: 111–119.

6 Plackett, R.L. and Burman, J.P. (1946). The design of optimum multifactorial experiments. *Biometrika* 33 (4): 305–325.

7 Kościelniak, P., Walas, S., and Parczewski, A. (1980). On the matrix effect in the determination of main components of dolomite by atomic flame emission and absorption spectrometry. *Fresenius' Zeitschrift für analytische Chemie* 302 (5): 402–406.

8 Kościelniak, P. and Parczewski, A. (1983). Non-linear transformation of factors in the design of experiments. *Analytica Chimica Acta* 153: 103–109.

9 Kościelniak, P. and Parczewski, A. (1985). Efficient application of non-linear transformation of factors in empirical modelling based on experimental designs. *Analytica Chimica Acta* 177: 197–202.

10 Kościelniak, P. and Parczewski, A. (1985). Modelling of the interference effect based on $k_1 \cdot k_2 \cdot \ldots \cdot k_n$ factorials. *Chemia Analityczna (Warsaw)* 30 (1): 49–54.

11 Kochana, J., Madej, K., and Parczewski, A. (1992). Determination of Fe(II)–Fe(III) and Fe(III)–Ni(II) by the two-component potentiometric complexometric titration. *Fresenius Journal of Analytical Chemistry* 342: 54–57.

12 Kościelniak, P. and Parczewski, A. (1988). Application of a combination of analytical techniques in the multicomponent analysis of complex materials. *Chemia Analityczna (Warsaw)* 33 (1): 55–60.

13 Kościelniak, P. and Herman, M. (2000). Flow system for calibration and interfrence examination in multicomponent analysis. *Laboratory Robotics and Automation* 12: 228–235.

14 Saxberg, B.E.H. and Kowalski, B.R. (1979). Generalized standard addition method. *Analytical Chemistry* 51 (7): 1031–1038.

15 Kościelniak, P. and Parczewski, A. (1984). The application of the theory of experimental designing in the generalized standard addition method. *Microchimica Acta* 83: 359–367.

16 Kościelniak, P. (1984). Non-linear transformations of factors in the generalized standard addition method. *Microchimica Acta* 83: 369–380.

12

Final Remarks

The issues addressed in this book are based on a new generalized approach to analytical calibration. It consists in adopting the thesis that the calibration process cannot be a natural operation of transforming the analytical signal into the type or amount of analyte in the sample, but only an imperfect and approximate representation of this operation.

This vision corresponds to the essence of analytical research. The goal of the analyst's work is to look deep into matter and determine the type or quantity of the constituents it contains. Every chemical analysis is thus, in a sense, a search for the objective truth that the reality around us conceals. This goal can never be fully achieved, but it can be approached as closely as possible with the tools created by the analyst. Chemical analysis is thus a confrontation of what is unattainable (true, theoretical) with what can be created and achieved in practice.

Analytical calibration as a part of chemical analysis is subject to this logic. This is expressed by the concepts of the model function and the standard as approximations of the real function and the object (sample) under study, respectively. Treating these concepts as immanently related to analytical calibration, one easily comes to the conclusion that, depending on the type of model function, analytical calibration can be empirical or theoretical, and the standard can be not only a chemical standard but also a biological, physical, or mathematical one. Practice shows that, in many cases, the empirical and theoretical modeling elements intersect and support each other, as is reflected in Figure 12.1, which illustrates the general division of analytical calibration.

It is not impossible that further development of analytical methods will demonstrate the possibility of performing analytical calibration with standards of yet another kind. It is unlikely, however, that analysts will soon be completely free of the calibration problem. It seems that direct determination of the type or amount of the examined components of a sample will be feasible only when the measuring instruments will be able to detect them on the atomic and molecular level.

Assuming now that a standard of any type can be used for analytical calibration, it must also be assumed that any analytical method is involved in this process. "Calibratability" of an analytical method cannot depend on either the type of standard or the type of model function and in particular on whether the model function is empirical or theoretical. Nor does it depend on whether the model is based on strictly

Calibration in Analytical Science: Methods and Procedures, First Edition. Paweł Kościelniak.

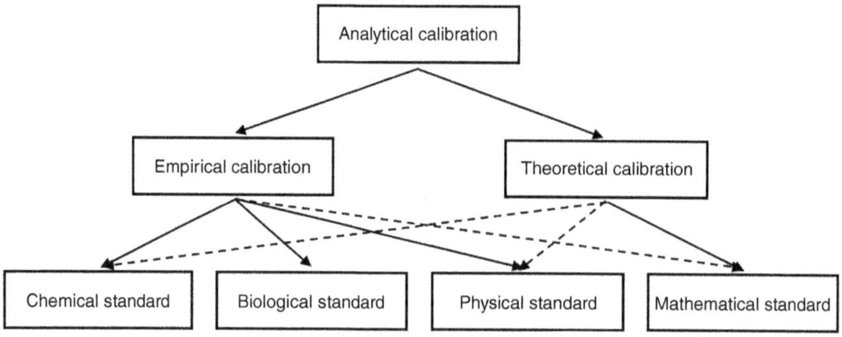

Figure 12.1 General structure of analytical calibration.

defined laws and rules, or whether it is formulated on the basis of a mathematical description of those processes and phenomena which the analyst considers most relevant and, in a given case.

The degree to which the analytical target is approached depends on the magnitude of the measurement errors made, that is, the random and systematic variations in the measurement signal, referred to in this paper as uncontrolled analytical effects. Calibration brings its own analytical errors, of course. This is due to the contribution of the standard, always inherently imperfect, introducing laboratory and measurement errors (in empirical calibration) or mathematical approximations and corrections (in theoretical calibration). The imperfection of the calibration standard results in the imperfection of the model function, and this in turn results in lower precision and accuracy of the analytical results.

On the other hand, as already mentioned and demonstrated, the calibration process can contribute to improving the precision and accuracy of analytical results through the process of compensating for uncontrolled analytical effects. Compensation occurs through the broadly defined similarity of standard and sample, which can be achieved in various ways through specific preparation of the standard and sample and appropriate handling during the measurement and interpretation stages. In other words, analytical errors caused by specific uncontrolled effects can be minimized by selecting an appropriate calibration procedure or method.

The book shows that the analyst can use quite a variety of calibration methods in both qualitative and quantitative analysis. Figures 3.1 and 4.1 provide a breakdown of the empirical calibration methods based on a chemical standard that is most commonly used in both cases. It may come as a surprise that the great similarity of these methods in both areas of chemical analysis, and consequently the possibility of giving them similar names and making a similar classification, has not yet been recognized. It is perfectly feasible, therefore, that in view of the unification of chemical analysis thus demonstrated, there will be a tendency in the course of years to further commonality between these methods and to take advantage of the analytical advantages which they offer.

How should individual calibration methods be viewed among the whole family of methods currently in use? Certainly, despite all the objections expressed earlier, the external standard method (ESM) should be considered as the basic calibration method. It is now and certainly will be in the future the most commonly used method both with chemical and other standards. This situation is favored primarily by its most natural and simple principle, which consists in the separate preparation (laboratory or mathematical) of standards in relation to the sample without (theoretically) having to support the calibration process by other laboratory or mathematical procedures. This gives the possibility to analyze multiple samples with the same standard (or series of standards).

The popularity of the external standard method (ESM) and other comparative calibration methods fosters a natural inclination to perform analysis as simply and quickly as possible. This is especially true when analyzing large sample series. It is worth noting that official analytical regulations (e.g. ISO standards) recommend that laboratories perform analyses using almost exclusively this calibration method. In the absence of uncontrolled effects or the presence of small effects easily eliminated by simple means, accurate analytical results can indeed be obtained quickly and efficiently by this method. In more difficult cases, other calibration methods could be of great help, even if used against established habits or recommendations. It seems that many problems with the accuracy of analytical results are due, among other things, to an overreliance on the ESM method and its routine use.

These problems are greater than they might appear and have serious social and material consequences. It has been revealed, for example, that as many as 30% of the analyses performed in industrial laboratories in the US are so erroneous that they need to be repeated, at a cost of up to $15 million per year [1]. In 2012, it was found that hundreds of innocent people may have been convicted in murder or rape cases in the US since the 1970s based on faulty laboratory analyses [2]. Other U.S. data indicate that most therapeutic misdiagnoses are due to improper selection, preparation, application, and interpretation of analytical procedures [3]. There is no doubt that similar phenomena occur in every other country, and their magnitude could be less if analytical procedures were more flexible in terms of the ability to select and apply different calibration methods.

In quantitative analysis, the interference effects (IEs) pose the greatest threat to the accuracy of analytical results. In principle, one can never be absolutely certain whether they occur in a given case and what their magnitude is since this depends not only on the type of analyte and interferents but also on their absolute and relative concentrations. Thus, an analytical method considered accurate for a given type of sample may fail if applied to samples of similar qualitative composition but different quantitative composition. In such a situation, the ESM method with simultaneous application of different ways to eliminate these effects may also fail. This is the reason that most other empirical calibration methods have arisen in quantitative analysis from the need to minimize errors due to IEs. Table 12.1 collects such approaches with the specific benefits that their use can provide. In the author's opinion, wider use of these calibration procedures could greatly improve the analytical error situation outlined above.

Undoubtedly, of particular importance among noncomparative calibration methods is the standard addition method in extrapolation version (E-SAM). Its

Table 12.1 Calibration methods for solving problems related to interference effects.

Task in relation to the interference effect (IE)	Method/procedure	Sections
Detection of IE	Test for general matrix effect	9.1
Detection of IE	Dilution method	4.1.1
Detection of IE	Integrated calibration method	8.3.1
Compensation for IE	Dilution method	4.1.1
Compensation for IE	Generalized calibration method	8.3.2
Compensation for proportional IE	Extrapolative standard addition method	6.1.1
Compensation for specific IE	Interpolative standard addition method	6.1.2
Compensation for specific IE	Indicative standard addition method	6.1.3
Compensation for additive IE	Multicalibration methods	9.2
Avoidance of IE through chemical reaction	Indirect method Titration	4.1.1 6.2
Avoidance of IE through the use of isotopes	Isotope dilution methods	6.3
General examination of IE	Experimental designing	11.1

usefulness is fully revealed in quantitative analysis. It provides compensation for uncontrolled effects provided that they are proportional in nature (and such are most often encountered in practice). These valuable properties can be given to other calibration methods by combining them with the E-SAM method in a single calibration procedure. The analytical value of the E-SAM method is best illustrated by the broad utility of various procedures involving this method (see Table 12.2).

The ability to combine calibration methods is certainly not getting the attention and use in analytical practice that it deserves. A good example of this is the mixed method, which can excellently help to minimize analytical errors for batch analyses. Integration of the E-SAM procedure into the determination of an analyte in samples of similar composition by the ESM method (i.e. analyses so often performed in different laboratories) could in many cases eliminate the need for reagents to eliminate interferences. Such a procedure is also more in line with the principles of green analytical chemistry.

In the author's opinion, more attention in everyday analytical work should also be paid to additive IEs in the sense of their detection and subsequent elimination or compensation. This is all the more important because, as we know, effects of this type cannot be compensated by any single calibration method, and their elimination by chemical means is also difficult and not always effective. In this context, the advantages of multicalibration methods, and the HPSAM method, in particular, cannot be overestimated. A serious limitation of this method is the stringent and difficult experimental conditions under which it demonstrates its effectiveness. It is

Table 12.2 The utility of procedures combining the E-SAM method with other calibration methods (or with it alone).

Multitask	Method/procedure	Sections
Compensation for proportional IE + avoidance of IE through chemical reaction	Extrapolative indirect method	8.1.1
Compensation for proportional IE + compensation for random signal fluctuations	Extrapolative internal standard method Standard dilution analysis	8.1.2 8.3.3
Compensation for proportional IE + compensation for IE of other types	Extrapolative dilution method Generalized calibration method	8.1.3 8.3.2
Compensation for proportional IE + compensation for additive effect	H-point standard addition method	9.2.2
Compensation for proportional IE + compensation for preparative effect	Surrogate recovery method	10.2
Compensation for proportional IE in a series of similar samples	Mixed method (E-SAM + external standard method)	8.2
Compensation for proportional IE in a series of similar samples + avoidance of IE through chemical reaction	Mixed method (E-SAM + indirect method)	8.2
Compensation for proportional IE in a series of similar samples + compensation for random signal fluctuations	Mixed method (E-SAM + internal standard method)	8.2

therefore necessary to develop an increasing number of analytical rules indicating the type of instrumental or chemical parameters to be taken into account in a given analytical method so that potential additive effects occurring in that method can be effectively compensated.

The question may be asked: in what direction will further developments in the science and practice related to analytical calibration occur? Certainly, further, probably increasingly fruitful efforts should be expected in the direction of theoretical calibration, in particular concerning quantitative analysis. Progress in this area will probably be parallel to the development of knowledge of the physicochemical phenomena and processes characteristic of individual measurement methods. Evidence of the usefulness of this type of work is provided by the increasingly excellent mathematical modeling of such complex processes as occur in the LIBS method. It should be assumed that a great help in these efforts may be the introduction of some elements of empirical calibration to the theoretical calibration to a limited extent.

In the field of empirical calibration, the author does not expect spectacular achievements in the sense of developing completely new calibration methods.

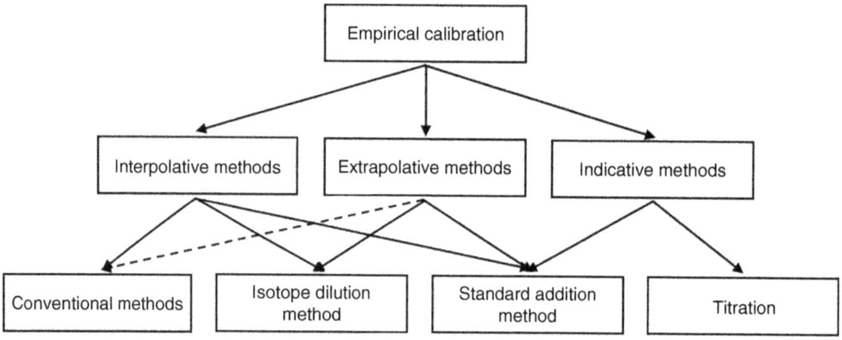

Figure 12.2 Classification of empirical calibration methods in quantitative analysis due to the method of transformation of measurement signals into analytical results.

One can suppose that new calibration approaches will rather be modifications or combinations of methods already developed. If one looks at empirical methods in qualitative analysis from the side of how the measurement signal is transformed into an analytical result (see Figure 12.2), one can see that in particular there is an opportunity to create calibration procedures that are indicative in nature. Taking the titration procedure as a model, one can think that such approaches should (at least theoretically) lead to results of increased precision and accuracy.

No less important than the development of new calibration approaches is the creation of conditions for performing empirical calibration as correctly as possible, leading to precise and accurate analytical results. In this context, therefore, any activity that increases the number and variety of certified reference materials should be considered particularly valuable. Increased availability of these materials will allow them to be used more frequently as standards similar to the analyzed samples in terms of chemical composition. With mass spectrometry playing an increasingly important role in quantitative analysis as a stand-alone and hyphenated detection system, it is also very important to have easier access to isotopic standards – isotopically enriched analogs and isotopically labeled analogs. In qualitative analysis, the creation of more and more extensive computer databases dedicated to various measurement methods is of paramount importance.

Modern analytical chemistry also requires care for the natural environment. In the case of analytical chemistry, however, there is a risk that excessive compliance with the rules of green chemistry may result in the diminishing attention of analysts to other, equally important features of analytical methods. Therefore, the sustainable development of analytical chemistry in all its areas, including analytical calibration, is very important. On the basis of this reasoning, the idea of "white analytical chemistry"(WAC) [4] has recently emerged, according to which the analytical method should be characterized – in addition to environmental friendliness – with an equally high reliability and analytical effectiveness.

Looking at the calibration methods from this point of view, it can be concluded that the approaches that are closest to the WAC idea are those that are implemented with a single standard solution, do not require the use of additional reagents, and resistance to uncontrolled effects is gained thanks to the operations of adding the standard to the sample and diluting it. best done in flow-through fish automatically. Examples of such calibration procedures are the extrapolative dilution method [5] or the standard dilution analysis [6]. The author hopes that this type of modern calibration methods will gain more and more recognition over time and will be permanently introduced into laboratory practice.

For this to happen, however, efforts must be made to increasingly promote flow techniques that provide mechanization and automation of sample and standard processing. The examples presented in the book show that, in addition to aspects important from the point of view of green analytical chemistry (reduced sample and reagent consumption, less waste produced, increased work safety), flow techniques allow calibrations to be performed in various original ways, reducing the time and cost of the calibration procedure, and contributing to improved precision and accuracy of analytical results. Thus, in themselves, they favor the development of the idea of WAC. It can be noted, however, that the scientific development in flow analysis, which has been noted for many years, does not translate – globally speaking – into adequate interest on the part of practicing analysts.

Above all, however, much more needs to be done to ensure that analytical calibration issues are seen by analysts in the proper light they deserve. They are, as has been tried to demonstrate, a fundamental and essential element of both qualitative and quantitative analysis. Therefore, in the author's opinion, calibration problems should enter – on an equal footing with other basic issues in analytical chemistry – into the compulsory canon of subjects at chemistry faculties of higher education, and also be an object of particular importance in other types of analytical courses. More scientific and popular scientific articles, as well as textbook and monographic studies should also be devoted to this topic. The author hopes that the present study is a convincing proof of such a rank and place of calibration in analytical science and practice.

References

1 Hertz, H.S. (1988). Are quality and productivity compatible in the analytical laboratory? *Analytical Chemistry* 60 (2): 75A–85A.

2 https://www.washingtonpost.com/local/crime/fbi-overstated-forensic-hair-matches-in-nearly-all-criminal-trials-for-decades/2015/04/18/39c8d8c6-e515-11e4-b510-962fcfabc310_story.html (accessed 27 September 2022).

3 McPherson, R. and Pincus, M. (2021). *Henry's Clinical Diagnosis and Management by Laboratory Methods*, 23e. Elsevier.

4 Nowak, P.M., Wietecha-Posłuszny, R., and Pawliszyn, J. (2021). White analytical chemistry: an approach to reconcile the principles of green analytical chemistry and functionality. *TrAC Trends in Analytical Chemistry* 138: 116223.

5 Kościelniak, P. and Kozak, J. (2002). Calibration by the gradient ratio-standard addition method in flow injection flame atomic absorption spectrometry. *Analytica Chimica Acta* 460 (2): 235–245.

6 Fortunato, F.M., Bechlin, M.A., Gomes Neto, J.A. et al. (2016). Standard dilution analysis in flow system: sodium determination by flame atomic emission spectrometry. *Microchemical Journal* 124: 662–667.

Index

Calibration in Analytical Science: Methods and Procedures, First Edition. Paweł Kościelniak.